Serono Symposia USA
Norwell, Massachusetts

Springer Science+Business Media, LLC

PROCEEDINGS IN THE SERONO SYMPOSIA USA SERIES

SEX-STEROID INTERACTIONS WITH GROWTH HORMONE
    Edited by Johannes D. Veldhuis and Andrea Giustina

EMBRYO IMPLANTATION: Molecular, Cellular and Clinical Aspects
    Edited by Daniel D. Carson

MALE STERILITY AND MOTILITY DISORDERS:
Etiological Factors and Treatment. A Serono Symposia S.A. Publication
    Edited by Samir Hamamah, Roger Mieusset, François Olivennes,
    and René Frydman

NUTRITIONAL ASPECTS OF OSTEOPOROSIS. A Serono Symposia S.A. Publication
    Edited by Peter Burckhardt, Bess Dawson-Hughes, and Robert P. Heaney

GERM CELL DEVELOPMENT, DIVISION, DISRUPTION AND DEATH
    Edited by Barry R. Zirkin

CELL DEATH IN REPRODUCTIVE PHYSIOLOGY
    Edited by Jonathan L. Tilly, Jerome F. Strauss III, and
    Martin Tenniswood

INHIBIN, ACTIVIN AND FOLLISTATIN: Regulatory Functions
in System and Cell Biology. A Serono Symposia S.A. Publication
    Edited by Toshihiro Aono, Hiromu Sugino, and Wylie W. Vale

PERIMENOPAUSE
    Edited by Rogerio A. Lobo

GROWTH FACTORS AND WOUND HEALING: Basic Science and
Potential Clinical Applications
    Edited by Thomas R. Ziegler, Glenn F. Pierce, and David N. Herndon

POLYCYSTIC OVARY SYNDROME
    Edited by R. Jeffrey Chang

IDEA TO PRODUCT: The Process
    Edited by Nancy J. Alexander and Anne Colston Wentz

BOVINE SPONGIFORM ENCEPHALOPATHY: The BSE Dilemma
    Edited by Clarence J. Gibbs, Jr.

GROWTH HORMONE SECRETAGOGUES
    Edited by Barry B. Bercu and Richard F. Walker

CELLULAR AND MOLECULAR REGULATION OF TESTICULAR CELLS
    Edited by Claude Desjardins

GENETIC MODELS OF IMMUNE AND INFLAMMATORY DISEASES
    Edited by Abul K. Abbas and Richard A. Flavell

MOLECULAR AND CELLULAR ASPECTS OF PERIIMPLANTATION PROCESSES
    Edited by S.K. Dey

THE SOMATOTROPHIC AXIS AND THE REPRODUCTIVE PROCESS IN HEALTH
AND DISEASE
    Edited by Eli Y. Adashi and Michael O. Thorner

Continued after Index

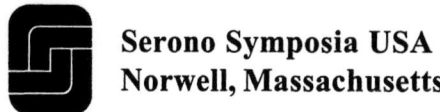

Serono Symposia USA
Norwell, Massachusetts

Johannes D. Veldhuis   Andrea Giustina
Editors

Sex-Steroid Interactions with Growth Hormone

With 94 Figures

Springer

Johannes D. Veldhuis, M.D.
Division of Endocrinology
Department of Internal Medicine
University of Virginia Health Sciences Center
Charlottesville, VA 22903
USA

Andrea Giustina, M.D.
Endocrine Section
Department of Internal Medicine
University of Brescia
25123 Brescia
Italy

Proceedings of the International Symposium on Sex-Steroid Interactions with Growth Hormone, sponsored by Serono Symposia USA, Inc., held October 22 to 25, 1998, in Naples, Florida.

For information on previous volumes, contact Serono Symposia USA, Inc.

Library of Congress Cataloging-in-Publication Data
Sex-steroid interactions with growth hormone / edited by Johannes D. Veldhuis, Andrea Giustina.
   p.   cm.
   "Serono Symposia USA"—Ser. t.p.
   Includes bibliographical references and index.
   ISBN 978-1-4612-7187-1   ISBN 978-1-4612-1546-2 (eBook)
   DOI 10.1007/978-1-4612-1546-2
   1. Hormones, Sex—Physiological effect Congresses.
   2. Somatotropin—Physiological effect Congresses. I. Veldhuis, Johannes D. II. Giustina, Andrea.
   [DNLM: 1. Sex Hormones—physiology Congresses. 2. Growth Substances—physiology Congresses. WK 900 S5173 1999]
   QP572.S4S496   1999
   612.6—dc21   99–13482
Printed on acid-free paper.
© 1999 Springer Science+Business Media New York
Originally published by Springer-Verlag New York, Inc. in 1999
Softcover reprint of the hardcover 1st edition 1999

All rights reserved. This work may not be translated or copied in whole or in part without the written permission of the publisher, Springer Science+Business Media, LLC , except for brief excerpts in connection with reviews or scholarly analysis. Use in connection with any form of information storage and retrieval, electronic adaptation, computer software, or by similar or dissimilar methodology now known or hereafter developed is forbidden.

The use of general descriptive names, trade names, trademarks, etc., in this publication, even if the former are not especially identified, is not to be taken as a sign that such names, as understood by the Trade Marks and Merchandise Marks Act, may accordingly be used freely by anyone.

While the advice and information in this book are believed to be true and accurate at the date of going to press, neither the authors, nor the editors, nor the publisher, nor Serono Symposia USA, Inc., nor Serono Laboratories, Inc., can accept any legal responsibility for any errors or omissions that may be made. The publisher makes no warranty, express or implied, with respect to the material contained herein.

Authorization to photocopy items for internal or personal use, or the internal or personal use of specific clients, is granted by Springer Science+Business Media, LLC , provided that the appropriate fee is paid directly to Copyright Clearance Center, 222 Rosewood Drive, Danvers, MA 01923, USA (Telephone: (508) 750-8400), stating the ISBN number, the volume title, and the first and last page numbers of each article copied. The copyright owner's consent does not include copying for general distribution, promotion, new works, or resale. In these cases, specific written permission must first be obtained from the publisher.

Production coordinated by Chernow Editorial Services, Inc., and managed by Francine McNeill; manufacturing supervised by Joe Quatela.
Typeset by KP Company, Brooklyn, NY.

9 8 7 6 5 4 3 2 1

ISBN 978-1-4612-7187-1

SEX-STEROID INTERACTIONS WITH GROWTH HORMONE

Scientific Committee

    Johannes D. Veldhuis, M.D., Co-Chair
    Andrea Giustina, M.D., Co-Chair
    Marc R. Blackman, M.D.
    Staffan Edén, M.D., Ph.D.
    Cyril Y. Bowers, M.D.
    Franco Sanchez-Franco, M.D., Ph.D.
    Gloria S. Tannenbaum, Ph.D.

Organizing Secretary

    Leslie Nies
    Serono Symposia USA, Inc.
    100 Longwater Circle
    Norwell, Massachusetts

Preface

Throughout the history of the experimental sciences, but perhaps in no others to the same extent as in the biological sciences, seemingly diverse research efforts examining different phenomena and separate problems often have merged into a unique focus of endeavor. A typical instance of such a unified approach in the past decade is the study of sex-steroid interactions with the growth hormone (GH) axis. Although in recent years the individual elements in this complex interaction have been stressed at an ever-increasing number of symposia, a major merit of this volume is to bring to the forefront fundamental sex-steroid–GH *interactions* as modulated by multiple relevant control mechanisms at both the neuroendocrine (output) and target tissue (end-organ) levels. To this end, a remarkably extensive compilation of scientific essays, each produced by highly competent investigative teams in distinct areas of research, is assembled within a single timely volume, devoted to the proceedings of a symposium held in Naples, Florida, in October 1998, entitled "Sex-Steroid Interactions with Growth Hormone."

The first section of the volume addresses recent developments arising from studies of the clinical impact of sex steroids on GH secretion in both adults and children. In this section, Cassorla and Roman and Houchin and Rogol discuss the mechanisms responsible for the interaction between estradiol or testosterone and growth in normal human puberty. Particularly, Cassorla emphasizes the achievements attained in this field with the advent of ultrasensitive estradiol assays, whereas Rogol poses the interesting question why GH secretion declines in young adults despite the continuing elevation in testosterone levels; Giustina et al. underscore the importance of longitudinal studies, in contrast to cross-sectional population studies, for investigating in depth the interaction between sex steroids and GH, while Metzger and Kerrigan reinforce the primary stimulatory role of endogenous estrogens in impacting GH secretion in normal male puberty based on androgen- and estrogen-receptor blockade studies. Fryburg et al. reestablish a primary role of androgens at the tissue level, in the presence of permissive amounts of GH. Blackman et al. and Ho and O'Sullivan report on the role of testosterone and estrogens in the regulation of GH secretion in the elderly. This theme is of continuing contemporary significance because important body-compositional alterations in aging have been attributed to the joint decline in sex hormone and GH secretion. Although both testosterone and estrogen replacement can increase GH secretion in males and females, respectively, Ho and O'Sullivan suggest that the route of administration of estrogen replacement may have differential effects on GH secretion. In relation to endogenous estrogen production, Jorgensen et al. show that spontaneous GH secretion is higher in young women in the estrogen-enriched preovulatory versus the estrogen-impoverished follicular phase, and Carr demonstrates that superovulation treatment with human menopausal gonadotropins further increases GH secretion. Boepple reports that treatment with GnRH agonistic analogs reduces otherwise elevated sex steroids and concomitantly heightened GH secretion in young children with precocious puberty. Finally, Veldhuis et al. attempt to summarize and highlight common features in this first section of the symposium and to articulate relevant unresolved issues concerning mechanisms underlying sex-steroid hormone neuromodulation of the human GH–IGF-I axis.

The second section is devoted to a thorough analysis of the mechanistic impact of sex steroids on the specific neural regulation of GH secretion, a topic of utmost interest. Tannenbaum reviews the mechanisms of neuroendocrine regulation of GH secretion in male and female rats, and Clifton and Steiner focus on the effects of testosterone and estrogen on the two main hypothalamic neuropeptides that govern GH secretion; that is, GHRH, which exerts a stimulatory action, and somatostatin, which imposes an inhibitory effect. Sanchez-Franco et al. further expound on the effects of sex steroids on GHRH and somatostatin secretion from cultured fetal neuronal hypothalamic cells. Finally, Copeland, Krieg et al., and Castaño et al. evaluate the possible actions of sex steroids directly on the pituitary gland, proposing different study models such as, respectively, primate somatotrophs, the perifused pituitary, and the heterogeneous population (low and high density) of porcine somatotrophs.

The third section examines the basic and clinical aspects of sex-steroidal modulation of GH actions on target tissues. Eden et al. emphasize the roles of the different patterns of GH secretion in male and female rats in determining the lipolytic effects of GH. Hoffman et al. analyze their respective experiences attesting to differential sensitivity of human males and females with GH deficiency to replacement GH therapy, all three groups showing a relative "GH resistance" in females. Finally, Neely and Martha review the rationale, the results, and the unresolved clinical issues of GH treatment in Turner syndrome.

The fourth section reviews the sex-based differences in the effects of GH secretagogues on GH secretion. Dieguez et al., Ghigo et al., Weltman et al., and Bowers and Granda-Ayala report that some classical and nonclassical GH secretagogues (GHRPs) can induce more marked GH secretory responses in females than in males. However, interestingly, the effects of GHRP seem to be less dependent on gender compared to some classical secretagogues.

The last section of the symposium appraises novel techniques for the investigation of the reproductive and GH axes. Frohman and Kineman describe the use of genetic and transgenic models; Friend updates knowledge on estrogen-receptor expression in the pituitary; Mauras investigates GH–sex-steroid anabolic interactions via specialized metabolic research methodologies based on tracer amino acid infusions and mass spectrometry; Johnson and Straume review innovative neuroendocrine analytical strategies, which allow quantification of pituitary hormone secretion and its feedback control, such as deconvolution analysis and approximate entropy; and Waxman and Gustafsson highlight sex-steroid hormone effects on the GH intracellular signaling pathway.

In ensemble, the book should provide an invaluable and timely forum in which outstanding scientists dissect, review, and discuss multiple new aspects of the biological nature of sex-steroid–GH axis interactions and their pathophysiological, diagnostic, and therapeutic implications from both basic science and clinical perspectives. This volume, therefore, should be of particular relevance to basic and clinical experimentalists, and also to neuroendocrinologists and to pediatric and reproductive endocrinologists.

The contributors to this volume are indebted to the outstanding resources of Serono Symposia USA, Inc., in providing a truly international scientific milieu and superb organization. We thank Leslie Nies for her patience in developing a complex program, and Judy Donahue for her assistance in preparing this volume for publication. We are grateful to the organizing committee for joining in the selection of program participants and orchestration of the flow of the symposium.

JOHANNES D. VELDHUIS
ANDREA GIUSTINA

Contents

Preface vii
Contributors xv

Part I. Clinical Impact of Sex Steroids on Growth Hormone Secretion

1. Sex-Steroid–Growth Hormone Axis Interactions in Normal Female Puberty 3
   FERNANDO CASSORLA AND ROSSANA ROMÁN

2. Sex-Steroid–Growth Hormone Interactions During Normal Male Puberty 11
   LISA D. HOUCHIN AND ALAN D. ROGOL

3. Impact of Testosterone Replacement on the Maturation of the Growth Hormone–IGF-I Axis 20
   ANDREA GIUSTINA, PAOLO DESENZANI, AND TIZIANO SCALVINI

4. Sex-Steroid–Hormone Modulation of the Somatotropic Axis During Normal Male Puberty 32
   DANIEL L. METZGER AND JAMES R. KERRIGAN

5. Influence of Testosterone on the GH–IGF-I Axis in Healthy Elderly Men 44
   MARC R. BLACKMAN, COLLEEN CHRISTMAS, THOMAS MÜNZER, KIERAN G. O'CONNOR, THOMAS E. STEVENS, MICHELE F. BELLANTONI, KATHERINE PABST, CAROL ST. CLAIR, H. BALLENTINE CARTER, E. JEFFREY METTER, AND S. MITCHELL HARMAN

6. Effect of Estrogen on GH Secretion and Action in
   Postmenopausal Women 54
   KEN K.Y. HO AND ANTHONY J. O'SULLIVAN

7. Menstrual Cycle Interaction with the Growth
   Hormone Axis 67
   PER OVESEN, NINA VAHL, JENS SANDAHL CHRISTIANSEN,
   JOHANNES D. VELDHUIS, AND JENS OTTO LUNDE JORGENSEN

8. Growth Hormone Axis Response to Superovulation 74
   BRUCE R. CARR

9. Androgenic Modulation of the Growth Hormone–IGF Axis
   and Its Impact on Metabolic Outcomes 82
   DAVID A. FRYBURG, ARTHUR L. WELTMAN, LINDA A. JAHN,
   JUDY Y. WELTMAN, EUGENE SAMOJLIK, RAYMOND L. HINTZ,
   AND JOHANNES D. VELDHUIS

10. Proposed Mechanisms of Sex-Steroid–Hormone Neuro-
    modulation of the Human GH–IGF-1 Axis 93
    JOHANNES D. VELDHUIS, WILLIAM S. EVANS, NIKHITA SHAH,
    E. SHANNON STORY, MEGAN J. BRAY, AND STACEY M. ANDERSON

11. The Growth Hormone Axis in Precocious Puberty 122
    PAUL A. BOEPPLE

Part II. Sex Steroids and GHRH-Somatostatin

12. Hypothalamic Control Mechanisms of Sexually
    Dimorphic Growth Hormone Secretory Patterns
    in the Rat 133
    GLORIA S. TANNENBAUM

13. Sex Differences in Growth Hormone-Releasing
    Hormone (GHRH) and Somatostatin Neurons 144
    DONALD K. CLIFTON AND ROBERT A. STEINER

14. Sex-Steroid Interaction with Somatostatin and GHRH
    in the Rat 153
    FRANCO SANCHEZ-FRANCO, FERNANDO PAZOS,
    GUMERSINDO FERNANDEZ, JUDITH LOPEZ, NURIA PALACIOS,
    AND LUCINDA CACICEDO

15. Sex-Steroid Effects on Primate Somatotrophs 160
    KENNETH C. COPELAND

16. Sex-Steroid Effects on Perifused Pituitary 172
RICHARD J. KRIEG, JR., PAUL M. MARTHA, JR., JAMES R. KERRIGAN,
JUDY M. BATSON, TIMOTHY E. SAYLES, STEVEN J. KRAUS,
DENNIS W. MATT, AND WILLIAM S. EVANS

17. Somatotrope Heterogeneity and Its Involvement
in Growth Hormone (GH) Regulation 183
JUSTO PASTOR CASTAÑO, JOSÉ LUIS RAMÍREZ,
JOSÉ CARLOS GARRIDO-GRACIA, AND FRANCISCO GRACIA-NAVARRO

Part III. How Sex Steroids Modulate GH Action on Target Tissues

18. Sex-Steroid and GH Interactions in the Regulation
of Lipid Metabolism 195
STAFFAN EDÉN, JAN OSCARSSON, AND MALIN OTTOSSON

19. Estrogen Replacement Therapy and the Response
to Human Growth Hormone 202
GIAN PAOLO CEDA, GIORGIO VALENTI, AND ANDREW R. HOFFMAN

20. Influence of Gender on Response to Growth Hormone
Substitution Therapy in Adults with Growth
Hormone Deficiency 209
FERDINAND ROELFSEMA AND YVONNE JOHANNA HENRICA JANSSEN

21. How Sex Steroids Modulate GH Action on Target Tissues:
Long-Term Follow-up in GH-Deficient Adults 219
GUDMUNDUR JOHANNSSON AND BENGT-ÅKE BENGTSSON

22. Growth Hormone Treatment in Turner Syndrome:
Rationale for Therapy 227
E. KIRK NEELY

23. Practical Issues of Clinical Growth Hormone Therapy
in Turner Syndrome 235
PAUL M. MARTHA, JR., AND KENNETH M. ATTIE

Part IV. GH Secretagogues: Differential Effects in Men and Women

24. Influence of Gonadal Function on GH Secretion 243
FRANCISCA LAGO, ANGELA PEÑALVA, EVA CARRO, ROSA MARIA SEÑARIS,
VERA POPOVIC, MANUEL POMBO, FELIPE F. CASANUEVA,
AND CARLOS DIEGUEZ

25. Endocrine Responses to GH Secretagogues in Relation to Sex and Age in Humans 249
EMANUELA ARVAT, ROBERTA GIORDANO, LAURA GIANOTTI, FABIO BROGLIO, GIAMPIERO MUCCIOLI, FRANCO CAMANNI, AND EZIO GHIGO

26. Gender Impact on the GH Response to Exercise 261
LAURIE WIDEMAN, JUDY Y. WELTMAN, NIKHITA SHAH, E. SHANNON STORY, CYRIL Y. BOWERS, JOHANNES D. VELDHUIS, AND ARTHUR L. WELTMAN

27. Stimulated Release of GH in Normal Younger and Older Men and Women 277
CYRIL Y. BOWERS AND RAMONA GRANDA-AYALA

Part V. Novel Techniques in Investigating the Reproductive and GH Axis

28. Genetic and Transgenic Models to Investigate the Growth Hormone Axis and Sexual Dimorphism 293
LAWRENCE A. FROHMAN AND RHONDA D. KINEMAN

29. Estrogen Receptor Expression in the Pituitary Gland 301
KEITH E. FRIEND AND IAN E. MCCUTCHEON

30. Metabolic Effects of GH/IGF-I, Testosterone, and Estrogen: Studies in Children and Adults 308
NELLY MAURAS

31. Innovative Quantitative Neuroendocrine Techniques 318
MICHAEL L. JOHNSON AND MARTIN STRAUME

32. Sexual Dimorphism of Liver Cytochrome P-450 Gene Expression: GH Pulse-Activated STAT Signaling Mechanisms 327
DAVID J. WAXMAN

33. Intracellular Signaling Networks 337
JAN-ÅKE GUSTAFSSON

Author Index 343

Subject Index 347

Contributors

STACEY M. ANDERSON, Division of Endocrinology, Department of Internal Medicine, University of Virginia Health Sciences Center, Charlottesville, Virginia, USA.

EMANUELA ARVAT, Division of Endocrinology, Department of Internal Medicine, Ospedale Molinette, Turin, Italy.

KENNETH M. ATTIE, Department of Medical Affairs, Genentech, Inc., South San Francisco, California, USA.

JUDY M. BATSON, Department of Biochemistry and Molecular Biology, Oklahoma State University, Stillwater, Oklahoma, USA.

MICHELE F. BELLANTONI, Department of Medicine, Johns Hopkins University School of Medicine, Baltimore, Maryland, USA.

BENGT-ÅKE BENGTSSON, Department of Medicine, Sahlgrenska University Hospital, Göteborg, Sweden.

MARC R. BLACKMAN, Department of Medicine, Johns Hopkins University School of Medicine, Baltimore, Maryland, USA.

PAUL A. BOEPPLE, Pediatric and Reproductive Endocrine Units, Massachusetts General Hospital and Harvard Medical School, Boston, Massachusetts, USA.

CYRIL Y. BOWERS, Division of Endocrinology and Metabolism, Department of Medicine, Tulane University Medical School, New Orleans, Louisiana, USA.

MEGAN J. BRAY, Department of Obstetrics and Gynecology, University of Virginia Health Sciences Center, Charlottesville, Virginia, USA.

FABIO BROGLIO, Division of Endocrinology, Department of Internal Medicine, Ospedale Molinette, Turin, Italy.

LUCINDA CACICEDO, Service of Endocrinology, Hospital Ramon y Cajal, Madrid, Spain.

FRANCO CAMANNI, Division of Endocrinology, Department of Internal Medicine, Ospedale Molinette, Turin, Italy.

BRUCE R. CARR, Department of Obstetrics and Gynecology, University of Texas Southwestern Medical Center, Dallas, Texas, USA.

EVA CARRO, Department of Physiology, University of Santiago de Compostela, Santiago de Compostela, Spain.

H. BALLENTINE CARTER, Department of Urology, Johns Hopkins University School of Medicine, Baltimore, Maryland, USA.

FELIPE F. CASANUEVA, Department of Medicine, University of Santiago de Compostela, Santiago de Compostela, Spain.

FERNANDO CASSORLA, Institute of Maternal and Child Research, University of Chile, Santiago, Chile.

JUSTO PASTOR CASTAÑO, Department of Cell Biology, University of Cordoba, Cordoba, Spain.

GIAN PAOLO CEDA, University of Parma, Parma, Italy.

JENS SANDAHL CHRISTIANSEN, Department of Medicine, Aarhus Kommunehospital, Aarhus, Denmark.

COLLEEN CHRISTMAS, Department of Medicine, Johns Hopkins University School of Medicine, Baltimore, Maryland, USA.

DONALD K. CLIFTON, Department of Obstetrics and Gynecology, University of Washington, Seattle, Washington, USA.

KENNETH C. COPELAND, Department of Pediatrics, Baylor College of Medicine, Houston, Texas, USA.

PAOLO DESENZANI, Endocrine Section, Department of Internal Medicine, University of Brescia, Brescia, Italy.

CARLOS DIEGUEZ, Department of Physiology, University of Santiago de Compostela, Santiago de Compostela, Spain.

STAFFAN EDÉN, Department of Physiology, Göteborg University, Göteborg, Sweden.

WILLIAM S. EVANS, Division of Endocrinology, Department of Internal Medicine, University of Virginia Health Sciences Center, Charlottesville, Virginia, USA.

GUMERSINDO FERNANDEZ, Service of Endocrinology, Center of Clinical Investigation, National Institute of Health Carlos III, Madrid, Spain.

KEITH E. FRIEND, Department of Endocrinology, University of Texas M.D. Anderson Cancer Center, Houston, Texas, USA.

LAWRENCE A. FROHMAN, Department of Medicine, University of Illinois at Chicago, Chicago, Illinois, USA.

DAVID A. FRYBURG, Division of Endocrinology and Metabolism, Department of Internal Medicine and the General Clinical Research Center, University of Virginia, Charlottesville, Virginia, USA.

JOSÉ CARLOS GARRIDO-GRACIA, Department of Cell Biology, University of Cordoba, Cordoba, Spain.

EZIO GHIGO, Division of Endocrinology, Department of Internal Medicine, Ospedale Molinette, Turin, Italy.

LAURA GIANOTTI, Division of Endocrinology, Department of Internal Medicine, Ospedale Molinette, Turin, Italy.

ROBERTA GIORDANO, Division of Endocrinology, Department of Internal Medicine, Ospedale Molinette, Turin, Italy.

ANDREA GIUSTINA, Endocrine Section, Department of Internal Medicine, University of Brescia, Brescia, Italy.

FRANCISCO GRACIA-NAVARRO, Department of Cell Biology, University of Cordoba, Cordoba, Spain.

RAMONA GRANDA-AYALA, Division of Endocrinology and Metabolism, Department of Medicine, Tulane University Medical School, New Orleans, Louisiana, USA.

JAN-ÅKE GUSTAFSSON, Departments of Medical Nutrition and Biosciences at Novum, Huddinge University Hospital, Karolinska Institute, Huddinge, Sweden.

S. MITCHELL HARMAN, Gerontology Research Center, National Institute on Aging, National Institutes of Health, Baltimore, Maryland, USA.

RAYMOND L. HINTZ, Division of Endocrinology, Department of Pediatrics, Stanford University Medical Center, Palo Alto, California, USA.

KEN K.Y. HO, Garvan Institute of Medical Research, St. Vincent's Hospital, Sydney, NSW, Australia.

ANDREW R. HOFFMAN, VA Palo Alto Health Care System, Stanford University School of Medicine, Palo Alto, California, USA.

LISA D. HOUCHIN, Department of Pediatrics, University of Virginia Health Sciences Center, Charlottesville, Virginia, USA.

LINDA A. JAHN, Division of Endocrinology and Metabolism, Department of Internal Medicine and the General Clinical Research Center, University of Virginia, Charlottesville, Virginia, USA.

YVONNE JOHANNA HENRICA JANSSEN, Department of Endocrinology, Leiden University Medical Center, Leiden, The Netherlands.

GUDMUNDUR JOHANNSSON, Research Center for Endocrinology and Metabolism, Sahlgrenska University Hospital, Göteborg, Sweden.

MICHAEL L. JOHNSON, Departments of Pharmacology and Internal Medicine, National Science Foundation Center for Biological Timing, University of Virginia Health Sciences Center, Charlottesville, Virginia, USA.

JENS OTTO LUNDE JORGENSEN, Department of Medicine, Aarhus Kommunehospital, Aarhus, Denmark.

JAMES R. KERRIGAN, Department of Pediatrics, East Tennessee State University, Johnson City, Tennessee, USA.

RHONDA D. KINEMAN, Department of Medicine, University of Illinois at Chicago, Chicago, Illinois, USA.

STEVEN J. KRAUS, Department of Radiology, Children's Hospital, Boston, Massachusetts, USA.

RICHARD J. KRIEG, JR., Departments of Anatomy and Pediatrics, Medical College of Virginia, Virginia Commonwealth University, Richmond, Virginia, USA.

FRANCISCA LAGO, Department of Physiology, University of Santiago de Compostela, Santiago de Compostela, Spain.

JUDITH LOPEZ, Department of Internal Medicine, Hospital Ramon y Cajal, Madrid, Spain.

PAUL M. MARTHA, JR., Endocrinology and Metabolism, PRAECIS Pharmaceuticals, Inc., Cambridge, Massachusetts, USA.

DENNIS W. MATT, Department of Obstetrics and Gynecology, Medical College of Virginia, Virginia Commonwealth University, Richmond, Virginia, USA.

NELLY MAURAS, Division of Endocrinology, Mayo Medical School, Nemours Children's Clinic, Jacksonville, Florida, USA.

IAN E. MCCUTCHEON, Department of Neurosurgery, University of Texas M.D. Anderson Cancer Center, Houston, Texas, USA.

E. JEFFREY METTER, Gerontology Research Center, National Institute on Aging, National Institutes of Health, Baltimore, Maryland, USA.

DANIEL L. METZGER, Department of Pediatrics, British Columbia Children's Hospital, Vancouver, British Columbia, Canada.

GIAMPIERO MUCCIOLI, Department of Anatomy, Pharmacology and Forensic Medicine, University of Turin, Turin, Italy.

THOMAS MÜNZER, Gerontology Research Center, National Institute on Aging, National Institutes of Health, Baltimore, Maryland, USA.

E. KIRK NEELY, Department of Pediatrics, Stanford University, Stanford, California, USA.

KIERAN G. O'CONNOR, Department of Geriatric Medicine, Louth Hospital, Dundalk, County Louth, Ireland.

JAN OSCARSSON, Department of Physiology, Göteborg University, Göteborg, Sweden.

ANTHONY J. O'SULLIVAN, Garvan Institute of Medical Research, St. Vincent's Hospital, Sydney, NSW, Australia.

MALIN OTTOSSON, Wallenberg Laboratory, Sahlgrenska Hospital, Göteborg University, Göteborg, Sweden.

PER OVESEN, Department of Gynecology and Obstetrics, Skejby Sygehus, Aarhus, Denmark.

KATHERINE PABST, Gerontology Research Center, National Institute on Aging, National Institutes of Health, Baltimore, Maryland, USA.

NURIA PALACIOS, Service of Endocrinology, Hospital Ramon y Cajal, Madrid, Spain.

FERNANDO PAZOS, Section of Endocrinology, Hospital Ramon y Cajal, Madrid, Spain.

ANGELA PEÑALVA, Department of Medicine, University of Santiago de Compostela, Santiago de Compostela, Spain.

MANUEL POMBO, Department of Pediatrics, University of Santiago de Compostela, Santiago de Compostela, Spain.

VERA POPOVIC, Institute of Endocrinology, Belgrade, Yugoslavia.

JOSÉ LUIS RAMÍREZ, Department of Cell Biology, University of Cordoba, Cordoba, Spain.

FERDINAND ROELFSEMA, Department of Endocrinology, Leiden University Medical Center, Leiden, The Netherlands.

ALAN D. ROGOL, Departments of Pediatrics and Pharmacology, University of Virginia Health Sciences Center, Charlottesville, Virginia, USA.

ROSSANA ROMÁN, Institute of Maternal and Child Research, University of Chile, Santiago, Chile.

EUGENE SAMOJLIK, Endocrine Laboratory, Newark Beth Israel Medical Center, United Medical and Dentistry–New Jersey Medical School, Newark, New Jersey, USA.

FRANCO SANCHEZ-FRANCO, Service of Endocrinology, Center of Clinical Investigation, National Institute of Health Carlos III, Madrid, Spain.

TIMOTHY E. SAYLES, Department of Obstetrics and Gynecology, Portsmouth Naval Hospital, Portsmouth, Virginia, USA.

TIZIANO SCALVINI, Endocrine Section, Department of Internal Medicine, University of Brescia, Brescia, Italy.

ROSA MARIA SEÑARIS, Department of Physiology, University of Santiago de Compostela, Santiago de Compostela, Spain.

NIKHITA SHAH, Omni Healthcare, Melbourne, Florida, USA.

CAROL ST. CLAIR, Department of Medicine, Johns Hopkins University School of Medicine, Baltimore, Maryland, USA.

ROBERT A. STEINER, Departments of Obstetrics and Gynecology, and Physiology and Biophysics, University of Washington, Seattle, Washington, USA.

THOMAS E. STEVENS, Department of Medicine, University of South Alabama, Mobile, Alabama, USA.

E. SHANNON STORY, Department of Endocrinology and Metabolism, Presbyterian Hospital, Charlotte, North Carolina, USA.

MARTIN STRAUME, Department of Internal Medicine, National Science Foundation Center for Biological Timing, University of Virginia Health Sciences Center, Charlottesville, Virginia, USA.

GLORIA S. TANNENBAUM, Departments of Pediatrics, and Neurology and Neurosurgery, McGill University, Montreal, Quebec, Canada.

NINA VAHL, Department of Medicine, Aarhus Kommunehospital, Aarhus, Denmark.

GIORGIO VALENTI, G. Stuard Hospital, University of Parma, Parma, Italy.

JOHANNES D. VELDHUIS, Division of Endocrinology, Departments of Internal Medicine and Obstetrics and Gynecology, National Science Foundation Center for Biological Timing, University of Virginia Health Sciences Center, Charlottesville, Virginia, USA.

DAVID J. WAXMAN, Division of Cell and Molecular Biology, Department of Biology, Boston University, Boston, Massachusetts, USA.

ARTHUR L. WELTMAN, Division of Endocrinology and Metabolism, Department of Internal Medicine and the General Clinical Research Center, and Department of Human Services, University of Virginia, Charlottesville, Virginia, USA.

JUDY Y. WELTMAN, Division of Endocrinology and Metabolism, Department of Internal Medicine and the General Clinical Research Center, University of Virginia, Charlottesville, Virginia, USA.

LAURIE WIDEMAN, General Clinical Research Center, Department of Medicine, University of Virginia, Charlottesville, Virginia, USA.

Part I

Clinical Impact of Sex Steroids on Growth Hormone Secretion

1

Sex-Steroid–Growth Hormone Axis Interactions in Normal Female Puberty

FERNANDO CASSORLA AND ROSSANA ROMÁN

During human infancy, the ovary shows signs of activity as evidenced by the development of follicles and by the production of sex steroids. During the first year of postnatal life there is evidence of active ovarian steroidogenesis, which leads to the secretion of estradiol by the infantile gonad. Circulating concentrations of estradiol may reach midpubertal levels during this period. Subsequently, the ovary enters a more quiescent period, with relatively low activity, even though there is evidence of residual follicular development during childhood (1). The ovary, however, remains steroidogenically active during this period, and produces small amounts of estradiol that can be detected with special ultrasensitive assays (2). These assays demonstrate that circulating concentrations of estradiol are higher in prepubertal girls compared to prepubertal boys and show a gradual increment with advancing age.

Serum growth hormone (GH) concentrations also increase during childhood, and are significantly higher in older compared to younger prepubertal girls. This pattern has been documented by analyzing baseline growth hormone concentrations in normal girls, determined in blood samples obtained every 20 min during 24 h (3). Thus, both serum estradiol and growth hormone begin to rise in normal girls before puberty, and continue to rise during puberty. The pubertal growth spurt in girls is produced by the synergistic effects of these hormones (4). Growth begins to accelerate in girls during late childhood at the time when serum concentrations of estradiol and growth hormone are beginning to increase. Thus, one of the earliest signs of impending puberty in girls appears to be acceleration of linear growth. The combined effects of estradiol and growth hormone produce an increase in the secretion of several growth factors, among them insulin-like growth factor (IGF-I), which also contribute to the development of the pubertal growth spurt.

The objective of this chapter is to review sex-steroid–growth hormone interactions during normal female puberty. In addition, we describe some human models that help us define possible cause-and-effect relationships between changes in sex steroids and growth hormone during this period.

Effects of Sex Steroids on Growth Hormone Secretion

Sex-steroid–growth hormone interactions during childhood and adolescence have been the subject of several reviews (5–8). Sex steroids act in synergy with growth hormone to induce the pubertal growth spurt in both sexes, and each one is responsible for approximately half the height gain during this period (4). This idea is supported by the observation that patients with isolated growth hormone deficiency experience a pubertal growth spurt that is approximately 50% of normal (9).

During childhood, mean serum growth hormone concentrations are similar in girls and boys and have a marked day–night rhythm (10). Growth hormone secretion reaches its peak during the early hours of sleep, and occurs throughout the night in small episodic bursts. Growth hormone concentrations are lower during the day than during the night during this period (3). Growth hormone production between secretory episodes is quite low, and serum concentrations often fall to the detection limit of most earlier assays.

Puberty is characterized by a dramatic activation of the growth hormone/IGF-I axis (3). The rise in mean 24-h growth hormone levels results from an increase in pulse amplitude, and in the mass of growth hormone secreted per burst (11). Using the method of deconvolution analysis, no significant changes in growth hormone pulse frequency, half-life, or duration have been documented (12). Girls show a significant increase in circulating growth hormone levels during late childhood and early puberty, with the highest levels observed during Tanner breast stage III to IV (10). Maximal growth hormone levels are observed in males at approximately Tanner genital stage IV. During midpuberty, the fractional day–night rhythm is attenuated by the greater rise in growth hormone secretion during the day (10). After sexual maturation is completed, growth hormone and IGF-I concentrations fall to adult and prepubertal values in both sexes.

The mechanisms whereby sex steroids stimulate growth hormone secretion by the pituitary appear to include increasing the production of hypothalamic growth hormone-releasing hormone (GHRH). In support of this hypothesis, data obtained in orchidectomized adult male rats shows that estradiol administration increases the content of hypothalamic GHRH (13). Testosterone and estradiol have been shown to increase GHRH messenger ribonucleic acid production (14). In addition, GHRH-containing hypothalamic neurons concentrate estradiol to a greater degree than somatostatin-containing neurons (15) and, in vitro, both estradiol and testosterone promote the survival of fetal hypothalamic neurons (16).

Both testosterone and estradiol can stimulate growth hormone secretion. The effects of testosterone on growth hormone secretion appear to be mediated through its conversion to estradiol (17,18). The available data show that sex steroids mostly increase growth hormone pulse amplitude. Weissberger et al. (17) treated normal and hypogonadal men with testosterone enanthate alone or associated with the estrogen blocker tamoxifen. In the normal men, tamoxifen produced a significant reduction in mean serum growth hormone concentrations and in growth hormone pulse amplitude. Eakman et al. (18) administered testosterone, or the nonaromatizable androgen dihydrotestosterone, to boys with constitutional delay. The patients that received testosterone showed an increase in growth hormone concentrations, but no significant differences in growth hormone levels were observed in the boys who received dihydrotestosterone. Metzger et al. (19) treated a group of late pubertal males with the androgen-receptor blocker flutamide. These investigators observed that during androgen blockade, there was an increase in serum estradiol concentrations that was associated with an increase in growth hormone production rate (19). These observations suggest that testosterone stimulates growth hormone secretion predominantly via its aromatization to estradiol.

In girls, significant relationships have been demonstrated between serum estradiol and growth hormone concentrations during both normal (20) and precocious puberty (21). In central precocious puberty, there is a significant concomitant rise in gonadal sex steroid production and growth hormone secretion, which is associated with an increase in growth velocity. Following suppression of the hypothalamic–pituitary–gonadal axis with gonadotropin-releasing hormone (GnRH) analog, a significant decline in growth velocity, skeletal maturation, growth hormone secretion, and IGF-I concentrations can be observed (21). Treatment with luteinizing hormone-releasing hormone (LHRH) analog for 6 months in patients with central precocious puberty reduces their growth rate by about 40%, whereas serum IGF-I decreases by only 10% and remains elevated for age (22). Growth rate during LHRH analog treatment fails to correlate with serum IGF-I but does correlate with serum estradiol (22). Circulating levels of IGF-I and IGF-binding protein-3 (IGF-BP-3) are GH dependent and tend to vary in a parallel fashion, but some studies suggest that sex steroids may regulate IGF-I and IGF-BP-3 differentially (23). Juul et al. (24) found that LHRH analog treatment in girls with precocious puberty had no effect on IGF-I levels, whereas serum IGF-BP-3 concentrations increased during the first 12 months of treatment.

Additional support for a relationship between sex steroids and growth hormone is provided by data indicating that growth hormone concentrations vary during the menstrual cycle in normal women: mean growth hormone concentrations increase during the late follicular phase, at the time when estrogen levels reach a peak (25). In addition, Ho et al. (26) observed that mean growth hormone levels correlate positively with free serum estradiol, but not with free serum testosterone, in both men and women.

At physiological concentrations, sex steroids stimulate the secretion of IGF-I during puberty and in hypogonadal patients treated with replacement doses of estradiol (27). At higher concentrations, however, estradiol can inhibit the production of IGF-I, as observed in acromegalic patients (28) and in girls with constitutionally tall stature. In male-to-female transsexuals, however, pharmacological doses of estradiol do not appear to alter serum IGF-I concentrations (29). To complicate the issue even further, acute administration of estradiol at both physiological and pharmacological doses to castrated female baboons increases serum IGF-I concentrations (30). Thus, depending on the period of exposure and dose of estradiol administered, serum IGF-I concentrations may vary in opposite directions. In general terms, estradiol at physiological concentrations stimulates IGF-I production, whereas estradiol at pharmacological concentrations appears to inhibit IGF-I production.

Relationships Between Estradiol and Growth Hormone in Turner Syndrome

Several years ago, we demonstrated that low doses of ethinyl estradiol (100 ng kg^{-1} day^{-1}) administered to patients with Turner syndrome increase growth velocity first and subsequently stimulate breast development (27,31). Thus, linear growth appears to be a more sensitive index of estrogen effect than breast development. These small doses of ethinyl estradiol can also increase spontaneous growth hormone concentrations determined in samples obtained every 20 min during 24 h (32). Therefore, in girls with Turner syndrome treatment with low doses of estradiol increases growth velocity (27), probably in part by stimulating the secretion of growth hormone by the pituitary. This effect, however, is not sustained, and growth velocity subsequently declines if treatment with low doses of estradiol is maintained (31).

These data are compatible with the acceleration of growth velocity that is observed during early puberty in normal girls. As mentioned earlier, the first sign of impending puberty in girls is an increase in growth velocity, which occurs at a time when serum estradiol levels are beginning to rise. Thus, both in normal girls (2) and in patients with Turner syndrome (27), low concentrations of estradiol are associated with acceleration in growth velocity. In contrast, the pubertal growth spurt in boys is a relatively late event in adolescence. During this period, circulating concentrations of androgen are high but circulating concentrations of estrogen are low. Therefore, the pubertal growth spurt of both normal girls and boys occurs at a time when serum concentrations of estradiol are relatively low (5). In support of an effect of low concentrations of estradiol stimulating growth velocity, our group has demonstrated that administration of estradiol at low doses can accelerate ulnar growth in boys with delayed puberty (33). It is therefore likely that low concentrations of estradiol may help promote the pubertal growth spurt in both genders.

Additional data in favor of a relationship between serum estradiol and growth hormone in patients with Turner syndrome derives from the measurement of growth hormone concentrations in these patients. Several studies have shown that serum growth hormone levels are normal in girls with Turner syndrome up to approximately 8 years of age. Beyond that age, growth hormone concentrations are reduced in these patients (34), probably as a result of their inappropriately low levels of sex steroids. Estrogen replacement therapy restores integrated serum concentrations of growth hormone to levels observed in normally maturing girls (35). Serum IGF-I levels, however, may rise during the second decade of life in girls with Turner syndrome, even in the absence of estradiol (35).

Androgen and Estrogen Insensitivity: Models to Assess the Isolated Effects of Estradiol and Testosterone on Growth Hormone Secretion

Patients with complete androgen insensitivity undergo a relatively normal genetic female growth spurt in terms of both magnitude and timing. This observation suggests that the pubertal growth spurt of normal girls is mostly dependent upon estrogen and that androgens do not appear to play a significant role in this process (36). If patients with androgen insensitivity are gonadectomized during puberty, a reduction in their mean serum nocturnal growth hormone and IGF-I levels is observed (37). Thus, in the absence of androgen action, estrogen alone can stimulate growth hormone release by the pituitary and promote linear growth.

The marked effects of estrogen on skeletal maturation have become evident in the single patient with estrogen insensitivity reported to date (38). This patient had a markedly delayed bone age and continued to grow during the third decade of life. This case suggests that estrogen plays a significant role in promoting skeletal maturation during puberty and that lack of estrogen allows continued growth, which can produce pathological tall stature.

A related estrogen deficiency state is caused by aromatase deficiency. Morishima et al. (39) reported male and female siblings with this condition. At the age of 14 years the female patient was tall, had delayed puberty, and showed evidence of virilization. Treatment with estrogen resulted in growth acceleration, breast development, and advancement of bone age. Another case of aromatase deficiency was reported by Conte et al. (40). This girl had normal growth and development during childhood, but did not undergo a pubertal growth spurt or develop secondary sexual characteristics during the second decade of life. Estrogen replacement therapy resulted in growth acceleration and breast development in this patient.

These findings in patients without biological androgen or estrogen effects demonstrate that estrogen plays a key role in determining the timing and magnitude of the pubertal growth spurt in females and the rate of skeletal maturation in males. It is evident that a complex interplay between estrogens, androgens, growth hormone, and IGF-I is necessary for normal linear growth and skeletal maturation during puberty in both sexes. However, androgens seem to play a lesser role in the female pubertal growth spurt than do estrogens in the male pubertal growth spurt. This role is exemplified by the relatively normal female pubertal growth spurt experienced by patients with androgen insensitivity, and the pathological continued growth caused by the delay in skeletal maturation observed in the patient with estrogen insensitivity.

References

1. Venturoli S, Porcu E, Macrelli S, Cavallari C, Flamigni C. L'Ovaire pubertaire. Contracept Fertil Sex 1994;22:469–74.
2. Oerter Klein K, Baron J, Colli M, McDonnell D, Cutler GB Jr. Estrogen levels in childhood determined by ultrasensitive recombinant cell bioassay. J Clin Invest 1994;94:2475–80.
3. Rose S, Municchi G, Barnes K, Kamp G, Uriarte M, Ross J, et al. Spontaneous growth hormone secretion increases during puberty in normal girls and boys. J Clin Endocrinol Metab 1991;73:428–35.
4. Brook CG, Hindmarsh PC. The somatotropic axis in puberty. Endocrinol Metab Clin North Am 1992;21(4):767–82.
5. Cutler GB Jr, Cassorla F, Ross J, Pescovitz O, Barnes K, Comité F, et al. Pubertal growth: physiology and pathophysiology. Rec Prog Horm Res 1986;42:443–70.
6. Kerrigan J, Rogol A. The impact of gonadal steroid hormone action on growth hormone secretion during childhood and adolescence. Endocr Rev 1992;13:281–98.
7. Clark P, Rogol A. Growth hormones and sex steroid interactions at puberty. Endocrinol Metab Clin North Am 1996;25:665–81.
8. Lee P, Witchel S. The influence of estrogen on growth. Curr Opin Pediatr 1997;9: 431–6.
9. Zachman M. Interrelations between growth hormone and sex hormones: physiologic and therapeutic consequences. Horm Res (Basel) 1992;38(suppl 1):1–8.
10. Albertsson-Wikland K, Rosberg S, Karlberg J, Groth T. Analysis of 24-hour growth hormone profiles in healthy boys and girls of normal stature. Relation to puberty. J Clin Endocrinol Metab 1994;78:1195–201.
11. Martha PM Jr, Gorman KM, Blizzard RM, Rogol AD, Veldhuis JD. Endogenous growth hormone secretion and clearance rates in normal boys, as determined by deconvolution analysis. Relationship to age pubertal status, and body mass. J Clin Endocrinol Metab 1992;74:336–44.
12. Veldhuis JD, Johnson ML. Deconvolution analysis of hormone data. Methods Enzymol 1992;201:539–75.
13. Gabriel SM, Millard WJ, Koening JL, Badger TM, Russell WE, Maiter DM, et al. Sexual and developmental differences in peptides regulating growth hormone secretion in the rat. Neuroendocrinology 1989;50:299–307.
14. Zeitler P, Argente J, Chowen-Breed JA, Clifton DK, Steiner RA. Growth hormone releasing hormone messenger ribonucleic acid in the hypothalamus of the adult male rat is increased by testosterone. Endocrinology 1990;127:1362–8.
15. Shirasu K, Stumpf WE, Sar M. Evidence for direct action of estradiol on growth hormone-releasing factor (GRF) in rat hypothalamus: localization of estradiol in GRF neurons. Endocrinology 1990;127:344–9.

16. Chowen JA, Torres Aleman I, García-Segura LM. Trophic effects of estradiol on fetal hypothalamic neurons. Neuroendocrinology 1992;56:895–901.
17. Weissberger AJ, Ho KKY, Activation of the somatotropic axis in adult males: evidence for the role of aromatization. J Clin Endocrinol Metab 1993;76:1407–12.
18. Eakman G, Dallas J, Ponder S, Keenan B. The effects of testosterone and dihydrotestosterone on hypothalamic regulation of growth hormone secretion. J Clin Endocrinol Metab 1996;81:1217–23.
19. Metzger D, Kerrigan J. Androgen receptor blockade with flutamide enhances growth hormone secretion in late pubertal males: evidence for independent actions of estrogen and androgen. J Clin Endocrinol Metab 1993;76:1147–52.
20. Wennink JMB, Delemarre-van de Waal HA, Shcemaker R, Blaauw G, Van den Braken C, Shoemaker J. Growth hormone secretion patterns in relation to LH and estradiol secretion throughout normal female puberty. Acta Endocrinol 1991;123:129–35.
21. Mansfield MJ, Rudlin CR, Crigler JF, Karol KA, Crawford JD, Boepple PA, et al. Changes in growth and serum, growth hormone and plasma somatomedin-C during suppression of gonadal sex steroid secretion in girls with central precocious puberty. J Clin Endocrinol Metab 1988;66:3–9.
22. Pescovitz OH, Rosenfeld RG, Hintz RL, Barnes K, Comité F, Loriaux DL, et al. Somatomedin-C in accelerated growth of children with precocious puberty. J Pediatr 1985;107:20–5.
23. Juul A, Dalgaard P, Blum WF, Bang P, Hall K, Michaelsen KF, et al. Serum levels of insulin growth factor (IGF)-binding protein-3 (IGFBP-3) in healthy infants, children and adolescents: the relation to IGF-I, IGF-II, IGFBP-1, IGFBP-2, age, sex, body mass index and pubertal maturation. J Clin Endocrinol Metab 1995;80:2534–42.
24. Juul A, Scheike T, Nielsen C, Krabbe S, Muller J, Skakkebaek N. Serum insulin-like growth factor (IGF-I) and IGF-binding protein-3 levels are increased in central precocious puberty: effects of low different treatment regimens with gonadotropin-releasing hormone agonist, without or in combination with an antiandrogen (cyproterone acetate). J Clin Endocrinol Metab 1995;80:3059–67.
25. Faria ACS, Bekenstein LW, Booth RA Jr, Vaccaro VA, Asplin CM, Veldhuis JD, et al. Pulsatile growth hormone release in normal women during the menstrual cycle. Clin Endocrinol 1992;36:591–6.
26. Ho KY, Evans WS, Blizzard RM, Veldhuis JD, Merriam GR, Samojlik E, et al. Effects of sex and age on the 24-hour profile of growth hormone secretion in man: importance of endogenous estradiol concentrations. J Clin Endocrinol Metab 1987;64:51–8.
27. Ross JL, Cassorla FG, Skerda MC, Valk IM, Loriaux DL, Cutler GB Jr. A preliminary study of the effect of estrogen dose on growth in Turner's syndrome. N Engl J Med 1983;309:1104–6.
28. Clemmons DR, Underwood LE, Ridgway EC, Kliman B, Kljelberg RN, Van Wyk JJ. Estradiol treatment of acromegaly, reduction of immunoreactive somatomedin-C and improvement in metabolic status. Am J Med 1980;69:571–5.
29. Meyer WJ, Furlanetto RW, Walker PA. The effects of sex steroids on radioimmunoassayable plasma somatomedin-C concentrations. J Clin Endocrinol Metab 1982;55:1184–7.
30. Copeland K, Johnson D, Kuehl T, Castracane D. Estrogen stimulates growth hormone and somatomedin-C in castrate and intact female baboons. J Clin Endocrinol Metab 1984;58:698–702.

31. Martínez A, Heinrich JJ, Domené H, Escobar ME, Jasper H, Montuori E, et al. Growth in Turner's syndrome: long term treatment with low dose ethinyl estradiol. J Clin Endocrinol Metab 1987;65:253–7.
32. Mauras N, Rogol A, Veldhuis J. Specific time-dependent actions of low-dose ethinyl estradiol administration on the episodic release of growth hormone, follicle-stimulating hormone, and luteinizing hormone in prepubertal girls with Turner's syndrome. J Clin Endocrinol Metab 1989;69:1053–8.
33. Caruso-Nicoletti M, Cassorla F, Skerda M, Ross JL, Loriaux DL, Cutler GB Jr. Short term, low dose estradiol accelerates ulnar growth in boys. J Clin Endocrinol Metab 1985;61:896–8.
34. Ross JL, Long LM, Loriaux DL, Cutler GB Jr. Growth hormone secretory dynamics in Turner's syndrome. J Pediatr 1985;106:202–6.
35. Zadik Z, Landau H, Chen M, Altman Y, Lieberman E. Assessment of growth hormone (GH) axis in Turner's syndrome using 24-hour integrated concentrations of GH, insulin-like growth factor-I, plasma GH-binding activity, GH binding to IM9 cells, and GH response to pharmacological stimulation. J Clin Endocrinol Metab 1992;75(2):412–6.
36. Zachmann M, Prader A, Sobol EM, Crigler JF Jr, Ritzen EM, Atarés M, et al. Pubertal growth in patients with androgen insensitivity: indirect evidence for the importance of estrogens in pubertal growth of girls. J Pediatr 1986;108(5):694–7.
37. Cicognani A, Cacciari E, Tacconi M, Pascucci MG, Tonioli S, Pirazzoli F, et al. Effect of gonadectomy on growth hormone, IGF-I and sex steroids in children with incomplete androgen insensitivity. Acta Endocrinol 1989;121:777–83.
38. Smith EP, Boyd J, Frank GR, Takahashi H, Cohen RM, Specker B, et al. Estrogen resistance caused by a mutation of the estrogen-receptor gene in a man. N Engl J Med 1994;331:1056–61.
39. Morishima A, Grumbach M, Simpson E, Fisher C, Qin K. Aromatase deficiency in male and female siblings caused by a novel mutation and the physiological role of estrogens. J Clin Endocrinol Metab 1995;80:3689–98.
40. Conte F, Grumbach M, Ito Y, Fisher R, Simpson E. A syndrome of female pseudohermaphrodism, hypergonadotropic, hypogonadism, and multicystic ovaries associated with missense mutations in the gene encoding aromatase (P450 arom). J Clin Endocrinol Metab 1994;78:1287–92.

2

Sex-Steroid–Growth Hormone Interactions During Normal Male Puberty

LISA D. HOUCHIN AND ALAN D. ROGOL

The neuroendocrine mechanisms involved in the initiation of the pubertal growth spurt epitomize the complexity of physiological interactions occurring during this stage of development. While secondary sexual characteristics develop as a result of the awakening of the hypothalamic–pituitary–gonadal (HPG) axis, the integrity of both the HPG and growth hormone–insulin-like growth factor-I (GH-IGF-I) axes is essential for the acceleration of height velocity during mid-puberty. Knowledge of the endocrine interactions involved during the growth spurt has come primarily from clinical models in which the integrity of one or both of the neuroendocrine axes is compromised. Although these interactions are still incompletely understood, many significant strides have been made recently in their characterization. This review summarizes some of these findings.

Physiological Changes During Normal Male Puberty

The development of secondary sexual characteristics is an easily recognizable physical indicator of the onset of puberty, and its underlying endocrinological mechanisms are well defined. The physical changes occurring in boys result from increased adrenal androgen production (adrenarche) in concert with augmented testosterone production from the testes. Although often occurring contemporaneously, adrenarche is physiologically distinct from the awakening of the HPG axis or gonadarche and usually precedes the other hormonal changes of puberty by 2 or more years. Physical signs of adrenarche, such as pubic hair development, however, are usually not present initially and generally follow the first visible signs of gonadarche in boys, namely, testicular enlargement and thinning of the scrotal sac (1). Typically between the ages of 11.5 and 12 years, the testicles begin to enlarge from prepubertal volumes of less than 4 ml to reach postpubertal volumes of 15 to 25 ml (2). The more commonly recognized markers of puberty in males—deepening of the voice, acne, and beard growth—appear later (3).

Accompanying the development of secondary sexual characteristics is the pronounced acceleration of linear growth known as the pubertal growth spurt. Longitudinal growth proceeds rapidly during infancy and then slows at approximately age 3–5 years to a constant rate of 5–6 cm per year until puberty (4). At puberty, the growth rate accelerates significantly to reach a peak velocity during mid- to late adolescence. This peak height velocity occurs later in pubertal development in boys (Tanner genital stage IV–V) than in girls (Tanner breast stage II–III) (3,5). Girls average a pubertal peak height velocity of 9 cm per year at age 12 (5), whereas boys reach a pubertal peak height velocity of 10.3 cm per year at an average age of 14 years (3). The linear growth rate then slows until it ceases at the time of epiphyseal fusion.

The physical changes of puberty become manifest during a period of maximal activation of both the HPG and somatotrophic axes. With the availability of more sensitive gonadotropin assays (6), it is clear that the HPG axis is active throughout childhood. Before puberty, gonadotropin concentrations remain low, presumably because of the exquisite sensitivity of the axis to negative feedback from the low levels of sex steroids produced by the prepubertal gonads (7). At least 2 years before the clinical onset of puberty, apparent disinhibition at the hypothalamic–pituitary level allows mean luteinizing hormone (LH) concentrations to rise, initially only at nighttime (8–10). Early-morning testosterone levels rise (11) and testicular size increases (Fig. 2.1). As the HPG axis continues to mature, LH secretion increases throughout the

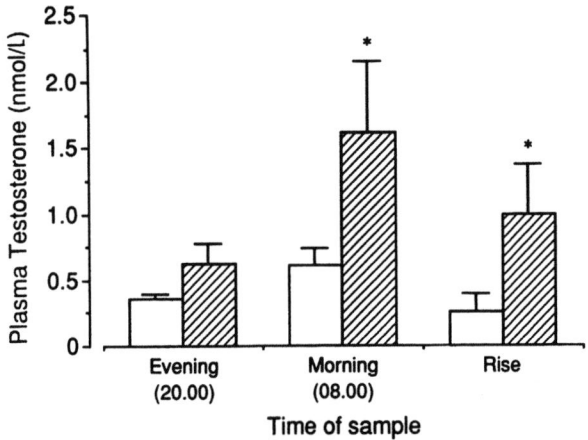

FIGURE 2.1. Initial plasma testosterone (T) concentration at 2000 and 0800 and overnight T increase (morning–evening concentration) in 45 prepubertal boys with constitutional delayed growth or puberty (defined as the attainment of testicular volume ≥ 4 ml) (*hatched columns*; n = 14) and those who remained clinically prepubertal (*open columns*; n = 31) during the subsequent 12 months of follow-up. *Asterisk*, $p < 0.005$ pubertal vs. prepubertal, Student's two-tailed unpaired *t*-test. Reproduced with permission from Wu et al. (11).

day as well, resulting in a more sustained serum concentration of testosterone that eventually reaches the concentrations found in young adult males (12).

Like the HPG axis, the GH–IGF-I axis also undergoes activation at the onset of puberty. Prepubertally, secretion patterns of GH in both genders show a marked diurnal rhythm with maximal secretion during sleep (13). At puberty, GH pulse amplitude increases nocturnally (14–17). Clinically, mean GH concentrations rise in concert with the onset of the acceleration of height velocity (15,18), suggesting GH as a major mediator of the growth spurt. After reaching peak height velocity, GH concentrations decrease gradually to levels found prepubertally. Approximate entropy (ApEn) serves as indirect estimate of feedback control within an axis. Mean ApEn of GH secretion increases in boys evaluated cross-sectionally during the growth spurt to a maximum in mid-to-late puberty, coincident with the peak height velocity, presumably because of activation of the somatotrophic axis in the milieu of increasing IGF-I and androgen concentrations (19) (Fig. 2.2). Due to their cross-sectional nature,

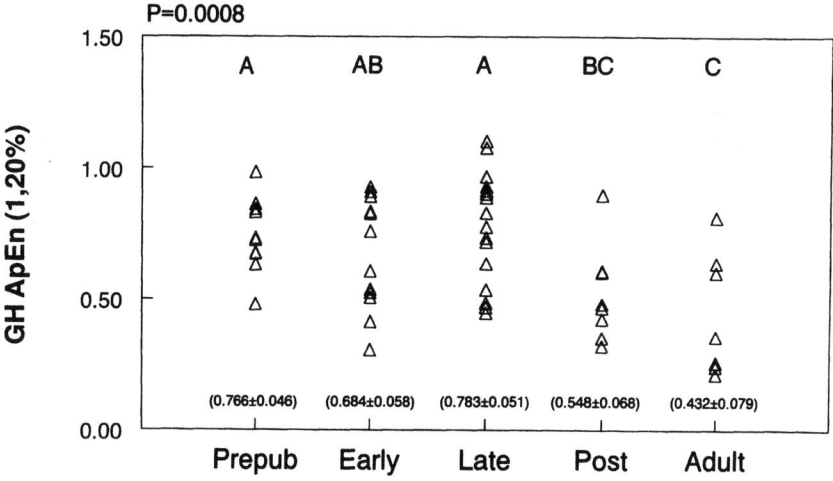

FIGURE 2.2. Approximate entropy (*ApEn*) quantification of the orderliness or pattern regularity within 24-h serum GH concentration time series (measured by IRMA) obtained in a total of 53 boys and young men sampled every 20 min for 24 h. A scale-invariant and model-free statistic, normalized ApEn (1, 20%), was calculated here at an m (window length) value of 1 and an r (tolerance) value of 20% of the SD of each subject GH series. Higher ApEn values denote greater disorderliness or irregularity of GH release over time. By ANOVA and Duncan's multiple-comparison test, significant group mean differences are denoted by nonidentical alphabetic superscripts ($p = 0.008$). Mean ± SEM ApEn values are given in parentheses below the individual data. Estrogen and testosterone, but not a nonaromatizable androgen, direct network integration of the hypothalamo-somatotrope (growth hormone)–insulin-like growth factor I axis in the human: evidence from pubertal pathophysiology and sex-steroid hormone replacement. Reproduced with permission from Veldhuis et al. (38).

these results do not permit distinction between the independent and synergistic roles of the GH–IGF-I or HPG axis in the pubertal growth spurt, nor do they reveal information on the mechanisms of interaction. They do, however, strongly imply that testosterone, GH, and their respective neuroendocrine axes are integral to the normal pubertal growth spurt.

Interactions of the Growth Hormone and Sex–Steroid Axes in the Pubertal Growth Spurt

An intact somatotrophic axis is a requirement for a normal pubertal growth spurt. Children with isolated GH deficiency undergo pubertal development after attaining an appropriate stage of skeletal maturation (19), although at a delayed chronological age compared to their normal peers. Once begun, sexual development progresses relatively normally (20); however, adult height is compromised without adequate GH replacement (21). Pubertal growth rates in these children are slowed if adequate exogenous GH replacement is not provided (22) or is withheld for short periods of time (23), despite normal serum gonadal steroid concentrations. Likewise, children with complete GH insensitivity (e.g., the Laron syndrome) or GH gene deletion demonstrate a pubertal growth spurt concurrent with rising testosterone levels; they do not achieve normal pubertal growth rates, even though they attain full sexual maturation (24,25). Thus, a functional GH–IGF-I axis must be present to ensure a completely normal growth spurt during puberty.

The sexual dimorphism of growth during adolescence implies a role for gonadal steroids in determining the initiation as well as character of the growth spurt. As stated earlier, boys reach their peak height velocity at a later chronological age and maturational stage than girls (3,5). Interestingly, genetic males with complete androgen-receptor insensitivity demonstrate a growth spurt resembling the female pattern in timing and magnitude (26). On serial sampling, girls have a greater sum of GH peak heights than do boys of the same maturational stage, and the dimorphism in GH release most closely correlates with bone age and gender (14). Similarly, premenopausal women have greater GH release than do men (27,28), presumably secondary to their higher estrogen level. Circulating estrogen concentrations differ significantly between prepubertal girls and boys when measured using ultrasensitive bioassay techniques (29), although the difference was previously undetectable using conventional assays. These findings further support the role of sex steroids in determining the timing and character of the pubertal growth spurt.

Gonadal steroids appear to mediate the pubertal growth spurt through augmentation of GH and IGF-I release and action. Administration of low-dose androgens or estrogen before GH stimulation testing is well known to increase GH secretion (30,31). The increases in GH and IGF-I that occur during the course of normal male puberty (13,32,33) appear to depend directly on testosterone, because GH and IGF-I concentrations in boys with constitutional delay increase from prepubertal levels to those more appropriate for chronological age after testosterone administration (34,35). In addition, testosterone therapy does not significantly increase somatomedin C levels in prepubertal boys with isolated GH deficiency unless they are also receiving GH replacement (34). The concerted activation of the somatotrophic and HPG axes at puberty is transient, because GH and IGF-I levels decline during late puberty and adulthood (13), despite sustained elevated testosterone levels.

Testosterone is postulated to stimulate the GH–IGF-I axis through estrogenic receptors following its peripheral aromatization. Supporting this hypothesis, tamoxifen (an estrogen-receptor blocker) decreases mean 24-h GH concentrations without altering LH or testosterone levels when given to normal late-pubertal boys (36). Similarly, flutamide (an androgen-receptor blocker) increases LH and free testosterone levels, in addition to increasing mean 24-h GH concentrations in normal boys in late puberty (37). Administration of the nonaromatizable androgen, dihydrotestosterone (DHT), fails to increase ApEn in the somatotrophic system in boys with constitutional delay of puberty (CDGA) (38), further supporting the hypothesis that testosterone mediates GH release during male puberty through an estrogenic mechanism. These issues are discussed in greater detail elsewhere in this volume (see Chapter 10).

In contrast, some data suggest that androgens mediate growth independently of estrogenic receptors. Dihydrotestosterone, when given short term to boys with CDGA, does not significantly increase serum GH or IGF-I concentrations, although height velocity increases to the same extent as with testosterone therapy (32,39), presumably through stimulation of long-bone growth via androgenic receptors. Both DHT and testosterone stimulate human osteoblastic proliferation and differentiation in vitro, and androgen receptor-blocking agents inhibit the mitogenic effects of DHT (40). One longitudinal study of boys with idiopathic hypogonadotropic hypogonadism (IHH) showed that small therapeutically induced increases in circulating testosterone concentrations, mimicking those seen in early puberty, increase GH secretion in the absence of concomitant elevations in estrogen levels as measured by conventional assays (41). With gradual, incremental increases in serum testosterone concentrations, the GH secretory burst mass substantially increased without significant changes in burst frequency or duration, mirroring the changes seen during normal male puberty (15). ApEn of GH secretion also increased (Fig. 2.3).

Additionally, in the foregoing study (41), treatment with testosterone did not enhance the GH response to stimulation by GH-releasing hormone (GHRH) when administered alone; however, GH release was enhanced when GHRH was administered with L-arginine, a presumed inhibitor of endogenous somatostatin secretion (42). To explain this observation, the investigators hypothesized that testosterone stimulates both GHRH and somatostatin release from the hypothalamus in addition to stimulating somatotropes and that these additive effects are evident only when somatostatin release is suppressed by

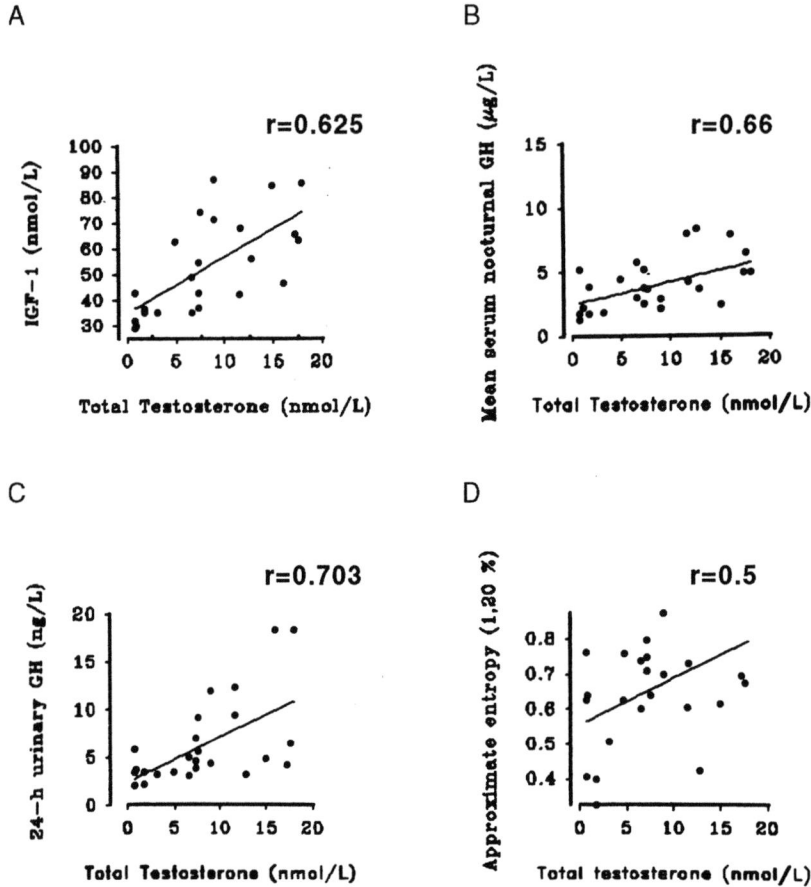

FIGURE 2.3A–D. Significant ($p < 0.05$) linear relationships between serum total testosterone concentration measured at each stage of treatment and serum IGF-I level (*A*), mean nocturnal GH concentration (*B*), 24-h urinary GH (*C*), and ApEn of nocturnal GH secretion (*D*). The *r* values are the correlation coefficients. (Reproduced with permission from Giustina et al. (41).

L-arginine. Thus, increases in circulating testosterone concentrations comparable to those found during early puberty are capable of modulating GH pulse amplitude and mass without measurable changes in circulating estrogen. While these findings are consistent with an androgenic mechanism of GH activation by testosterone, the authors acknowledge that the ability to detect biologically relevant changes in estrogen concentration may have been limited by the sensitivity of the estrogen assay used.

In summary, puberty is a unique period of development, marking the transition between childhood and adulthood. Puberty entails dramatic changes in linear growth and sexual differentiation that occur in the midst of increasing activation of the HPG and somatotrophic axes. A normal pubertal growth spurt requires these axes to be functionally intact and finely integrated. The synergism between GH and testosterone during puberty is transient as GH concentrations decrease after epiphyseal fusion, even though circulating testosterone levels remain elevated. A major site of interaction between GH and testosterone during the growth spurt appears to be the hypothalamus and pituitary, where increasing testosterone concentrations augment GH secretion. Whether this activation occurs solely through estrogenic receptors or in concert with androgenic receptors remains to be determined.

Acknowledgments. This work was supported in part by the following grants: General Clinical Research Center at the University of Virginia M01 RR00847 (A.D.R.), National Science Foundation Center for Biological Timing (A.D.R.), Diabetes and Hormone Action Training Grant T32 DK07320 (L.D.H.), and the Genentech Foundation Fellowship Grant (L.D.H.).

References

1. Parker LN. Adrenarche. Endocrinol Metab Clin 1991;20(1):71–81.
2. Marshall WA. Growth and sexual maturity in normal puberty. Clin Endocrinol Metab 1975;4:3–25.
3. Marshall WA, Tanner JM. Variations in patterns of pubertal changes in boys. Arch Dis Child 1970;45:13–23.
4. Tanner JM. Foetus into man: physical growth from conception to maturity. Cambridge: Harvard University Press, 1978:6.
5. Marshall WA, Tanner JM. Variations in patterns of pubertal changes in girls. Arch Dis Child 1969;44:291–303.
6. Pandian MR, Odell WD, Carlton E, Fisher DA. Development of third-generation immunochemiluminometric assays of follitropin and lutropin and clinical application in determining pediatric reference ranges. Clin Chem 1993;39:1815–9.
7. Beitins IZ, Padmanabhan V. Bioactivity of gonadotropins. Endocrinol Metab Clin 1991;20(1):85–120.
8. Wu FCW, Butler GE, Kelnar CJH, Sellar RE. Patterns of pulsatile luteinizing hormone secretion before and during the onset of puberty in boys: a study using an immunoradiometric assay. J Clin Endocrinol Metab 1990;70:629–37.
9. Jakacki RI, Kelch RD, Sounder Lloyd JS, Hopwood JN, Marshall JC. Pulsatile secretion of luteinizing hormone in children. J Clin Endocrinol Metab 1982:55:453–8.
10. Burr IM, Sizonenko PC, Kaplan SL, Grumbach MM. Hormonal changes in puberty. I. Correlation of serum luteinizing hormone and follicle stimulating hormone with stages of puberty, testicular size, and bone age in normal boys. Pediatr Res 1970:4:25–35.
11. Wu FC, Brown DC, Butler GE, Stirling HF, Kelnar CJ. Early morning plasma testosterone is an accurate predictor of imminent pubertal development in prepubertal boys. J Clin Endocrinol Metab 1993;76:26–31.
12. Styne DM. Puberty and its disorder in boys. Endocrinol Metab Clin 1991;20(1):43–69.

13. Martha PM Jr, Rogol AD, Veldhuis JD, Kerrigan JR, Goodman DW, Blizzard RM. Alterations in the pulsatile properties of circulating growth hormone concentrations during puberty in boys. J Clin Endocrinol Metab 1989;69:563–70.
14. Roemmich JN, Clark PA, Mai V, Berr SS, Weltman A, Veldhuis JD, et. al. Alterations in growth and body composition during puberty: III. Influence of maturation, gender, body composition, fat distribution, aerobic fitness, and energy expenditure on nocturnal growth hormone release. J Clin Endocrinol Metab 1998;83:1440–7.
15. Martha PM Jr, Gorman KM Jr, Blizzard RM, Rogol AD, Veldhuis JD. Endogenous growth hormone secretion and clearance rates in normal boys, as determined by deconvolution analysis: relationship to age, pubertal status and body mass. J Clin Endocrinol Metab 1992;74:336–44.
16. Costin G, Kaufman FR, Brasel JA. Growth hormone secretory dynamics in subjects with normal stature. J Pediatr 1989;115:537–44.
17. Rose SR, Municchi G, Barnes KM, Cutler GB Jr. Spontaneous growth hormone secretion increases during puberty in normal girls and boys. J Clin Endocrinol Metab 1991;73:428–35.
18. Blizzard RM, Thompson RG, Baghdassarian A, Kowarski A, Migeon CJ, Rodriquez A. The interrelationship of steroids, growth hormone and other hormones on pubertal growth. In: Grumbach MM, Grave GD, Mayer FE, eds. Control of the onset of puberty. New York: Wiley, 1974:342.
19. Tanner JM, Whitehouse RH. A note on the bone age at which patients with true isolated growth hormone deficiency enter puberty. J Clin Endocrinol Metab 1975;41:788–90.
20. Burns EC, Tanner JM, Preece MA, Cameron N. Final height and pubertal development in 55 children with idiopathic growth hormone deficiency treated for between 2 and 15 years with human growth hormone. Eur J Pediatr 1981;137:155–64.
21. Rimoin DL, Merimee TJ, Rabinowitz D, McKusick VA. Genetic aspects of clinical endocrinology. Rec Prog Horm Res 1968;24:365–437.
22. Brasel JA, Wright JC, Wilkins L, Blizzard RM. Evaluation of seventy-five patients with hypopituitarism beginning in childhood. Am J Med 1965;38:484–98.
23. Tanner JM, Whitehouse RH, Hughes PC, Carter BS. Relative importance of growth hormone and sex steroids for the growth at puberty of trunk length, limb length and muscle width in growth hormone-deficient children. J Pediatr 1976;89:1000–8.
24. Laron Z, Sarel R, Pertzelan A. Puberty in Laron type dwarfism. Eur J Pediatr 1980;134:79–83.
25. Rivarola MA, Phillips JA III, Migeon CJ, Heinrich JJ, Hjelle BJ. Phenotypic heterogeneity in familial isolated growth hormone deficiency type I-A. J Clin Endocrinol Metab 1984;59:34–40.
26. Zachmann M, Prader A, Sobel EH, Crigler JF Jr, Ritzen EM, Atares M. Pubertal growth in patients with androgen insensitivity: indirect evidence for the importance of estrogens in pubertal growth in girls. J Pediatr 1986;108:694–7.
27. Ho KY, Evans WS, Blizzard RM, Veldhuis JD, Merriam GR, Samojlik E, et al. Effects of sex and age on the 24-hour profile of growth hormone secretion in man: importance of endogenous estradiol concentrations. J Clin Endocrinol Metab 1987;64:51–8.
28. Van den Berg G, Veldhuis JD, Frolich M, Roelfsema F. An amplitude-specific divergence in the pulsatile mode of GH secretion underlies the gender difference in mean GH concentrations in men and premenopausal women. J Clin Endocrinol Metab 1996;81:2460–6.

29. Klein KO, Baron J, Colli MJ, McDonnell DP, Cutler GB Jr. Estrogen levels in childhood determined by an ultrasensitive recombinant cell bioassay. J Clin Invest 1994;94:2475–80.
30. Moll GW Jr, Rosenfield RL, Fang VS. Administration of low dose estrogen rapidly and directly stimulates growth hormone production. Am J Dis Child 1986;140:124–7.
31. Martin LG, Clark JW, Connor TB. Growth hormone secretion enhanced by androgens. J Clin Endocrinol Metab 1968;28:425–8.
32. Keenan BS, Richards GE, Ponder SW, Dallas JS, Nagamani M, Smith ER. Androgen-stimulated pubertal growth; the effects of testosterone and dihydrotestosterone on growth hormone and insulin-like growth factor-I in the treatment of short stature and delayed puberty. J Clin Endocrinol Metab 1993;76:996–1001.
33. Cara JF, Rosenfield RL, Furlanetto RW. A longitudinal study of the relationship of plasma somatomedin-C concentration to the pubertal growth spurt. Am J Dis Child 1987;141:562–4.
34. Parker MW, Johanson AJ, Rogol AD, Kaiser DL, Blizzard RM. Effect of testosterone on somatomedin-C concentrations in pubertal boys. J Clin Endocrinol Metab 1984;58:87–90.
35. Rosenfield RL, Furlanetto RW. Physiologic testosterone or estradiol induction of puberty increases somatomedin-C. J Pediatr 1985;107:415–7.
36. Metzger DL, Kerrigan JR. Estrogen receptor blockade with tamoxifen diminishes growth hormone secretion in boys: evidence for a stimulatory role of endogenous estrogens during male adolescence. J Clin Endocrinol Metab 1994;79:513–8.
37. Metzger DL, Kerrigan JR. Androgen receptor blockade with flutamide enhances growth hormone secretion in late pubertal males: evidence for independent actions of estrogen and androgen. J Clin Endocrinol Metab 1993;76:1147–52.
38. Veldhuis JD, Metzger DL, Martha PM Jr, Mauras N, Kerrigan JR, Keenan B, et al. Estrogen and testosterone, but not a nonaromatizable androgen, direct network integration of the hypothalamo-somatrope (growth hormone)-insulin-like growth factor I axis in the human: evidence from pubertal pathophysiology and sex-steroid hormone replacement. J Clin Endocrinol Metab 1997;82:3414–20.
39. Link K, Blizzard RM, Evans WS, Kaiser DK, Parker MW, Rogol AD. The effect of androgens on the pulsatile release and the twenty-four-hour mean concentration of growth hormone in peripubertal males. J Clin Endocrinol Metab 1986;62:159–64.
40. Kasperk CH, Wergedal JE, Farley JR, Linkhart TA, Turner RT, et al. Androgens directly stimulate proliferation of bone cells in vitro. Endocrinology 1989;124:1576–8.
41. Giustina A, Scalvini T, Tassi C, Desenzani P, Poiesi C, Wehrenberg WB, et al. Maturation of the regulation of growth hormone secretion in young males with hypogonadotropic hypogonadism pharmacologically exposed to progressive increments in serum testosterone. J Clin Endocrinol Metab 1997;82:1210–9.
42. Giustina A, Veldhuis JD. Pathophysiology of the neuroregulation of growth hormone secretion in experimental animals and the human. Endocr Rev 1998;19:717–97.

3

Impact of Testosterone Replacement on the Maturation of the Growth Hormone–IGF-I Axis

ANDREA GIUSTINA, PAOLO DESENZANI, AND TIZIANO SCALVINI

The regulated mode of growth hormone (GH) secretion is sexually dimorphic in both animals (1) and humans (2–6). During human pubertal development, there is preferential augmentation of the amplitude of spontaneous GH pulses, with a subsequent return to or a fall below prepubertal values in early adulthood (2). Some clinical data indicate a sex difference in the timing of these physiological changes in activity of the somatotropic axis (3). Moreover, a large body of evidence has been accumulated showing significant sex differences in GH responses to various pharmacological stimuli in young adults (4). However, the majority of data on sex hormone-mediated regulation of spontaneous and stimulated GH secretion in the human has been derived from cross-sectional population studies including healthy subjects of normal stature representing one or more stages of pubertal development (2,5,6). In relation to the specific actions of testosterone on GH secretion, these previous experiments have suggested an augmentation of the calculated maximal rate of GH release attained per secretory episode, resulting in a greater mass of GH released per secretory burst (6). A positive correlation between the magnitude of plasma GH responses to some provocative stimuli and androgen concentrations in human subjects (6) has also been suggested by cross-sectional studies.

The time course and role of androgens in the heightened GH response to physiological and pharmacological stimuli remain to be clarified on a longitudinal basis in the same male subject during puberty, as does the nature of the specific actions of testosterone on the development of GH responsiveness to hypothalamic growth hormone-releasing factor (GHRH) and somatostatin acting alone or in combination. To address these aims, we performed a longitudinal study of nocturnal pulsatile GH secretion and GH secretory responses to GHRH alone or combined with L-arginine (a functional somatostatin antagonist) (7) in previously untreated young male patients with isolated hypogonadotropic hypogonadism who were exposed to progressively higher testosterone levels designed to mimic the androgen environment recognized during the early stages of puberty.

Patients and Methods

Patients

Six previously untreated boys were studied, all of whom were referred to our Endocrine Section for delayed sexual development and were subsequently diagnosed with isolated hypogonadotropic hypogonadism (IHH) (failure of pubertal onset by age 15 years; low serum immunoreactive luteinizing hormone [LH] and follicle-stimulating hormone [FSH] concentrations in the presence of low serum testosterone levels), normal pituitary-thyroid and pituitary-adrenal function, and normal magnetic resonance imaging of the sella and parasellar structures. Furthermore, during the 6 months following the end of the study, none of the subjects apparently entered spontaneous puberty (7).

Experimental Design

All patients received three incremental doses of testosterone enanthate (Testoenant, Geymonat, Italy), namely, 25, 50, and 100 mg i.m. injections; each dose was administered three times during a period of 4 weeks (1 injection every 2 weeks). Studies of GH secretion (nocturnal and stimulated) were performed four times: before testosterone administration (T1), and on the sixth day after the third administration of each dose of testosterone (T2, T3, and T4) (1).

Nocturnal GH Secretion

The study was performed according to a protocol previously described (8). Briefly, the first blood sample was withdrawn for GH assay at 2200. Subsequent blood samples were obtained at 20-min intervals until 0600 on the following day. Lights were turned off at 2230 and turned on at 0600. Sleep was evaluated by visual inspection at 20-min intervals. Baseline blood samples for serum total and free testosterone, 17-β-estradiol, sex hormone-binding protein (SHBG), IGF-I, and IGF-BP-3 assays were also drawn.

Stimulated GH Secretion

GHRH and L-arginine stimulation studies (7) were also repeated four times (before testosterone administration and the seventh and eighth days after the third administration of each testosterone dose) in a single-blind crossover design. Briefly, all subjects underwent the following two tests in random order and in consecutive days: (1) an infusion of L-arginine hydrochloride (Damor, Naples, Italy; 30 g, i.v. in 100 ml saline), or (2) an infusion of saline (100 ml, i.v.) from −30 to 0 min. On both occasions, at 0 min the subjects received an i.v. bolus injection of human GHRH-(1,29)NH_2 (Geref, Serono, Italy; 100 µg in 1 ml saline). Blood samples for GH assays were withdrawn at −45, −30 (time of the start of saline or arginine infusion), 0 (time of GHRH injection), 15, 30, 45, 60, 90, and 120 min.

Assays

A commercial immunoradiometric assay (IRMA) was used for estimation of serum GH concentrations (Allegro hGH: Nichols Institute, San Juan Capistrano, CA; inter- and intraassay coefficients of variation, 5.4% and 2.3%, respectively; sensitivity limit of the assay, 0.06 µg/l). Commercial radioimmunoassays were used for measurement of total IGF-1 (Nichols Institute; after acid–ethanol extraction: inter- and intraassay coefficients of variation 5.2% and 9.4%, respectively; sensitivity limit of the assay 0.013 nmol/l), IGF-BP-3 (Nichols Institute: inter- and intraassay coefficients of variation, 5.8% and 5.6%, respectively; sensitivity limit of the assay, 0.06 µg/ml), total testosterone (Diagnostic Product Corporation [DPC], Los Angeles, CA; inter- and intraassay coefficients of variation, 7.3% and 5%, respectively; sensitivity limit of the assay, 0.14 nmol/l), free-testosterone (DPC; inter- and intraassay coefficients of variation, 3.2% and 3.4%, respectively; sensitivity limit of the assay 0.52 pmol/l), 17-β-estradiol (DPC; inter- and intraassay coefficients of variation, 4.9% and 4.5%, respectively; sensitivity limit of the assay, 18.4 pmol/l), SHBG (Radim, Liege, Belgium; inter- and intraassay coefficients of variation, 4.6% and 5.6%, respectively; sensitivity limit of the assay, 2.5 nmol/l).

Data Analysis

Quantitative characteristics of pulsatile GH secretion basally and following administration of secretagogues were assessed assuming a simple burst model of hormone secretion using multiparameter deconvolution analysis (9,10). The serum GH concentration responses to saline plus GHRH and L-arginine plus GHRH also were deconvolved (11), and additionally expressed as mean concentrations (µg/l) and peak (maximum) serum values (µg/l). Differences among basal and stimulated mean GH pulse properties within and between subjects were detected by analysis of variance (ANOVA) with a repeated-measures design. Correlations were assessed by applying univariate linear regression analysis. Data are expressed as means ± SEM. $p < 0.05$ was considered statistically significant. Approximate entropy is a recently introduced model-independent regularity statistic, which is designed to estimate the relative degree of orderliness of serial observations (12). This analysis is designed to quantify the logarithmic likelihood that runs of patterns (consecutive data values) of length m will be repeated within a tolerance of r. Here, we used $m = 1$ and $r = 0.20$ times the within-series (8-h GH profile) SD to normalize approximate entropy. This normalized version is scale-invariant and is designated as ApEn (1,20%).

Results

Serum Testosterone

Before treatment (T1), serum testosterone concentrations (mean, 1.2 ± 0.13 nmol/l) were low in all subjects. Serum testosterone levels increased in all subjects during therapy. The mean of all total testosterone measurements during the treatment periods were 4.0 ± 0.87 nmol/l, T2 ($p < 0.05$ vs. baseline); 7.1 ± 1.1 nmol/l, T3 ($p < 0.05$ vs. T2); and 11.6 ± 0.58 nmol/l, T4 ($p < 0.05$ vs. T3). The highest total serum testosterone concentrations recorded during the various phases of treatment were 5.8 ± 1.04 nmol/l, T2 ($p < 0.05$ vs. baseline); 9.5 ± 1.0 nmol/dl, T3 ($p < 0.05$ vs. T2); and 1.6 ± 1.0 nmol/l, T4 ($p < 0.05$ vs. T3) (Fig. 3.1). The highest free serum testosterone levels measured during the phases of treatment were 16 ± 3.6 pmol/l, T2 ($p < 0.05$ vs. baseline); 25 ± 5.8 pmol/l, T3 ($p < 0.05$ vs. T2); and 46 ± 4.1 pmol/l, T4 ($p < 0.05$ vs. T3).

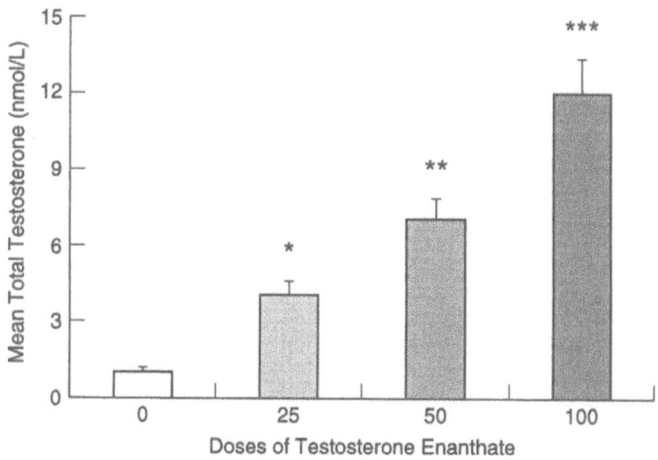

FIGURE 3.1. Serum levels of mean total testosterone measured throughout the graded testosterone-repletion study. *, $p < 0.05$ vs. baseline (zero dose); **, $p < 0.05$ vs. 25 mg testosterone enanthate; ***, $p < 0.05$ vs. 50 mg testosterone enanthate.

17-β Estradiol and SHBG

Before treatment (T1), and after the 25-mg dose of testosterone replacement (T2), serum estradiol was undetectable (< 18.4 pmol/l) in all subjects. Serum 17-β estradiol levels became detectable only in two of the subjects after 50 mg testosterone administration (T3) (29 and 33 pmol/l, respectively). Estradiol levels were detectable in all but one of the patients after the last stage of treatment (T4): mean 42 ± 9.5 pmol/l ($p < 0.05$ vs. T3, T2, and T1). T1 SHBG levels were 27 ± 13 nmol/l. A progressive decrease in SHBG levels was observed during the various phases of treatment, which was not statistically significant at T2 (27 ± 9 nmol/l), but was significant at T3 (20 ± 10 nmol/l, $p < 0.05$ vs. T1 and T2) and at T4 (17 ± 9.5 nmol/l, $p < 0.05$ vs. T1 and T2). In three of the subjects, SHBG levels were undetectable at the end of the study.

Spontaneous GH Secretion

Mean nocturnal serum GH concentrations were higher after the 25-mg dose of testosterone compared to pretreatment values (2.64 ± 0.6 vs. 3.72 ± 0.4 µg/l, $p < 0.05$). The further increases in the mean nocturnal serum GH concentration values (µg/l) observed at the other treatment steps (T3, 4.19 ± 0.9; T4, 5.15 ± 0.77) were significantly higher than T1 but did not achieve statistical significance compared to T2 (Fig. 3.2). Deconvolution analysis revealed that the 8-h (overnight) GH secretion rate was increased by testosterone treatment. In particular, the mass of GH secretory bursts was clearly greater during the first stage of treatment (T2) compared with the values before treatment (T1) ($p < 0.05$); no further statistical increases in burst mass were observed at T3 and T4 stages of treatment. Also, the mean amplitude (maximal rate) of GH

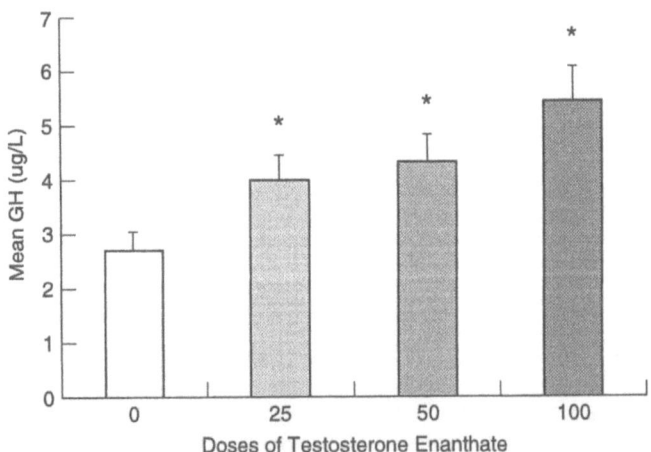

FIGURE 3.2. Serum levels of mean nocturnal spontaneous GH release throughout the study. *, $p < 0.05$ vs. baseline (zero dose).

secretory pulses was greater in the T3 and T4 stages compared to T1 and T2 ($p < 0.05$).

A small but significant rise in the mean number of detectable GH secretory pulses occurred concurrently with the increase in pulse amplitude. These amplitude and mass changes were specific, because GH half-life and the estimated secretory half-duration (duration at half-maximal amplitude) did not change. The total mass of pulsatile GH secretion rose significantly (versus pretreatment) during stage T2 of treatment without further changes at higher treatment doses. Mean values of approximate entropy rose significantly ($p < 0.05$), indicating increased disorderliness or irregularity of the GH release process at T2 (25 mg testosterone) and higher doses of testosterone (50 and 100 mg) versus T1 (preandrogen treatment). Mean values were not significantly different from one another at T2, T3, and T4 stages of treatment.

Stimulated GH Secretion

The mean pre-GHRH injection serum GH levels were not significantly different at the four stages of the study. In fact, infusion of L-arginine alone did not increase mean time zero serum GH levels in a similar fashion across the four different stages of testosterone treatment.

The mean preandrogen-treatment (T1) GH response to saline plus GHRH (GH peak, 13 ± 5.1 µg/l) was not significantly different compared to that observed during any of the three doses of testosterone treatment (GH peaks: T2, 16 ± 1.8 µg/l; T3, 16 ± 4.2 µg/l; T4, 11 ± 1.2 µg/l). Peak serum GH concentrations pretreatment (T1) occurred 15–60 min after GHRH injection in all patients. Testosterone administration did not change the timing of GH peaks significantly or alter the mean absolute serum GH levels or deconvolution-estimated GH secretion after GHRH. L-arginine pretreatment induced a significant increase in the GH responses to GHRH (compared to saline pretreatment), both at baseline (T1: GH peak, 32 ± 4.7 µg/l; $p < 0.05$ vs. saline plus GHRH) and during treatment. Specifically, testosterone significantly ($p < 0.05$) increased the GH responses to L-arginine plus GHRH compared to T1 during stage T3 (GH peak, 49 ± 5.7 µg/l) and T4 (GH peak, 49 ± 8.3 µg/l) of treatment. Deconvolution analysis revealed greatest calculated total mass of GH secreted after L-arginine plus GHRH at T3 and T4 compared to T1 stages. Mean absolute serum GH concentrations at 15, 30, and 45 min after L-arginine plus GHRH at stages T3 and T4 were also significantly higher than those observed at T1. The ratio between GH peaks after L-arginine plus GHRH versus saline plus GHRH was significantly higher at T3 and T4 as compared to the T1 stage. All subjects experienced facial flushing after GHRH injection. No side effects were observed during L-arginine infusion.

IGF-I and IGF-I BP-3

The plasma IGF-I concentrations for the six subjects before treatment (T1) and at the end of the study are summarized in Figure 3.3. Values during treatment were always greater than those before treatment (T1; 35 ± 6 nmol/l) and progressively

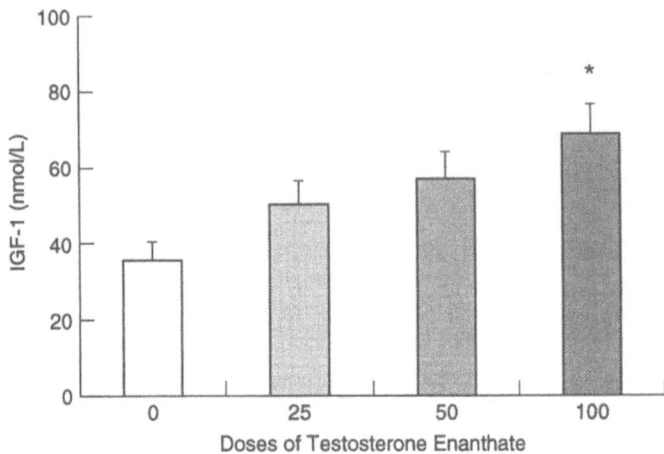

FIGURE 3.3. Serum levels of IGF-I measured during incremental testosterone treatment. *, $p < 0.05$ vs. baseline (zero dose).

increased to reach maximal values in all the subjects at the last stage of treatment (T2, 49 ± 6.67 nmol/l; T3, 58 ± 4.7 nmol/l; T4, 69 ± 6 nmol/l). Conversely, no significant changes in IGF BP-3 levels were observed during treatment: T1, 3.3 ± 0.2 µg/ml; T2, 3.6 ± 0.4 µg/ml; T3, 3.2 ± 0.3 µg/ml; T4, 3.5 ± 0.31 µg/ml.

Correlations

Significant ($p < 0.05$) positive linear correlations were found between mean nocturnal GH concentration ($r = 0.682$), GH secretory burst mass ($r = 0.504$), number of GH peaks ($r = 0.5$), GH approximate entropy ($r = 0.484$), 24-h urinary GH ($r = 0.703$), plasma IGF-I ($r = 0.459$), and free serum testosterone recorded for each subject at each of the stages of treatment. Serum IGF-I concentrations correlated significantly with mean 24-h serum GH levels and total mass of GH secreted in pulses ($r = 0.86$; $p < 0.05$). No significant correlation was found between serum testosterone levels and any parameters of saline plus GHRH-stimulated GH release in our study population. Peak serum GH concentrations achieved after L-arginine plus GHRH were positively correlated with peak total testosterone at a level approaching statistical significance ($r = 0.38$; $p = 0.067$). No significant correlations were found between any of the parameters of spontaneous or stimulated GH secretion and circulating 17-β estradiol levels or other measured sex steroids.

Discussion

We have demonstrated specific, prominent, and rapid actions of testosterone on the human somatotropic axis in a clinical model of male isolated hypogonadotropic hypogonadism previously unreplaced with androgen. A small (approximately threefold) increase in serum testosterone concentrations for only 6 weeks induced by the lowest dose (25 mg biweekly) of testosterone amplified GH secretory burst mass significantly without altering detectable GH secretory frequency, half-duration, or GH half-life, thus closely mimicking the GH response during normal puberty in boys (7). Mechanistically, testosterone also augmented pituitary GH secretion significantly after presumptive somatostatin withdrawal combined with GHRH (L-arginine plus GHRH) (13).

Sex-steroid hormones profoundly influence multiple endocrine axes and the GH axis in both the rat (13) and humans (6). In the human, previous reports have described an increase of 24-h plasma GH levels in both sexes during spontaneous puberty or exogenous sex-hormone treatment (2,3,5,6). As earlier studies were cross-sectional, large interindividual variations in baseline plasma GH concentrations could have obscured the influence of androgens or estrogens, and the studies could not clarify the time course of GH axis responses. Mechanistic investigations by deconvolution techniques demonstrated that the increase in peak serum GH concentrations resulted from an amplified and maximal GH secretory rate attained within each discrete release episode (5).

A previous study investigated the effects of parenterally injected testosterone in boys with constitutionally delayed puberty. In these subjects, testosterone administration doubled the height and area of serum GH concentration peaks because of a doubling or tripling of the mass of GH secreted per burst (14). These experimental observations in boys are consistent with the inference that androgen can positively regulate GH secretory burst mass or amplitude. Analogously, in a group of previously untreated young men with isolated gonadotropin deficiency (15), when mean plasma testosterone levels were normalized, there was a twofold rise in mean 24-h serum GH concentrations and GH pulse amplitude.

Our study gives insights into the maturation of mechanisms that control spontaneous pulsatile GH secretion in the young male because we modulated the GH axis by progressively increasing serum testosterone levels within the physiologically observed range in early puberty. Our data, obtained by frequent blood sampling and studied by means of deconvolution analysis, confirm that testosterone is able to significantly increase the amplitude and mass of spontaneous GH secretory events. Of considerable interest is our observation of augmented GH peak amplitude following 6 weeks of the lowest dose of testosterone (25 mg i.m. every 2 weeks). Indeed, blood concentrations of testosterone attained after this dose approximated only one-fifth to one-seventh those anticipated in healthy young men (14). Therefore, we hypothesized that GH secretory burst mass is a highly regulated fundamental unit of spontaneous GH secretion and serves as primary mechanism activated during the early stages of male puberty. Interestingly, the total mass of GH secreted was significantly increased compared to untreated levels even after the 4-week regimen of 25-mg testosterone injections.

There are conflicting reports (6) in the literature concerning the effects of testosterone on the frequency of spontaneous GH peaks. Our 20-min data suggest that both the 50- and the 100-mg dose of testosterone may increase the frequency of detectable spontaneous nocturnal GH peaks. Indeed, previous studies comparing different subjects in various pubertal stages may have been unable to detect transient changes in GH peak frequency in early puberty. Independently of possible changes in GH burst frequency, the present clinical model of low-dose testosterone administration unmasked a positive correlation between circulating total and free testosterone concentrations and total and individual GH secretory burst mass in the male.

Our data show, for the first time, that testosterone acts specifically to increase the disorderliness of the GH secretory process in humans. It could be hypothesized that this effect is simply a manifestation of an increase in GH secretion rather than a isolated or independent event. However, this hypothesis is unlikely because this parameter is not dependent on the hormone concentration per se; in fact, the approximate entropy statistic as normalized is independent of scale (12). Thus, loss of regularity of GH secretion can be hypothesized to be a distinct feature of pubertal development in males occurring in the very early stages of the process, possibly reflecting the maturation, in the sense of an increased activity, of the hypothalamic (GHRH–somatostatin interplay) drive on GH secretion determined by testosterone.

Our data show that while testosterone potently and specifically increases IGF-I levels, it does not change IGF-BP-3 levels over the same time course, even at the highest androgen dose used. Therefore, a testosterone-mediated increase in (predicted) biologically active IGF-I may be hypothesized to play a role in the synergism between testosterone and GH in the growth process (16), although we cannot exclude possible concomitant testosterone-mediated changes in other IGF-BPs, which might in turn influence the levels of circulating free IGF-I.

Based on available experimental data in the rat (13), one can hypothesize that the increase in GH peak amplitude is likely to be caused either by a testosterone-mediated increase in effective GHRH release and action (17) or by a direct sex-steroid effect at the pituitary level (18). In the human, evidence is available that sex steroids are among the factors that enhance GH release following stimulation by numerous pharmacological agents. Because L-arginine blocks glucocorticoid-induced GH suppression (19) and enhances the GH response to a maximal dose of GHRH (20), it has been hypothesized to stimulate GH secretion by inhibiting endogenous somatostatin secretion (19).

No data are available on changes in GH responses to either direct pituitary (GHRH) or hypothalamic (e.g., L-arginine) stimuli longitudinally during pubertal development. Our observations indicate that the GH response to GHRH is not significantly influenced by the increases in testosterone levels that concomitantly stimulate pulsatile GH secretion. In fact, both the amplitude of the peaks and the total mass of GH secreted following GHRH stimulation were similar in our group of hypogonadal males at the beginning and at the end of the study, when serum levels of testosterone typical of late puberty were attained. Conversely, when the GHRH test was performed after pretreatment with L-arginine, the higher doses of testosterone (50 and 100 mg) significantly enhanced the GH responses, suggesting that testosterone can modulate somatostatinergic inhibitory tone. In humans, the GH response to GHRH is maximal during infancy and becomes less pronounced with advancing adult age (21). This decreased responsiveness may be caused by an increase in somatostatinergic tone with age because a restoration of the GH response to GHRH can be achieved by cholinergic agonist pretreatment (22), which is thought to block somatostatin activity at the hypothalamic level (23).

In contrast, available indirect in vivo studies in the rat strongly suggest that testosterone may have a stimulating effect on hypothalamic somatostatin tone (24). Therefore, to reconcile available data in the rat with our clinical findings, we can hypothesize that testosterone does not change the GH response to GHRH alone in our patients because its direct stimulating actions at the pituitary level are obscured by an increase in hypothalamic somatostatin tone (19). To test this hypothesis, we performed an experiment investigating the effects of changes in circulating testosterone levels on a combined hypothalamic-pituitary stimulus of GH secretion. Our data clearly demonstrate that testosterone enhances the GH response to GHRH plus L-arginine in boys, and this effect became evident even with a 50-mg replacement dose of this hormone. The observation that, in the presence of L-arginine, an amino acid thought to decrease somatostatin tone (19), the GH response to GHRH is increased significantly during testosterone replacement, supports the overall hypothesis of a dual action of testosterone at the hypothalamic level (increase in GHRH and somatostatin), and does not exclude any additional direct GH-stimulating action at the pituitary level.

Aromatization to estrogen has been hypothesized to be important in testosterone-mediated stimulation of the GH axis in the human (25). In our study, the first changes in GH secretion were evident in the presence of very small increases in serum testosterone concentrations concomitantly with undectable circulating estrogen levels. Moreover, the various parameters of GH release correlated only with circulating testosterone. These findings argue for the unexpected thesis that the GH-stimulating action of testosterone occurs via the androgen receptor (26). Moreover, these data cannot rule out significant biological effects of small changes in circulating 17-β-estradiol concentrations undectable with the assay used by us or of intrahypothalamopituitary aromatization of testosterone. Nonetheless, testosterone-mediated changes in GH secretion occurred in the presence of decreasing serum levels of SHBG, which can be considered a reliable biological marker of peripheral activity of testosterone via the androgen receptor (27).

In conclusion, our longitudinal human studies in androgen-deficient boys suggested that the small increases in testosterone occurring during early puberty are able to promote an increase in the mass of GH secretory peaks, plasma IGF-I levels, and the irregularity of the GH release process. Mechanistically, our pharmacological data further suggested that testosterone may act directly or indirectly at the pituitary level to increase the GH secretory capacity of somatotropes, and also alone or via its metabolities at the hypothalamic level to augment the activity of both GHRH and somatostatin (8).

References

1. Eden S. Age- and sex-related differences in episodic growth hormone secretion in the rat. Endocrinology 1979;105:555–60.
2. Martha PM Jr, Rogol AD, Veldhuis JD, Kerrigan JR, Goodman DW, Blizzard RM. Alterations in the pulsatile properties of circulating growth hormone concentrations during puberty in boys. J Clin Endocrinol Metab 1989;69:563–70.
3. Rose SR, Municchi G, Barnes KM, Kamp GA, Uriate MM, Ross JL, et al. Spontaneous growth hormone secretion increases during puberty in normal girls and boys. J Clin Endocrinol Metab 1991;73:428–35.
4. Giustina A, Veldhuis JD. Pathophysiology of the neuroregulation of growth hormone secretion in experimental animals and the human. Endocr Rev 1998;19:717–97.
5. Martha PM Jr, Goorman KM, Blizzard RM, Rogol AD, Veldhuis JD. Endogenous growth hormone secretion and clearance rates in normal boys as determined by deconvolution analysis: relationship to age, pubertal status and body mass. J Clin Endocrinol Metab 1992;74:336–44.
6. Kerrigan JR, Rogol AD. The impact of gonadal steroid hormone action on growth hormone secretion during childhood and adolescence. Endocr Rev 1992;13:281–98.
7. Giustina A, Bossoni S, Bodini C, Girelli A, Balestrieri GP, Pizzocolo G, et al. Arginine normalizes the growth hormone (GH) response to GH-releasing hormone in adult patients receiving chronic daily immunosuppressive glucocorticoid therapy. J Clin Endocrinol Metab 1992;74:1301–5.
8. Giustina A, Scalvini T, Tassi C, Desenzani P, Poiesi C, Wehrenberg WB, et al. Maturation of the regulation of growth hormone secretion in young males with hypogonadotropic hypogonadism pharmacologically exposed to progressive increments in serum testosterone. J Clin Endocrinol Metab 1997;82:1210–9.
9. Giustina A, Girelli A, Alberti D, Bossoni S, Buzzi F, Doga M, et al. Effects of pyridostigmine on spontaneous and growth hormone-releasing hormone stimulated growth hormone secretion in children on daily glucocorticoid therapy after liver transplantation. Clin Endocrinol 1991;35:491–8.
10. Veldhuis JD, Carlson ML, Johnson ML. The pituitary gland secretes in bursts: appraising the nature of glandular secretory impulses by simultaneous multiple-parameter deconvolution of plasma hormone concentrations. Proc Natl Acad Sci USA 1987;84:7686–90.
11. Giustina A, Bresciani E, Bossoni S, Chiesa L, Misitano V, Wehrenberg WB, et al. Reciprocal relationship between the level of circulating cortisol and growth hormone secretion in response to growth hormone-releasing hormone in man: studies in patients with adrenal insufficiency. J Clin Endocrinol Metab 1994;79:1266–72.
12. Pincus SM, Keefe DL. Quantification of hormone pulsatility via an approximate entropy algorithm. Am J Physiol 1992;262:E741-54.
13. Wehrenberg WB, Giustina A. Basic counterpoint: mechanisms and pathways of gonadal steroid modulation of growth hormone secretion. Endocr Rev 1992;13:299–308.

14. Ulloa-Aguirre A, Blizzard RM, Garcia-Rubi E, Rogol AD, Link K, Christie CM, et al. Testosterone and oxandrolone, a non-aromatizable androgen, specifically amplify the mass and rate of growth hormone (GH) secreted per burst without altering GH secretory burst duration or frequency or the GH half-life. J Clin Endocrinol Metab 1990;71:846–54.
15. Liu L, Merriam GR, Sherins RJ. Chronic sex steroid exposure increases mean plasma growth hormone concentration and pulse amplitude in men with isolated hypogonadotropic hypogonadism. J Clin Endocrinol Metab 1987;64:651–6.
16. Tanner JM, Witehouse RH, Huhges PCR, Carter BS. Relative importance of growth hormone and sex steroids for the growth at puberty of trunk length, limb length, and muscle width in growth hormone-deficent children. J Pediatr 1976;89:1000–8.
17. Zeitler P, Argente J, Chowen-Breed JA, Clifton DK, Steiner RA. Growth hormone releasing-hormone messenger ribonucleic acid in the hypothalamus of the adult male rate is increased by testosterone. Endocrinology 1990;127:1362–8.
18. Wehrenberg WB, Baird A, Ying SY, Ling N. The effects of testosterone and estrogen on the pituitary growth hormone response to growth hormone-releasing factor. Biol Reprod 1985;32:369–75.
19. Giustina A, Wehrenberg WB. The role of glucocorticoids in the regulation of growth hormone secretion. Mechanisms and clinical significance. Trends Endocrinol Metab 1992;3:306–11.
20. Giustina A, Bossoni S, Bodini C, Doga M, Girelli A, Buffoli MG, et al. The role of cholinergic tone in modulating the growth hormone response to growth hormone-releasing hormone in normal man. Metabolism 1991;40:519–23.
21. Shibasaki T, Shizume K, Nakahara M, Masuda A, Jibiki K, Demura H, et al. Age-related changes in plasma GH response to GHRF in man. J Clin Endocrinol Metab 1984;58:212–4.
22. Giustina A, Bussi AR, Conti C, Doga M, Legati F, Macca C, et al. Comparative effect of galanin and pyridostigmine on the growth hormone (GH) response to GH-releasing hormone in normal aged subjects. Horm Res 1992;37:165–70.
23. Wehrenberg WB, Wiviott SD, Voltz DM, Giustina A. Pyridostigmine mediated growth hormone release: evidence for somatostatin involvement. Endocrinology 1992;130:1445.
24. Argente J, Chowen-Breed JA, Steiner RA, Clifton DK. Somatostatin messenger RNA in hypothalamic neurons is increased by testosterone through activation of androgen receptors and not by aromatization to estradiol. Neuroendocrinology 1990;52:342–9.
25. Veldhuis JD, Iranmanesh A, Rogol AD, Urban RJ. Regulatory action of testosterone on pulsatile growth hormone secretion in the human: studies using deconvolution analysis. In: Thorner MO, Adashi E, eds. The somatotrophic axis and the reproductive process in health and disease. New York: Raven Press, 1994:40–57.
26. Millard WJ, Politch JA, Martin JB, Fox TO. Growth hormone secretory patterns in androgen-resistant (testicular feminized) rats. Endocrinology 1986;119:2655–60.
27. Petra PH. The plasma sex steroid binding protein (SBP or SHBG): a critical review of recent developments on the structure, molecular biology and function. J Steroid Biochem Mol Biol 1991;40:735–53.

4

Sex-Steroid–Hormone Modulation of the Somatotropic Axis During Normal Male Puberty

DANIEL L. METZGER AND JAMES R. KERRIGAN

The growth spurt of puberty is a manifestation of the effects of elevated growth hormone (GH) and sex-steroid hormone levels (1–5). Elevated serum concentrations of both testosterone and 17-β-estradiol are observed during the growth spurt in late pubertal males. Although GH and gonadal steroid hormones act independently to facilitate pubertal growth, available evidence also supports a physiological role of androgens and estrogens in the neuroendocrine control of the somatotropic axis. Currently available data support both a stimulatory and inhibitory role of androgen in GH release. Estrogen appears to alter GH secretion primarily by a facilitative action.

To better understand the mechanism(s) by which androgens and estrogens alter the somatotropic axis, we employed the paradigms of sex-steroid hormone receptor blockade in cohorts of normal pubertal males (6,7). Flutamide, a potent, nonsteroidal, specific competitive inhibitor of androgen binding to its receptor in the central nervous system and peripheral tissues (8), was administered to a group of 6 late pubertal male volunteers. In a separate experiment, the nonsteroidal competitive inhibitor of the estrogen receptor, tamoxifen (9), was administered to 10 late pubertal males. Frequent venous blood sampling and deconvolution analysis (4,10,11) were employed to appraise alterations in GH release and elimination kinetics.

Subjects and Methods

Study Subjects

All study subjects had Tanner stage IV or V pubertal development; chronological ages ranged from 13.1 to 17.0 years. Heights and weights were between the 5th and 95th percentiles for each study participant (based on National Center for Health Statistics). Body mass index [BMI: weight/height^2

(12)] values were 0.2 ± 0.4 SD for skeletal age. Skeletal ages were within 2 SD for chronological age. All participants had normal screening blood chemistry profiles, complete blood cell counts, and thyroid function tests. Written parental consent and subject assent were obtained for each of the study subjects. For the flutamide protocol, 6 late pubertal males participated; 10 late pubertal males were enrolled in the tamoxifen protocol.

Assays

All serum GH and luteinizing hormone (LH) concentrations for a given subject were determined in duplicate in single-run assays. GH concentratous were determined by the Allegro human GH immunoradiometric assay (Nichols Institute, San Juan Capistrano, CA) with a detection limit of 0.2 µg/l. LH concentrations were measured using an immunofluorometric assay (Delfia human LH; Pharacia ENI Diagnostics, Columbia, MD) with a lower detection limit of 0.3 IU/l. All other assays were as previously described (6,7).

Experimental Design

Flutamide Protocol

All subjects were admitted on two occasions in randomized order. For the baseline (BASELINE) admission, no medications were taken before the study. For the treatment (FLUTAMIDE) admission, subjects were administered flutamide (Eulexin; Schering, Kenilworth, NJ) 250 mg orally four times a day for 3 days before and throughout the admission.

Tamoxifen Protocol

All subjects were admitted on two occasions, separated in time by at least 4 weeks, in randomized order. Subjects were administered either placebo (PLACEBO) or tamoxifen (TAMOXIFEN) citrate (Nolvadex; ICI Pharma, Wilmington, DE) 10 mg orally every 12 h for 72 h before and throughout the admission. Subjects were blinded to the treatment protocol.

Both Protocols

During each admission, a heparin lock was placed in a forearm vein approximately 1 h before venous sampling. Blood samples were obtained at 10-min intervals for 24 h beginning at 0800 for measurement of serum GH and LH concentrations. All other serum hormone concentrations were measured in a single blood sample obtained at 0800. Normal physical activity was permitted. Three standard meals and a bedtime snack were provided; a caffeine-free diet was followed. Subjects were required to be in bed with the room lights turned out from 2300 to 0600. No adverse effects of flutamide or tamoxifen were reported.

Deconvolution Analysis

Deconvolution analysis is a mathematical technique that, when applied to hormone concentration-versus-time series, estimates subject-specific measures of hormone production and half-life (10,11). Among the pertinent measures are (1) the number and positions in time of significant secretory episodes (burst frequency), (2) secretory burst half-duration (duration of the calculated secretory episode at half-maximal amplitude), (3) amplitude (maximal secretory rate in each secretory burst), and (4) hormone half-life corresponding to a disappearance rate constant. The elimination process here was described by a monoexponential model; a Gaussian distribution of hormone secretory rates was used. The mass of hormone released per secretory episode is determined as the analytical area of the secretory rate-versus-time curve in a given burst. No tonic (nonpulsatile) secretion was necessary to model the present experimental data. Total daily pulsatile GH production rates were derived as the product of secretory burst frequency and the mass of hormone released per secretory event. The integrated product of the secretory impulse and elimination function, the convolution integral, depicts the best-fit nonlinear least-squares function of the actual concentration-versus-time profile of the hormone.

Statistical Analysis

Comparisons between the baseline and flutamide-treated states or the placebo- and tamoxifen-treated states were made using the paired two-tailed Student's t-test. The Wilk-Shapiro statistic was applied to all variables to evaluate the hormality of the distribution; those not normally distributed were logarithmically transformed before analysis. Data analysis and management were accomplished using the CLINFO Data Reduction System at the University of Virginia General Clinical Research Center. $p < 0.05$ was used to define statistical significance. Results are expressed as the mean ± SEM.

Results

Flutamide Protocol

Effects of Androgen-Receptor Blockade on Serum Concentrations of LH, Sex-Steroid Hormones, and GH-Related Proteins

Mean serum LH concentrations increased significantly (< 0.005) from 3.2 ± 0.4 at baseline to 4.8 ± 0.6 IU/l during flutamide administration (refer to Table 4.1). Statistically significant increases in concentrations of free serum testosterone (T) and total serum estradiol (E_2) were observed during androgen-receptor blockade. Serum concentrations of total T demonstrated a small, but nonsignificant, increase during flutamide treatment. No statistically significant changes in serum concentrations of insulin-like growth factor-I (IGF-I) or GH-binding protein were observed.

TABLE 4.1. Effects of androgen receptor blockade on serum concentrations of LH, sex-steroid hormones, and GH-related proteins.

Hormone	Baseline	Flutamide	p
24-h mean LH (IU/l)	3.2 ± 0.4	4.8 ± 0.6	<0.005
Total T (nmol/l)	25 ± 2.7	34 ± 6.8	NS
Free T (pmol/l)	94 ± 8	122 ± 6	<0.01
Total E_2 (pmol/l)	54 ± 8	83 ± 16	<0.05
IGF-I (nmol/l)	66 ± 4.5	59 ± 5.1	NS
GHBP (pmol/l)	140 ± 22	147 ± 12	NS

NS, not statistically significant.
Adapted from Metzger et al. (6).

Effects of Androgen-Receptor Blockade on GH Secretory and Elimination Dynamics

Graphical results of deconvolution analysis for two representative subjects are illustrated in Figure 4.1.

The effects of androgen-receptor blockade on GH secretory and elimination parameters are depicted graphically in Figure 4.2 and are summarized in Table 4.2. The 24-h mean serum GH concentration increased significantly [2.9 ± 0.3 (flutamide) versus 1.8 ± 0.3 (baseline); $p < 0.01$] in response to androgen-receptor antagonism. GH production rate increased significantly ($p < 0.001$) from 93 ± 16 to 152 ± 15 µg/l^{-1} 24 h^{-1} during flutamide administration. Contributing to this change were demonstrable increases in the mass of GH secreted per burst [12.0 ± 1.4 (flutamide) versus 8.4 ± 1.0 µg/Lv (baseline); $p < 0.05$] and the maximal GH secretory rate [0.39 ± 0.04 (flutamide) versus 0.30 ± 0.03 µg^{-1}min^{-1} (baseline); $p < 0.05$]. A small, but statistically significant, increase in the GH secretory burst frequency [12 ± 1 (flutamide) versus 10 ± 1 per 24 h (baseline); $p < 0.05$] was observed during androgen-receptor antagonism. Administration of flutamide had no demonstrable effects on the GH elimination half-life or GH burst half-duration.

Tamoxifen Protocol

Effects of Estrogen Receptor Blockade on Serum Concentrations of LH, Sex-Steroid Hormones and GH-Related Proteins

Mean 24-h serum LH concentrations were not significantly altered during tamoxifen administration (Table 4.3). Serum IGF-I concentrations (nmol/l) decreased significantly [59 ± 3.5 (placebo) versus 48 ± 4.3 (tamoxifen); $p < 0.05$] during estrogen-receptor blockade. No demonstrable alterations in the serum concentrations of total T, free T, total E_2, and IGF-binding protein-3 (IGF-BP-3) were observed during antagonism of the estrogen receptor.

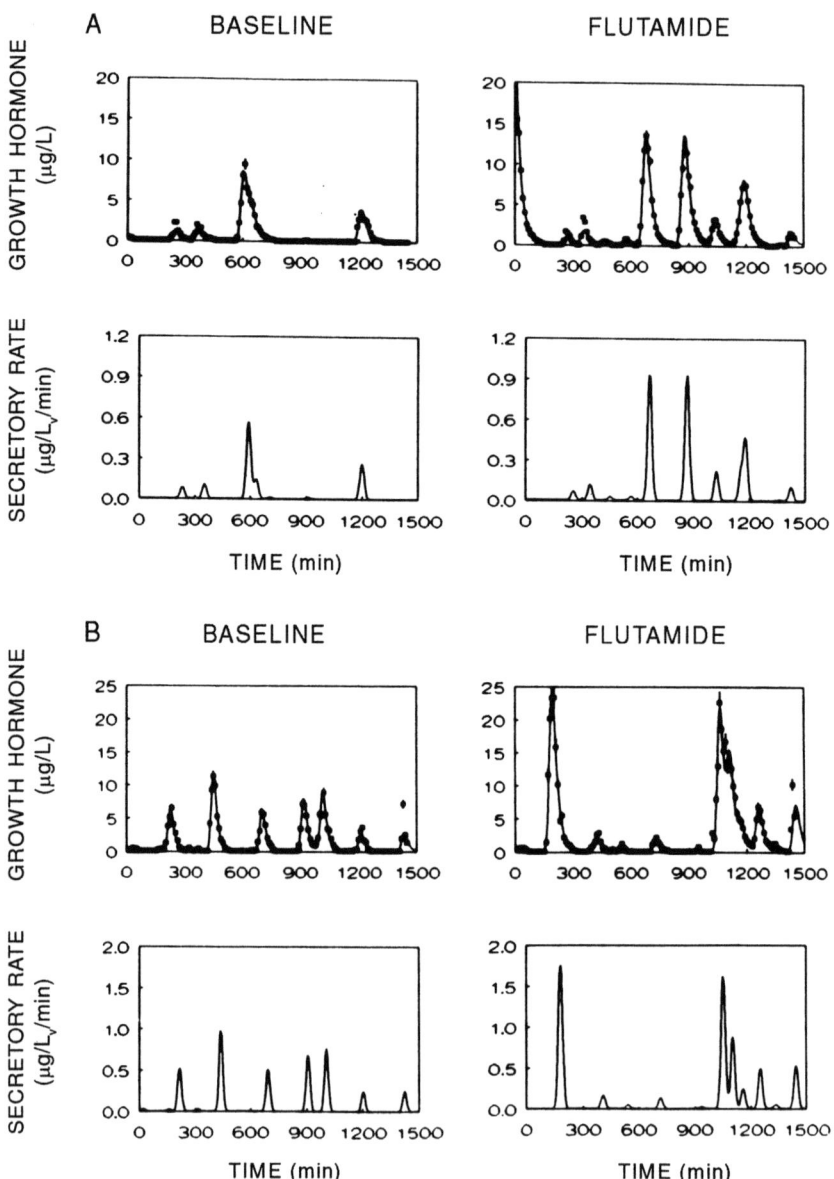

FIGURE 4.1A,B. Representative deconvolution analysis of GH profiles in placebo versus flutamide treatment sessions in two subjects (*A* and *B*). The *upper subpanels* for each subject depict serum GH concentrations versus time (min) beginning at 0800. The *lower subpanels* illustrate the deconvolution-resolved GH secretory rate versus time curves. Lv, liters of GH distribution volumes. Reproduced with permission from Metzger et al. (6).

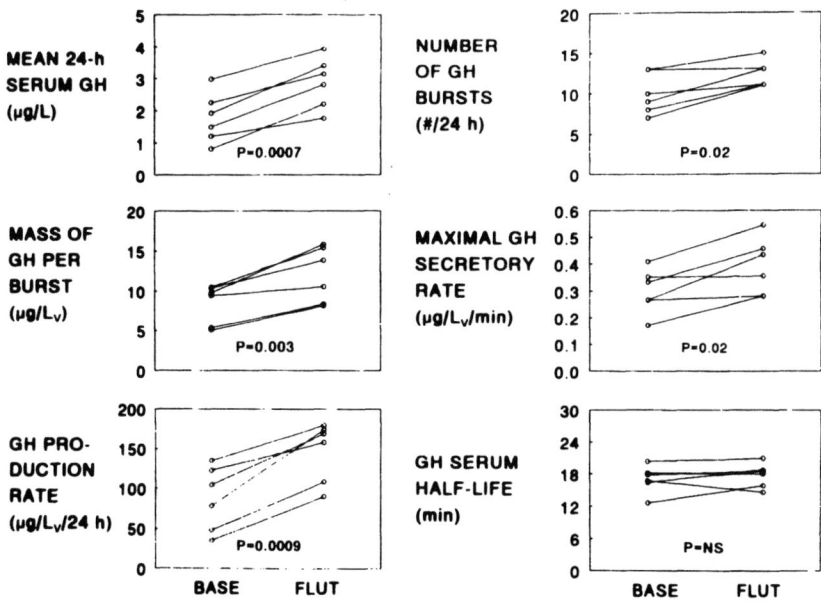

FIGURE 4.2. Combined graphical results of flutamide treatment for all study subjects. *Tie bars* connect the results of both experimental conditions for individual study subjects. NS, not statistically significant; *BASE*, basal (untreated) state; *FLUT*, flutamide administration. L_v, liters of GH distribution volumes. Reproduced with permission from Metzger et al. (6).

TABLE 4.2. Effects of androgen-receptor blockade on GH secretory and elimination dynamics.

GH parameter	Baseline	Flutamide	p
24-h mean serum GH (µg/l)	1.8 ± 0.3	2.9 ± 0.3	<0.001
GH burst frequency (per/24 h)	10 ± 1	12 ± 1	<0.05
Mass of GH per burst (µg/l)	8.4 ± 1.0	12.0 ± 1.4	<0.005
Maximal GH secretory rate (µg l^{-1} · min^{-1})	0.30 ± 0.03	0.39 ± 0.04	<0.05
GH burst half-duration (min)	27 ± 2	29 ± 1	NS
GH production rate (µg l^{-1} · 24 h^{-1})	93 ± 16	152 ± 15	<0.001
GH elimination half-life (min)	17 ± 1	18 ± 1	NS

l, liters of distribution volume; NS, not statistically significant.
Adapted from Metzger et al. (6).

TABLE 4.3. Effects of estrogen-receptor blockade on serum concentrations of LH, sex-steroid hormones, and GH-related proteins.

Hormone	Placebo	Tamoxifen	p
24-h mean LH (IU/l)	3.1 ± 0.3	3.4 ± 0.3	NS
Total T (nmol/l)	19.4 ± 1.6	21.5 ± 1.7	NS
Free T (pmol/l)	60 ± 8.3	63 ± 9.0	NS
Total E_2 (pmol/l)	55 ± 8	57 ± 6	NS
IGF-I (nmol/l)	59 ± 3.5	48 ± 4.3	<0.05
IGF-BP-3 (mg/l)	3.3 ± 0.2	3.8 ± 0.3	NS

NS, not statistically significant.
Adapted from Metzger et al. (7).

Effects of Estrogen-Receptor Blockade on GH Secretory and Elimination Dynamics

Graphical results of deconvolution analysis for two representative subjects are illustrated in Figure 4.3.

The effects of estrogen-receptor blockade on GH secretory and elimination parameters are depicted graphically in Figure 4.4 and summarized in Table 4.4. The 24-h mean serum GH concentration decreased significantly [3.9 ± 1.0 µg/l (placebo) versus 2.7 ± 0.6 (tamoxifen); $p < 0.05$] in response to estrogen-receptor antagonism. GH production rate decreased significantly ($p < 0.01$) from 237 ± 55 to 155 ± 33 µg l^{-1} 24 h^{-1} during tamoxifen administration. Contributing to this change were a significant decrease in the maximal GH secretory rate [0.46 ± 0.08 (placebo) versus 0.34 ± 0.06 µg l^{-1} min^{-1} (tamoxifen); $p < 0.01$] and a small, but statistically significant, decrease in the detectable GH secretory burst frequency [16 ± 1 (placebo) versus 14 ± 1 per 24 h (tamoxifen); $p < 0.05$]. During tamoxifen administration there was a decrease, although not statistically significant, in the mass of GH secreted per burst [13.3 ± 2.5 (placebo) versus 10.3 ± 2.0 (tamoxifen); $p = 0.06$]. Administration of tamoxifen had no demonstrable effects on the GH elimination half-life or GH burst half-duration.

Discussion

The potential mechanisms whereby sex-steroid hormones impact GH secretion and elimination were investigated in late pubertal males using selective antagonists of androgen (flutamide) and estrogen (tamoxifen) receptors, frequent venous blood sampling, and deconvolution analysis.

Antagonism of androgen action during flutamide administration resulted in increased 24-h mean serum GH concentrations caused by augmented GH production without demonstrable alterations in GH elimination kinetics. Appreciable rises in both the mass of GH secreted per burst and the maximal

4. Somatotropic Axis Modulation in Normal Male Puberty   39

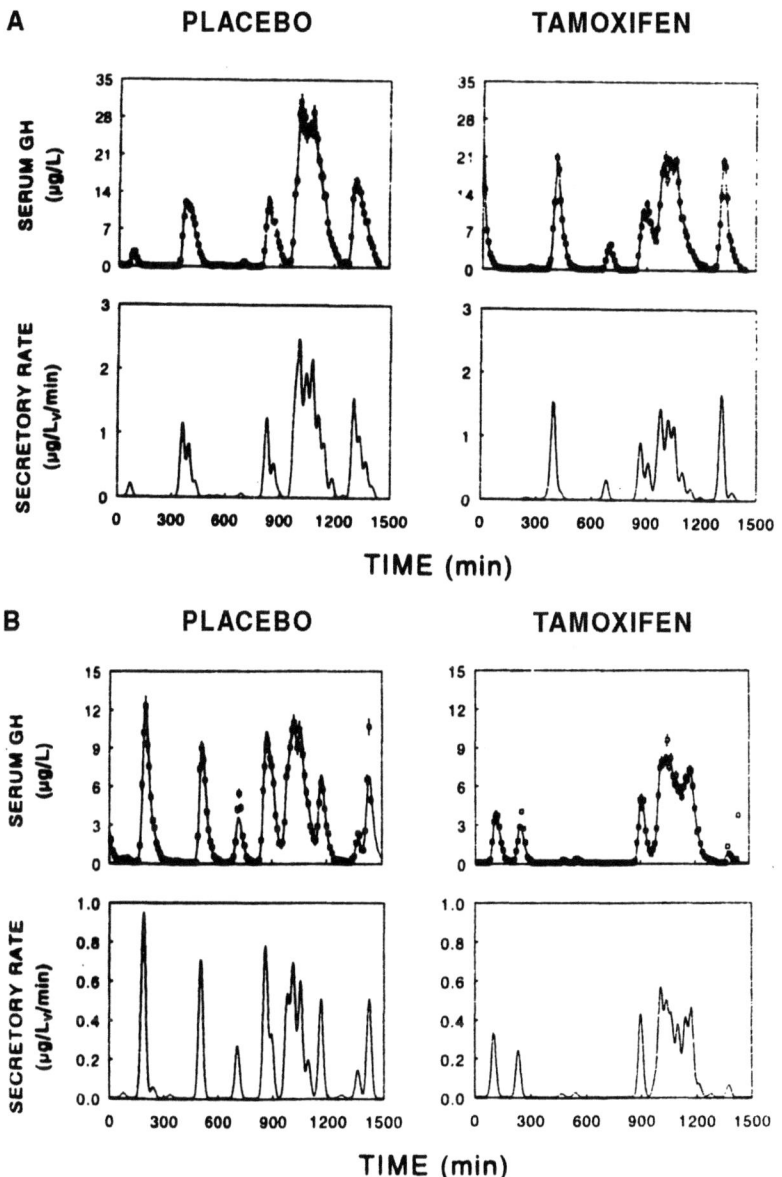

FIGURE 4.3 A,B. Representative deconvolution analysis for two subjects (A and B) treated with placebo or tamoxifen. The *upper subpanels* for each subject depict serum GH concentrations versus time. The *lower subpanels* illustrate the deconvolution-resolved GH secretory rate versus time curves. Reproduced with permission from Metzger et al. (7).

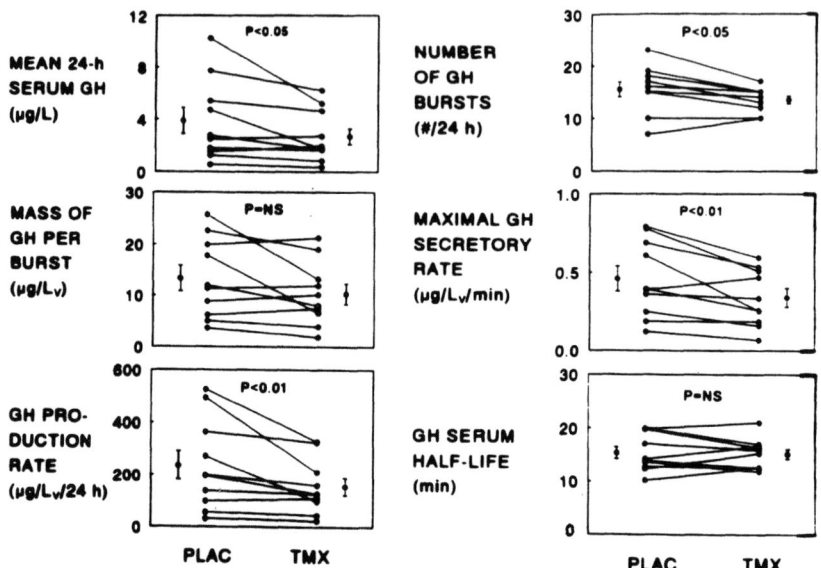

FIGURE 4.4. Combined graphical results for all subjects in the tamoxifen study. *Tie bars* connect the results of both experimental conditions for individual study subjects. NS, not statistically significant; PLAC, placebo (untreated) state; TMX, tamoxifen administration. L_v, liters of GH distribution volumes. Reproduced with permission from Metzger et al. (7).

GH secretory rate, as well as a small increase in the number of detectable GH secretory episodes, contributed to increased GH production rate. No change occurred in the GH elimination rate. Concomitant increases in serum concentrations of mean LH, free testosterone, and total estradiol were observed during androgen-receptor antagonism. These findings are compatible with either enhanced estrogen action or diminished androgen action, or both, as the

TABLE 4.4. Effects of estrogen receptor blockade on GH secretory and elimination dynamics.

GH parameter	Placebo	Tamoxifen	p
24-h mean serum GH (μg/l)	3.9 ± 1.0	2.7 ± 0.6	<0.05
GH burst frequency (per/24 h)	16 ± 1	17 ± 1	<0.05
Mass of GH per burst (μg/l)	13 ± 2.5	10 ± 2.0	NS
Maximal GH secretory rate ($\mu g \; l^{-1} \cdot min^{-1}$)	0.46 ± 0.08	0.34 ± 0.06	<0.01
GH burst half-duration (min)	27 ± 1	28 ± 2	NS
GH production rate ($\mu g \; l^{-1} \cdot min^{-1}$)	237 ± 55	155 ± 33	<0.01
GH elimination half-life (min)	15 ± 1	15 ± 1	NS

l, liters of body distribution volume; NS, not statistically significant.
Adapted from Metzger et al. (7).

mechanism(s) subserving androgen-mediated alterations in GH secretory dynamics during normal male puberty.

Administration of tamoxifen facilitated the investigation of a potential role of estrogen in the physiological control of GH secretion. During antagonism of estrogen action in late pubertal males, a significant reduction in GH production rate, without demonstrable alterations in GH elimination, led to a diminution of 24-h mean serum GH concentrations. Contributing to decreased GH production were demonstrable declines in the maximal GH secretory rate, the number of detectable GH secretory events and a trend toward diminished mass of GH secreted per burst. These findings support a facilitative role of endogenous estrogens in the neuroendocrine control of the somatotropic axis during normal male puberty.

Previous investigations have studied the effect of sex-steroid hormones on GH secretion. Long-term testosterone administration has been ssociated with consistently increased GH release in young males with constitutionally delayed puberty (13–15) and adult males with hypogonadism (16,17). Administration of a nonaromatizable androgen, (5-α-dihydrotestosterone), in contrast to testosterone administration, was associated with diminished serum concentrations of GH and IGF-I (15). Treatment of normal adult male subjects with nandrolone, a nonaromatizable androgen, was not associated with altered serum concentrations of GH or IGF-I (18). In this same study, however, administration of testosterone was associated with increased serum levels of GH and IGF-I. These studies provide evidence that stimulation of the somatotropic axis by androgen is mediated via the estrogen receptor following aromatization of testosterone to estrogen. Androgen appears to exert an inhibitory effect on GH release via its action through the androgen receptor.

The role of estrogen in the control of the somatotropic axis has been investigated by numerous researchers. Administration of estrogen has resulted in augmented serum concentrations of GH but not usually IGF-I (19–23). Tamoxifen administration has been demonstrated to result in damped GH release (17,24,25). These findings provide data supportive of a facilitative role of estrogen in the neuroendocrine control of GH release.

Combining the findings in the present studies with those of previous investigations provides insight into the mechanisms underlying sex-steroid hormone modulation of GH secretion in pubertal males. Based on these observations, we speculate that any facilitative role of endogenous androgens in GH production is mediated via the estrogen receptor. Independently of such an effect, it is likely that androgen exerts an inhibitory action on GH secretion via androgen receptor-mediated pathways. Endogenous estrogens appear to impact the somatotropic axis primarily in a stimulatory manner.

Acknowledgments. This work was supported in part by Clinical Investigator Award HD00926 from the NICHD (to J.R.K.), NIH Training Grant T32-DK-07642 (to D.L.M.), USPHS Grant RR-00847 to the University of Virginia General Clinical Research Center, and NIH-supported CLINFO Data Reduction Systems.

References

1. Albertson-Wikland K, Rosberg S, Libre E, Lundberg L-O, Groth T. Growth hormone secretory rates as estimated by deconvolution analysis of 24-h plasma concentration profiles. Am J Physiol 1989;257:E809–14.
2. Rose SR, Municchi G, Barnes KM, Kamp GA, Uriarte MM, Ross JL, et al. Spontaneous growth hormone secretion increases during puberty in normal girls and boys. J Clin Endocrinol Metab 1991;73:428–35.
3. Wennink JMB, Delemarre-van de Waal HA, Schoemaker R, Blaauw G, van den Braken C, Schoemaker J. Growth hormone secretion patterns in relation to LH and estradiol secretion throughout normal female puberty. Acta Endocrinol (Copenh) 1991;124:129–35.
4. Martha PM Jr, Gorman KM, Blizzard RM, Rogol AD, Veldhuis JD. Endogenous growth hormone secretion and clearance rates in normal boys, as determined by deconvolution analysis: relationship to age, pubertal status, and body-mass. J Clin Endocrinol Metab 1992;74:336–44.
5. Kerrigan JR, Rogol AD. The impact of gonadal steroid hormone action on growth hormone secretion during childhood and adolescence. Endocr Rev 1992;13:281–98.
6. Metzger DL, Kerrigan JR. Androgen receptor blockade with flutamide enhances growth hormone secretion in late pubertal males: evidence for independent actions of estrogen and androgen. J Clin Endocrinol Metab 1993;76:1147–52.
7. Metzger DL, Kerrigan JR. Estrogen receptor blockade with tamoxifen diminishes growth hormone secretion in boys: evidence for a stimulatory role of endogenous estrogens during male adolescence. J Clin Endocrinol Metab 1994;79:513–8.
8. Quercia RA. Focus on flutamide: a nonsteroidal antiandrogenic chemotherapeutic agent. Hosp Formul 1989;24:421–8.
9. Buckley MM-T, Goa KL. Tamoxifen: a reappraisal of its pharmacodynamic and pharmacokinetic properties, and therapeutic use. Drugs 1989;37:451–90.
10. Veldhuis JD, Carlson ML, Johnson ML. The pituitary gland secretes in bursts: appraising the nature of glandular secretory impulses by simultaneous multiple-parameter deconvolution of plasma hormone concentrations. Proc Natl Acad Sci USA 1987;84:7686–90.
11. Veldhuis JD, Faria A, Vance ML, Evans WS, Thorner MO, Johnson ML. Contemporary tools for the analysis of episodic growth hormone secretion and clearance in vivo. Acta Paediatr Scand 1988;347(Suppl):63–82.
12. Hammer LD, Kraemer HC, Wilson DM, Ritter PL, Dornbusch SM. Standardized percentile curves of body mass index for children and adolescents. Am J Dis Child 1991;145:259–63.
13. Link K, Blizzard RM, Evans WS, Kaiser DL, Parker MW, Rogol AD. The effect of androgens on the pulsatile release and the twenty-four-hour mean concentration of growth hormone in peri-pubertal males. J Clin Endocrinol Metab 1986;62:159–64.
14. Ulloa-Aquirre A, Blizzard RM, Garcia-Rubi E, Rogol AD, Link K, Christie CM, Johnson ML, Veldhuis JD. Testosterone and oxandrolone, a non-aromatizable androgen, specifically amplify the mass and rate of growth hormone (GH) secreted per burst without altering GH secretory burst duration or frequency or the GH half-life. J Clin Endocrinol Metab 1990;71:846–54.

15. Keenan BS, Richards GE, Ponder SW, Dallas JS, Nagamani M, Smith ER. Androgen-stimulated pubertal growth: the effects of testosterone and dihydrotestosterone on growth hormone and insulin-like growth factor-I in the treatment of short stature and delayed puberty. J Clin Endocrinol Metab 1993;76:996–1001.
16. Liu L, Merriam GR, Sherins RJ. Chronic sex steroid exposure increases mean plasma growth hormone concentration and pulse amplitude in men with isolated hypogonadotropic hypogonadism. J Clin Endocrinol Metab 1987;64:651–6.
17. Weissberger AJ, Ho KKY. Activation of the somatotropic axis by testosterone in adult males: evidence for the role of aromatization. J Clin Endocrinol Metab 1993;76:1407–12.
18. Hobbs, CJ, Plymate SR, Rosen CJ, Adler RA. Testosterone administration increases insulin-like growth factor-I levels in normal men. J Clin Endocrinol Metab 1993;77:776–9.
19. Wiedemann E, Schwartz E, Frantz AG. Acute and chronic estrogen effects upon serous somatomedin activity, growth hormone, and prolactin in men. J Clin Endocrinol Metab 1976;42:942–52.
20. Caruso-Nicoletti M, Cassorla F, Skerda M, Ross JL, Loriaux DL, Cutler GB Jr. Short-term low dose estradiol accelerates ulnar growth in boys. J Clin Endocrinol Metab 1985;61:896–8.
21. Dawson-Hughes B, Stern D, Goldman J, Reichlin S. Regulation of growth hormone and somatomedin-C secretion in post-menopausal women: effect of physiological estrogen replacement. J Clin Endocrinol Metab 1986;63:424–32.
22. Moll GW Jr, Rosenfield RL, Fang VS. Administration of low-dose estrogen rapidly and directly stimulates growth hormone production. Am J Dis Child 1986;140:124–7.
23. Mauras N, Rogol AD, Valdhuis JD. Increased hGH production rate after low-dose estrogen therapy in prepubertal girls with Turner's syndrome. Pediatr Res 1990;28:626–30.
24. Pollak M, Constantino J, Polychronakes C, Blauer SA, Guyda H, Redmond C, et al. Effect of tamoxifen on serum insulin-like growth factor I levels in stage I breast cancer patients. J Natl Cancer Inst 1990;82:1693–7.
25. Lima L, Arce V, Lois N, Fraga C, Lechuga MJ, Tresquerres JA, et al. Growth hormone (GH) responsiveness to GHRH in normal adults is not affected by short-term gonadal blockade. Acta Endocrinol (Copenh) 1989;120:31–6.

5

Influence of Testosterone on the GH–IGF-I Axis in Healthy Elderly Men

Marc R. Blackman, Colleen Christmas, Thomas Münzer,
Kieran G. O'Connor, Thomas E. Stevens, Michele F. Bellantoni,
Katherine Pabst, Carol St. Clair, H. Ballentine Carter,
E. Jeffrey Metter, and S. Mitchell Harman

Relationship of Testosterone with the GH–IGF-I Axis in Experimental Animals

Numerous in vivo and in vitro animal studies attest to the complex interplay between gonadal steroids and the GHRH–somatostatin–GH–IGF-I axis in the male and female, a topic that has been excellently and extensively reviewed (1–3). In male rats, consensus exists that the daily secretion of growth hormone (GH) is severalfold greater than that of females; that males release GH in high-amplitude, regular pulses, with low interpulse GH concentrations, whereas females secrete GH in low-amplitude, irregular pulses, with higher interpulse GH levels; and that there are many gender-distinct differences in the central nervous system and peripheral regulatory components of the GH axis, both in the absence and presence of endogenous or exogenous testosterone (1–3).

Most data suggest that, in the male rat, testosterone exerts its effects on the GH axis primarily through the androgen, rather than the estrogen, receptor, whereas in humans the opposite appears to be true (see following). Moreover, the orderliness of GH secretion, as assessed by analysis of approximate entropy, is greater in male versus female rats (4), and is decreased visually in male rats after administration of antiestrogens, suggesting a possible role for estrogen in directing this aspect of GH release (5). GH-receptor expression, particularly in the liver, is sexually dimorphic and species specific (6,7). Correspondingly, in the rat, hepatic expression of GH-binding protein (GHBP), as well as of various cytochromes and intracellular pathways that activate the STAT5b signaling cascade, is differentially modulated by male versus female patterns of GH secretion (8–10).

Relationship of Testosterone with the GH–IGF-I Axis in Boys and in Adolescent, Young Adult, and Middle-Aged Men

In humans, gender differences in GH axis activity have been recognized for many years, with most reports indicating greater basal and gonadal-steroid-stimulated GH secretion in females versus males during childhood, adolescence, and young adulthood to middle age (3,11–19). In contrast, gender disparities in the GH–IGF-I axis diminish or disappear after the menopause and in the elderly (14). Testosterone appears to stimulate GH secretion in males primarily via estrogen receptor-mediated mechanisms (20,21), and exerts an effect on central GHRH–somatostatin–somatotropic networking leading to increased disorderliness of GH release (22). Recent findings emphasize that testosterone can exert its anabolic tissue effects independently, without stimulating GH secretion (23).

Testicular androgens, in concert with normal GH axis function, are required for the normal pubertal transition in boys (13,24), whereas normal testis function in the absence of GH secretion is associated with a several-year delay in pubertal development (25). Prior cross-sectional and longitudinal studies revealed a close correlation between pubertal rises in androgens and augmented GH section. Not surprisingly, use of the more contemporary technique of deconvolution analysis has revealed significant positive relationships between androgen levels in pubertal boys and daily pulsatile GH secretory rate and GH secretory burst mass, as well as serum IGF-I levels (26). The related observation that small and gradually increasing doses of testosterone given to boys with idiopathic hypogonadotropic hypogonadism leads to progressive increments in the number and amplitude of GH secretory peaks underscores the sensitivity of the GH axis to androgen exposure during the pubertal transition (27). In boys undergoing normal puberty and in those with idiopathic hypogonadotropic hypogonadism, testosterone treatment increases the disorderliness of GH secretion, as assessed by augmented scores for approximate entropy (28). In contrast, in pubertally delayed boys, administration of the nonaromatizable androgen, DHT, neither stimulates GH secretion nor alters the approximate entropy score (22).

In healthy men, as in pubertal boys, serum concentration of total or free testosterone are closely, and directly, correlated with mean serum GH levels, as well as with GH secretory burst mass and amplitude (27,29). Administration of a nonsteroidal antiandrogen to healthy young men augments pulsatile GH secretion (20), whereas antiestrogen treatment exerts an inhibitory effect (21). In nonelderly men with primary hypogonadism, GH secretion and IGF-I levels increase after testosterone treatment, an effect that is reversed after administration of antiestrogen (22,23,30). Taken together, these and the aforementioned findings suggest that in nonelderly men testosterone mediates its stimulatory effect on the amount and orderliness of pulsatile GH secretion via the estrogen, not the androgen, receptor.

Gender differences have also been reported for GH release following acute stimulation by GHRH (growth hormone-releasing hormone) of some, but not other, GH secretagogues. In general, healthy young and middle-aged men respond less well to provocative GHRH testing than do similarly aged women (12,31). Moreover, in men, maximal GH responsivity is closely related to serum levels of estradiol (12). In contrast, elderly women and men respond similarly to an acute bolus of GHRH, perhaps because their estrogen levels are similar (12). Arginine- and insulin-induced hypoglycemia (in some studies) elicit lower acute GH secretory responses in men versus women, and GH responses increase after administration of estrogen to men (32). Similarly, galanin-induced GH release is reduced in young men versus women (33), whereas GH secretory responses to clonidine (34) or the GH secretagogue GHRP-2 (35) are greater in men, and GH responsivity to glucagon does not differ by gender (36).

Relationship of Testosterone with the GH–IGF-I Axis in Elderly Men

Aging in healthy men is associated with gradual reductions in circulating levels of total or free testosterone and in GH secretion and serum concentrations of IGF-I and IGF-binding protein-3 (IGF-BP-3), the principal circulating IGF-binding protein (14,37,38). In older men who have age-related reductions in serum testosterone levels, significant positive correlations have been observed between serum testosterone levels and spontaneous (39) GH secretion, and with GHRH-stimulated GH release in some (40), but not other (41), studies. Thus, it can be hypothesized that the age-related decline in testosterone secretion in healthy men contributes to the concomitant physiological reduction in GH secretion with age, and that the latter may be reversed, at least in part, by testosterone administration.

To begin to assess the role of gonadal steroid in modulating GH secretion in elderly men, we performed a cross-sectional survey in which we evaluated the relationships among baseline morning serum values of testosterone (T), sex hormone-binding globulin (SHBG), free T index (T/SHBG), IGF-I, IGF-BP-3, and IGF-II in 203 healthy men aged 40–83 years. Linear regression analysis revealed nonsignificant relationships of serum T with levels of IGF-I ($r = 0.062, p > 0.3$), IGF-BP-3 ($r = -0.089, p > 0.2$) and IGF-II ($r = -0.130, p < 0.06$) (Fig. 5.1). In contrast, serum levels of SHBG were inversely related to circulating concentrations of IGF-I ($r = -0.235, p < 0.001$) and IGF-BP-3 ($r = -0.278, p < 0.001$), but not to IGF-II ($r = 0.070, p > 0.3$) (Fig. 5.2). Conse-

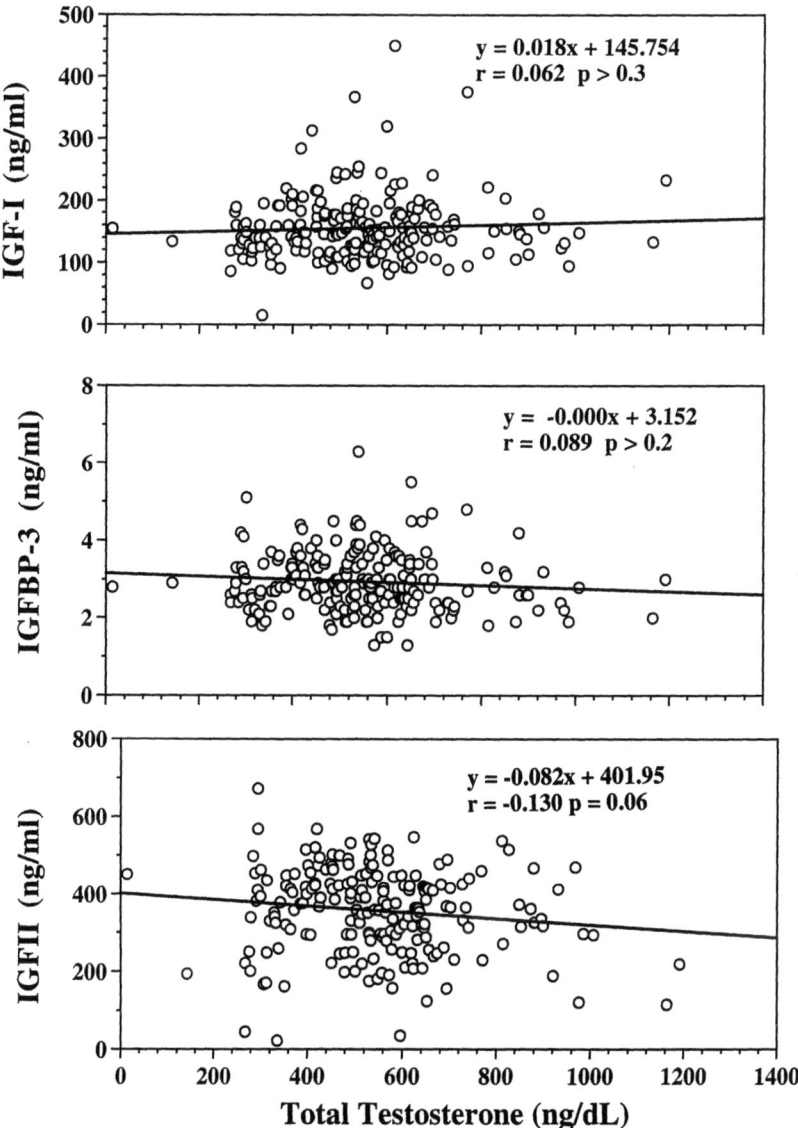

FIGURE 5.1. Bivariate plots of the relationships of serum concentrations of testosterone with circulating IGF-I, IGF-BP-3, and IGF-II in 203 healthy men aged 40–83 years.

quently, the free T index was directly related to levels of IGF-I ($r = 0.324$, $p > 0.0001$) and IGF-BP-3 ($r = 0.224$, $p < 0.002$), but not to IGF-II ($r = 0.065$, $p > 0.3$) (Fig. 5.3). Linear regression analysis also revealed that age was directly related to SHBG ($r = 0.487$, $p < 0.0001$), and inversely related to free T index

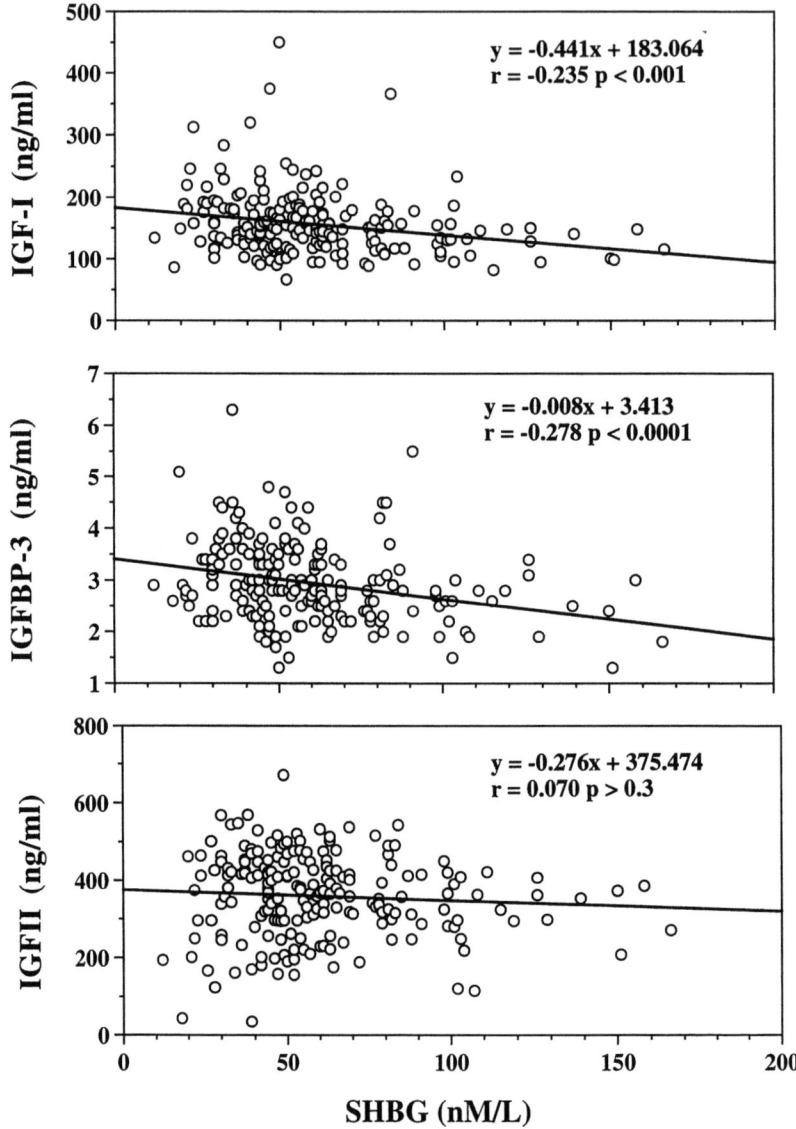

FIGURE 5.2. Bivariate plots of the relationships of serum levels of sex hormone-binding globulin (SHBG) with circulating concentrations of IGF-I, IGF-BP-3, and IGF-II in 203 healthy men aged 40–83 years.

($r = -0.544, p < 0.0001$), IGF-I ($r = -0.310, p < 0.0001$), IGF-BP-3 ($r = 0.194$, $p < 0.01$), and IGF-II ($r = -0.181, p < 0.01$) in this same group of men. Consequently, we used multiple regression analysis to assess the relative influences of age versus T, SHBG, or free T index on levels of IGF-I, IGF-BP-3, and IGF-II (Table 5.1). Although age was a strong independent predictor of serum IGF-I,

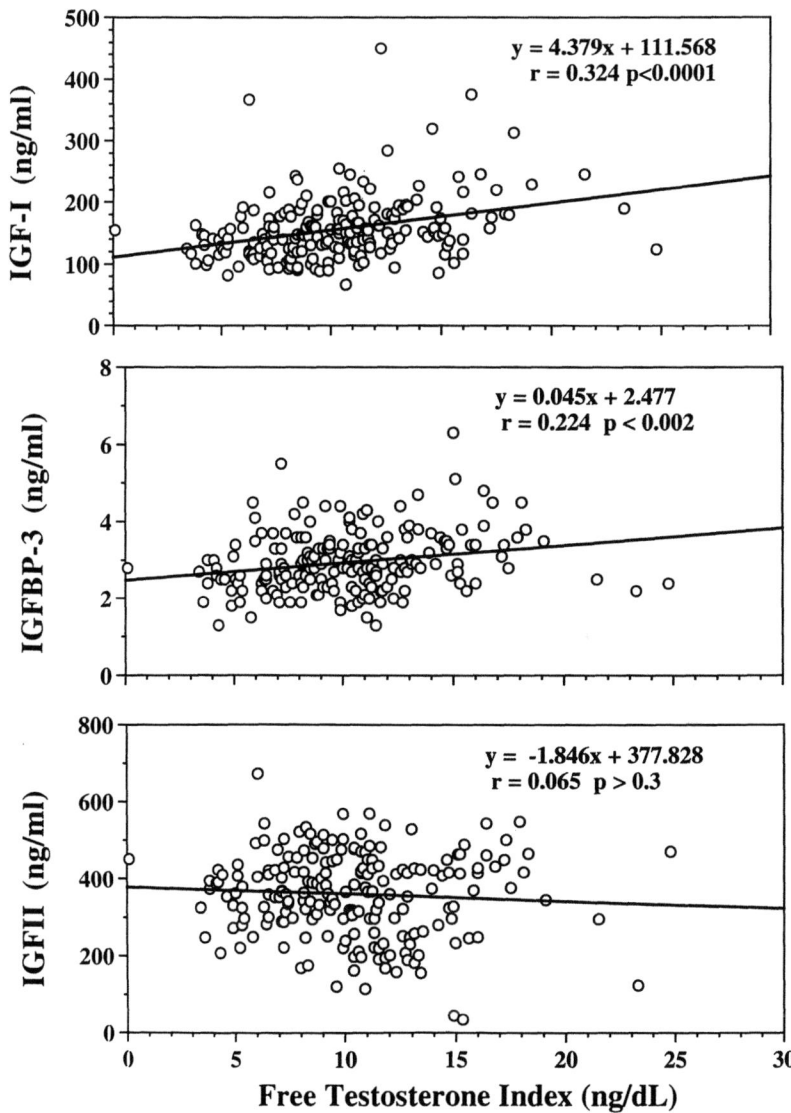

FIGURE 5.3. Bivariate plots of the relationships of serum values for the free testosterone index (free T index) with circulating concentrations of IGF-I, IGF-BP-3, and IGF-II in 203 healthy men aged 40–83 years.

IGF-BP-3, and IGF-II levels, T and free T index remained independently related to IGF-I; SHBG and free T index were independently related to IGF-BP-3; and free T index was independently related to IGF-II (Table 5.1).

We also assessed the relationships of baseline morning serum T levels with integrated overnight GH secretion (expressed as log AUPGH) and morning levels of IGF-I and IGF-BP-3 in a separate group of 74 healthy men, aged

TABLE 5.1. Relative influence of age vs. androgenic milieu on serum IGF-I, IGF-BP-3, and IGF-II levels in healthy men.

Independent variable	Dependent variables		
	IGF-I	IGF-BP-3	IGF-II
	p	p	p
Age	<0.0001	<0.005	<0.01
Testosterone	<0.05	NS	NS
Age	<0.01	NS	<0.05
SHBG	NS	<0.005	NS
Age	NS	NS	<0.001
Free T Index	<0.005	<0.05	<0.01

65–88 years, who exhibited age-related reductions in IGF-I levels and gonadal steroids. Blood sampling was performed every 20 min from 8:00 P.M. to 8:00 A.M.; GH was measured by immunoradiometric assay followed by PULSAR analysis of GH neurosecretory parameters. Linear regression analysis revealed no significant relationships of serum T with log AUPGH or IGF-I levels, and showed a positive relationship of serum T with levels of IGF-BP-3 that was of borderline significance ($r = 0.225$, $p < 0.07$). Finally, using a double-masked, placebo-controlled design, we are now evaluating the effects of 6 months of administration of T alone, GH alone, and T plus GH on various indices of GH secretion, including approximate entropy, as well as serum IGF-I and IGF-BP-3, in this latter group of healthy "somatopausal" and "andropausal" men.

This study is also designed to determine whether combined administration of T plus GH elicits synergistic changes in body composition and various endocrine-metabolic, musculoskeletal, cardiovascular, immunological, and psychobehavioral outcomes in this same population. The data obtained to date suggest that there is, at baseline, a direct relationship of free T levels with GH axis activity in healthy elderly men. However, the effects of long-term T administration in this population remain to be elucidated. Further research will be required to determine whether, in elderly men, as inferred from their younger counterparts, T activates the somatostatin–GHRH–somatotropic axis via the estrogen receptor, whether T influences the orderliness as well as the amount of pulsatile GH secretion, and whether androgen exerts effects on various target tissues independently of its actions on the GH axis.

References

1. Wehrenberg WB, Giustina A. Basic counterpoint: mechanisms and pathways of gonadal steriod modulation of growth hormone secretion. Endocr Rev 1992;13:299–308.
2. Jaffe CA, Ocampo-Lim B, Guo W, Krueger K, Sugahara I, Demott-Friberg R, Bermann M, Barkan AL. Regulatory mechanisms of growth hormone secretion are sexually dimorphic. J Clin Invest 1998;102:153–64.

3. Giustina A, Veldhuis JV. Pathophysiology of the neuroregulation of growth hormone secretion in experimental animals and the human. Endocr Rev 1998;19:717–97.
4. Pincus SM, Gevers E, Robinson ICAF, Van den Berg G, Roelfsema F, Hartman ML, Veldhuis JD. Females secrete growth hormone with more process irregularity than males in both human and rat. Am J Physiol 1996;270:E107–15.
5. Tannenbaum GS, Gurd W, Lapointe M, Pollak M. Tamoxifen attenuates pulsatile growth hormore secretion: mediation in part by somatostatin. Endocrinology 1992;130:3395–401.
6. Domene HM, Marin G, Sztein J, Yum YM, Baron J, Cassorla FG. Estradiol inhibits growth hormone receptor gene expression in rabbit liver. Mol Cell Endocrinol 1994;103:81–7.
7. Carmignac DF, Gabrielsson BG, Robinson IC. Growth hormone binding protein in the rat: effects of gonadal steroids. Endocrinology 1993;133:2445–52.
8. Waxman DJ, Ram PA, Pampori NA, Shapiro BH. Growth hormone regulation of male-specific rat liver P450s 2A2 and 3A2. Mol Pharmacol 1995;48:790–7.
9. Legraverend C, Mode A, Westin S, Strom A, Eguchi H, Zaphiropoulos PG, Gustafsson J-A. Transcriptional regulation of rat P450 2C gene subfamily members by the sexually dimorphic pattern of growth hormone secretion. Mol Endocrinol 1992;6:259–66.
10. Waxman DJ, Pampori NA, Ram PA, Agrawal AK, Shapiro BH. Interpulse interval in circulating growth hormone patterns regulates sexually dimorphic expression of hepatic cytochrome P450. Proc Natl Acad Sci USA 1991;88:6868–72.
11. Frantz AG, Rabkin MT. Effects of estrogen and sex difference of human growth hormone. J Clin Endocrinol Metab 1965;25:1470–80.
12. Lang I, Schernthaner G, Pietschmann P, Kurz R, Stephenson JM, Templ M. Effects of sex and age on growth hormone response to growth hormone releasing hormone in healthy individuals. J Clin Endocrinol Metab 1987;65:535–40.
13. Kerrigan JR, Rogol AD. The impact of gonadal steroid hormone action on growth hormone secretion during childhood and adolescence. Endocr Rev 1992;13:281–98.
14. Corpas E, Harman SM, Blackman MR. Human growth hormone and human aging. Endocr Rev 1993;14:20–39.
15. Ho KKY, O'Sullivan AJ, Weissberger AJ, Kelly JJ. Sex steroid regulation of growth hormone secretion and action. Horm Res 1996;45:67–73.
16. Veldhuis JV. New modalities for understanding dynamic regulation of the somatotropic (GH) axis: explication of gender differences in GH neuroregulation in the human. J Pediatr Endocrinol Metab 1996;9:237–53.
17. Veldhuis JD. Gender differences in secretory activity of the human somatotropic (growth hormone) axis. Eur J Endocrinol 1996;134:287–95.
18. Veldhuis JV. Neuroendocrine control of pulsatile growth-hormone release in the human: relationship with gender. Growth Horm IGF Res 1998;8:49–59.
19. Ho KY, Evans WS, Blizzard RM, et al. Effects of sex and age on 24-hour profile of growth hormone secretion in men. J Clin Endocrinol Metab 1987;64:518.
20. Metzger DL, Kerrigan JR. Androgen receptor blockage with flutamide enhances growth hormone secretion in late pubertal males: evidence for independent actions of estrogen and androgen. J Clin Endocrinol Metab 1993;76:1147–52.
21. Metzger DL, Kerrigan JR. Estrogen receptor blockade with tamoxifen diminishes growth hormone secretion in boys: evidence for a stimulatory role of endogenous estrogens during male adolescence. J Clin Endocrinol Metab 1994;79:513–8.

22. Veldhuis JD, Metzger DL, Martha PM Jr, Mauras N, Kerrigan JR, Keenan B, Rogol AD, Pincus SM. Estrogen and testosterone, but not a nonaromatizable androgen, direct network integration of the hypothalamo-somatotrope (growth hormone)-insulin-like growth factor I axis in the human: evidence from pubertal pathophysiology and sex-steroid replacement. J Clin Endocrinol Metab 1997;82:3414–20.
23. Fryburg DA, Weltman A, Jahn LA, Weltman JY, Samojlik E, Hintz RL, Veldhuis JD. Short-term modulation of the androgen milieu alters pulsatile, but not exercise- or growth hormone (GH)-releasing hormone-stimulated GH secretion in healthy men: impact of gonadal steroid and GH secretory changes on metabolic outcomes. J Clin Endocrinol Metab 1997;82:3710–9.
24. Mauras N, Rogol AD, Haymond MW, Veldhuis JD. Sex steroids, growth hormone, IGF-I: neuroendocrine and metabolic regulation in puberty. Horm Res 1996;45:74–80.
25. van der Werff ten Bosch JJ, Bot A. Growth of males with idiopathic hypopituitarism without growth hormone treatment. Clin Endocrinol (Oxf) 1990;32:707–17.
26. Martha PM Jr, Goorman KM, Blizzard RM, Rogol AD, Veldhuis JD. Endogenous growth hormone secretion and clearance rates in normal boys as determined by deconvolution analysis: relationship to age, pubertal status and body mass. J Clin Endocrinol Metab 1992;74:336–44.
27. Giustina A, Scalvini T, Tassi C, Desenzani P, Poiesi C, Wehrenberg W, Rogol A, Veldhuis JD. Maturation of the regulation of growth hormone secretion in young males with hypogonadotropic hypogonadism pharmacologically exposed to progressive increments in serum testosterone. J Clin Endocrinol Metab 1997;82:1210–9.
28. Pincus SM. Approximate entropy as a measure of system complexity. Proc Natl Acad Sci USA 1991;88:2297–301.
29. Iranmanesh A, Lizarralde G, Veldhuis JD. Age and relative adiposity and specific negative determinants of the frequency and amplitude of growth hormone (GH) secretory bursts and the half-life of endogenous GH in healthy men. J Clin Endocrinol Metab 1991;73:1081–8.
30. Weissberger AJ, Ho KK. Activation of the somatotropic axis by testosterone in adult males: evidence for the role of aromatiztion. J Clin Endocrinol Metab 1993;76:1407–12.
31. Gelato MC, Pescovitz OH, Cassorla F, Loriaux DL, Merriam GR. Dose-response relationships for the effects of growth hormone-releasing factor-(1-44)-NH2 in young adult men and women. J Clin Endocrinol Metab 1984;59:197–201.
32. Merimee TJ, Rabinowitz D, Fineberg SE. Arginine-initiated release of human growth hormone. Factors modifying the response in normal men. N Engl J Med 1969;280:1434–8.
33. Giustina A, Lincini M, Bussi AR, Girelli A, Pizzocolo G, Schettino M. Effects of sex and age on the growth hormone response to galanin in healthy human subjects. J Clin Endocrinol Metab 1993;76:1369–72.
34. Tulandi T, Lal S, Guyda H. Effect of estrogen on the growth hormone response to the alpha-adrenergic agonist clonidine in women with menopausal flushing. J Clin Endocrinol Metab 1987;65:6–10.
35. Wideman L, Weltman JY, Shah N, Story S, Bowers CY, Veldhuis JD, Weltman A. The effects of arginine and GHRP2 on resting and exercise induced growth hormone (GH) release. Program of the 79th Annual Meeting of the Endocrine Society, Minneapolis, MN 1997 (abstract P2-197).

36. Rao RH, Spathis GH. Intramuscular glucagon as a provocative stimulus for the assessment of pituitary function: growth hormone and cortisol responses. Metab Clin Exp 1987;36:658–63.
37. Corpas E, Harman SM, Blackman MR. Serum IGF-binding protein-3 is related to IGF-I, but not to spontaneous GH release, in healthy old men. Horm Metab Res 1992;24:543–5.
38. Blackman MR, Elahi D, Harman SM. Endocrinology and aging. In: DeGroot L, et al., eds. Endocrinology, 3rd ed. Philadelphia: Saunders, 1995:2702–30.
39. Veldhuis JV, Liem AY, South S, Weltman A, Weltman J, Clemmons DA, Abbott R, Mulligan T, Johnson ML, Pincus SM, Straume M, Iranmanesh A. Differential impact of age, sex-steroid hormones, and obesity on basal vs. pulsatile growth hormone secretion in men as assessed in an ultrasensitive chemiluminescence assay. J Clin Endocrinol Metab 1995;80:3209–22.
40. Corpas E, Harman SM, Piñeyro MA, Roberson R, Blackman MR. GHRH 1-29 twice daily reverses the deceased GH and IGF-I levels in old men. J Clin Endocrinol Metab 1992;75:530–5.
41. Ceda GP, Ceresini G, Denti L, Cortellini P, et al. Androgens do not regulate the growth hormone response to GHRH in elderly men. Horm Metab Res 1989;21:695–6.

6

Effect of Estrogen on GH Secretion and Action in Postmenopausal Women

KEN K.Y. HO AND ANTHONY J. O'SULLIVAN

There is evidence that estrogen regulates growth hormone (GH) secretion in the human female. Serum oestradiol and GH levels are significantly correlated in pubertal girls (1). GH secretion varies during the menstrual cycle, with mean concentrations highest during the late follicular phase when estrogen concentration is highest (2). Mean 24-h GH and IGF-I levels are lower in postmenopausal women than in premenopausal women, suggesting that reduced activity of the somatotrophic axis in the menopause may be secondary to estrogen deficiency (3). The collective observations suggest that gonadal steroid deficiency may be associated with a reduction in GH secretion, which can be restored by sex-steroid replacement.

In males, GH secretion is reduced in primary or prolonged hypogonadism and stimulated by androgen treatment, which also increases IGF-I levels (4,5). Recent data suggest that the stimulatory effect of testosterone may in part be dependent on aromatization to estrogen (6). The consequences of hypogonadism and of estrogen replacement on GH economy in women are less clear. Girls with Turner syndrome appear to have reduced spontaneous GH release, especially at the age of anticipated puberty (7). While estrogen treatment augments spontaneous GH secretion (8), effects on IGF-I appear to be inconsistent, with reports of either an increase (9) or no change (10) in IGF-I levels. Older studies utilizing higher doses of estrogens reported suppression of IGF-I activity in menopausal (11) and acromegalic women (12). Thus, the available data on estrogen effects in hypogonadal women appear confusing. In the last few years, we have undertaken studies to clarify the role of estrogen in the regulation of GH secretion and action using the postmenopausal state as a model of hypogonadism.

Estrogen Effect in the Menopause

We investigated the effects of unopposed estrogens using a dosage previously reported to increase serum IGF-I in Turner patients (13). GH secretion obtained from 20-min blood measurements and serum IGF-I levels were compared in weight-matched pre- and postmenopausal women. The effects of estrogen therapy on circulating levels of GH-binding protein, GHBP (14,15), which is believed to be derived from proteolytic cleavage of the extracellular domain of the GH receptor (16), were also investigated. GHBP alters the distribution and pharmacokinetics of GH and is likely to modulate GH action (17).

The route of administration may be an important consideration in studying the effects of estrogen replacement on circulating GH and on IGF-I, which is derived primarily from the liver (18). The route is important because orally administered estrogens exert nonphysiological effects on hepatic protein synthesis, effects that are avoided when the parenteral route is used (19,20). Accordingly, we evaluated two regimens for estrogen replacement, ethinyl estradiol administered orally and 17-β-estradiol administered transdermally (Estraderm TTS 100; Ciba Geigy) (13).

We confirmed previous observations that mean 24-h serum GH concentrations and IGF-I levels were significantly lower in postmenopausal compared to younger cycling women. Both replacement regimens resulted in significant and comparable reductions in circulating levels of LH (luteinizing hormone) and FSH (follicle-stimulating hormone). Administration of oral ethinyl oestradiol resulted in a threefold increase in mean 24-h GH concentrations (Fig. 6.1). In contrast, transdermal administration of 17-β estradiol did not result in a significant change in mean 24-h GH concentrations. Oral administration of ethinyl estradiol resulted in a uniform and significant reduction in mean IGF-I levels. In contrast, transdermal estrogen delivery resulted in a small but significant increase in mean IGF-I levels (Fig. 6.1). GHBP levels were not significantly different between pre- and postmenopausal women but were affected by the route of estrogen therapy (Fig. 6.1). There was a significant increase in binding activity following the oral route but not the transdermal route. The increase in binding activity arises from an increase in GHBP concentrations but not in affinity (21,22).

The results showed that estrogen treatment in the menopause has a significant route-dependent effect on the GH–IGF-I axis. The finding of IGF-I suppression by ethinyl estradiol administration in postmenopausal women is supported by the results of two other studies using a similar dosage (23,24). The mechanism by which orally administered estrogen reduces circulating IGF-I is not clear but may arise from inhibition of hepatic IGF-I mRNA generation, as demonstrated in the rat (25). Thus, it is likely that the IGF-I suppressive effect of oral ethinyl estradiol is a consequence of a first-pass hepatic effect. This proposal is in keeping with the well-recognized observation that orally administered estrogens alter the synthesis of proteins of hepatic origin such as angiotensinogen, coagulation factors, and lipoproteins, effects that are do not occur equivalently with the parenteral route. Another possible mechanism may involve GHBP, which has been shown in in vitro studies to blunt GH action (26). A first-pass mechanism is also likely to explain the

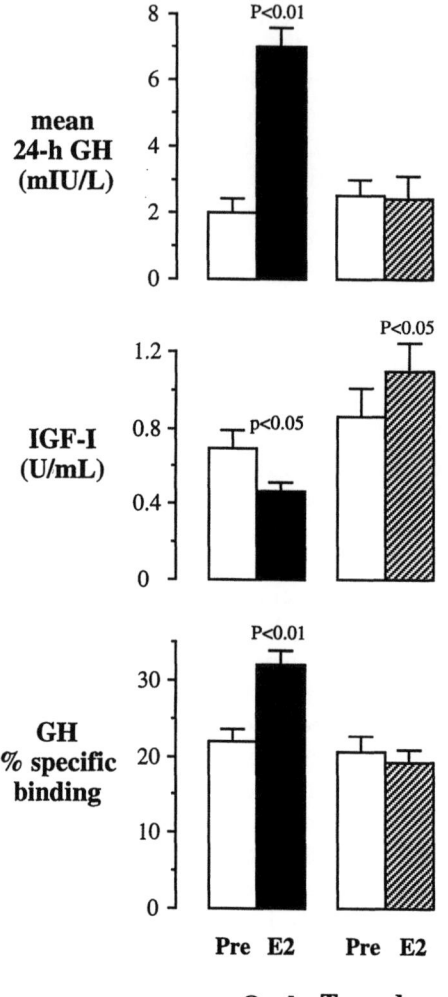

FIGURE 6.1. Mean (±SE) 24-h serum growth hormone (GH) (*top panel*), IGF-I concentrations (*middle panel*), and GH-binding protein (GHBP) activity (*lower panel*), in two groups of postmenopausal women ($n = 7$) before and after 12 weeks of oral ethinyl estradiol or transdermal 17-β-estradiol treatment. Reproduced with permission from Weissberger et al. (13).

differing effects of oral and transdermal estrogens on serum GHBP, because the liver is rich in GH receptors from which GHBP is thought to be derived (13,27). Because the fall in circulating IGF-I occurred in the setting of enhanced GH secretion, the increase in GH concentration can be explained by reduced negative feedback inhibition of IGF-I on GH secretion (13).

Different Oral Estrogen Formulations

The estrogen types used in our study were not identical. Consequently, the data do not totally exclude the possibility that the contrasting effects of ethinyl estradiol and 17-β estradiol reflect intrinsic chemical differences rather than the dissimilar routes of administration. It has been reported that induction of hepatic protein synthesis by ethinyl estradiol appears to be greater than its ability to suppress gonadotropin secretion when compared to other estrogen types. These hepatic effects were not entirely eliminated when ethinyl estradiol was administered parenterally via the vaginal route (28,29).

To address whether the reduction of IGF-I and elevations of GH and GHBP are specific properties of ethinyl estradiol or intrinsic to the oral route of administration, the effects of oral administration of ethinyl estradiol (20 µg), conjugated equine estrogen (Premarin, 1.25 mg) and estradiol ester (estradiol valerate, 2 mg) (30) were compared. The doses have been shown to have approximate systemic bioequivalence (28,31). The three estrogen formulations were administered to six postmenopausal women in a randomized crossover design for a period of 4 weeks each. Mean 24-h GH concentrations were measured before and at the end of each treatment cycle during the estrogen-only phase, immediately before coadministration of medroxyprogesterone acetate.

All three estrogen formulations resulted in significant suppression of LH and FSH and in elevation of the hepatic proteins SHBG and angiotensinogen. GHBP increased in parallel with these hepatic proteins. Each of the three estrogen formulations resulted in a significant reduction in IGF-I levels compared to baseline and corresponding elevations in mean 24-h GH and GHBP concentrations (Fig. 6.2). The percent increase in mean 24-h GH during treatment was significantly and inversely related to the percent decrease in IGF-I levels.

The uniform responses displayed by all three estrogen formulations administered by the oral route stand in contrast to those observed following transdermal delivery and strongly suggest that the reduction in IGF-I levels is an intrinsic effect of oral estrogens (30). The increase in GHBP concentration reflects another level of action of estrogen on the GH–IGF-I axis. All three estrogen formulations increased GHBP in parallel with elevations in SHBG and angiotensinogen, both recognized as estrogen-sensitive hepatic proteins. Together with our observations that transdermal estrogen delivery had no effect on GHBP, the data suggest that GHBP is an estrogen-sensitive hepatic protein and that the liver is a major source of GHBP in humans (21).

The inverse order of effect between treatment groups in IGF-I suppression and GH elevation provide further support for our proposal that the increases in GH concentration arises from reduced feedback inhibition by IGF-I on

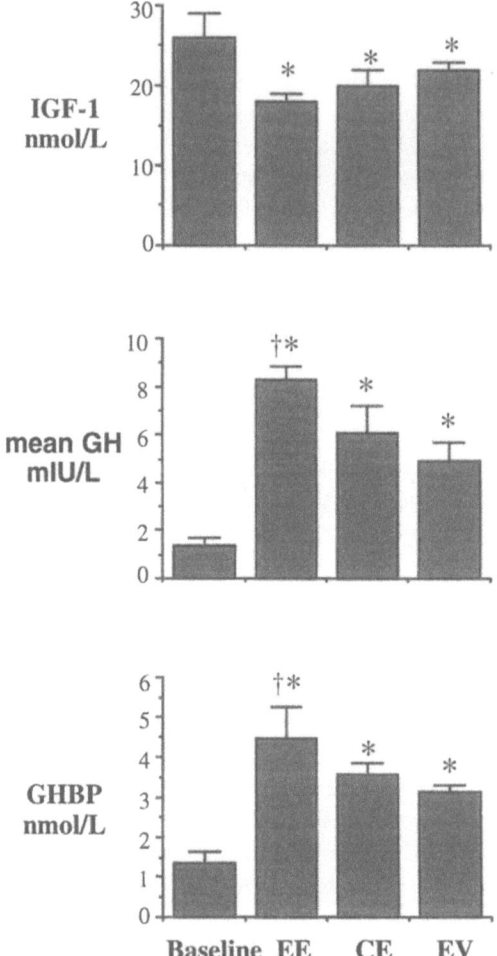

FIGURE 6.2. Mean (±SE) serum IGF-I (*top panel*), mean 24-h GH (*middle panel*) and GHBP (*lower panel*) concentrations in six postmenopausal women before and during treatment with ethinyl estradiol (*EE*), conjugated equine estrogen (*CE*), and estradiol valerate (*EV*). *, $p < 0.05$ vs. baseline; †, $p < 0.05$ vs. EV. Reproduced with permission from Kelly et al. (30).

pituitary GH release. Irrespective of the mechanisms involved, the perturbation of the GH–IGF-I axis suggests that this sex steroid may have significant biological and route-dependent effects on GH- and IGF-I responsive tissues.

Biological Effects

The anabolic actions of GH are mediated through IGF-I, whereas other metabolic actions such as stimulation of lipolysis and induction of insulin resistance are direct. Because three major components of the somatotropic axis (GH, IGF-I, and, GH-binding protein) are markedly affected by oral but not transdermal estrogen administration, we have investigated the possibility that estrogen may also exert other significant biological effects that are dependent on its route of administration.

Bone and Connective Tissue

We first investigated effects on bone and connective tissues, which are both GH- and IGF-I-responsive peripheral tissues (32). By measuring changes in levels of osteocalcin and procollagen I and III, we found that estrogen had distinct effects on connective and skeletal tissue metabolism that were dependent on the route of administration, with the changes in these biochemical markers of osteoblast and fibroblast function occurring in parallel with changes in circulating IGF-I. Thus levels of osteocalcin and procollagen I and III increased during transdermal compared to oral estrogen treatment (32). This finding provided the first evidence that perturbation of the GH–IGF-I axis by estrogen had a significant effect on peripheral GH-responsive tissues.

Glucose Metabolism

The stimulation of GH secretion induced by oral estrogen therapy may affect carbohydrate metabolism because GH impairs insulin action (33,34). We have investigated the effects of the route of estrogen replacement on glucose tolerance and insulin sensitivity in nine postmenopausal women after 12 weeks of treatment (35). Peak glucose and insulin levels during the glucose tolerance test were slightly higher with oral than with transdermal estrogen treatment, although the difference did not reach statistical significance. When compared to the transdermal route, the mean glucose infusion rates required to maintain euglycemia during the hyperinsulinemic euglycemic clamp were slightly lower with oral treatment but failed to reach significance. However, during the transdermal estrogen phase, mean nonesterified free fatty acid concentration was suppressed to a significantly lower level during the hyperinsulinemic-euglycemic clamp than during the oral estrogen phase. Thus, when compared to the oral route, transdermal estrogen therapy is associated with a slight but significant improvement of insulin action on lipid metabolism and trends toward improved action on glucose metabolism (35).

Substrate Oxidation

Recent studies of GH replacement in GH deficient adults provide unequivocal evidence that GH plays a major role in regulating substrate oxidation and body composition (36–38). Stimulation of lipid oxidation and protein synthesis during GH treatment leads to a progressive fall in body fat and increase in lean body mass in GH-deficient adults. These metabolic effects of GH are imparted by a complex interplay of IGF-I mediated actions and direct IGF-I independent actions of GH. In light of divergent effects on GH, GHBP, and IGF-I, we have investigated the metabolic impact of the route of estrogen administration on substrate oxidation and on body composition (39). The effects of 24 weeks each of oral and transdermal estrogen on energy metabolism and body composition in 18 postmenopausal women were studied in an open-label randomized crossover study. Energy expenditure, lipid oxidation, and carbohydrate oxidation were measured by indirect calorimetry in the fasted and fed state before and after 2 and 6 months treatment. Lean body mass, fat mass, and total body bone mineral density were measured by dual X-ray absorptiometry before and after 6 months of treatment. Mean LH levels fell to comparable levels during oral and transdermal treatments whereas mean IGF-I levels were significantly lower during oral estrogen treatment, as previously observed.

No significant difference in resting energy expenditure or basal lipid oxidation was observed between the two routes of estrogen therapy. Ingestion of a standardized mixed meal acutely suppressed lipid oxidation during each treatment phase. However, when compared to the transdermal route, oral estrogen administration resulted in a greater suppression of lipid oxidation (36 ± 5 vs. 54 ± 5 mg/min, $p < 0.01$) during the first hour after ingestion of the standardized meal (39). This greater but transient suppression of lipid oxidation in the early postprandial phase induced by the oral route was accompanied by a reciprocal stimulation of carbohydrate oxidation (147 ± 13 vs. 109 ± 12 mg/min, $p < 0.05$). Thus, when compared to transdermal estrogen, oral estrogen therapy resulted in an early but transient suppression of lipid oxidation and a reciprocal elevation of carbohydrate oxidation after a nutrient load.

The finding of a suppressive effect of oral estrogen on fat oxidation extends the observations made in a young girl treated with high doses of ethinyl estradiol (40). In this case report, we observed that oral ethinyl estradiol treatment with doses of 60, 100, and 200 µg/day produced a reversible, dose-dependent suppression of lipid oxidation associated with a reversible increase in fat mass (Fig. 6.3). Lipid oxidation was reduced throughout the basal and postprandial states during ethinyl estradiol treatment in contrast to significant suppression occurring only during the postprandial phase in the present study. As the potency of ethinyl estradiol in the doses used is considerably higher than that of conjugated estrogen, the observations suggests that the inhibitory effect of oral estrogen on lipid oxidation is dose dependent.

Body Composition

In the crossover study of postmenopausal women, no significant changes in body weight were observed between or for either route of estrogen therapy after 6 months (Fig. 6.4) (39). Mean bone mineral density (BMD) increased during both oral and transdermal estrogen therapy, with the increases not being significantly different between the two routes. Significantly different effects on fat mass and

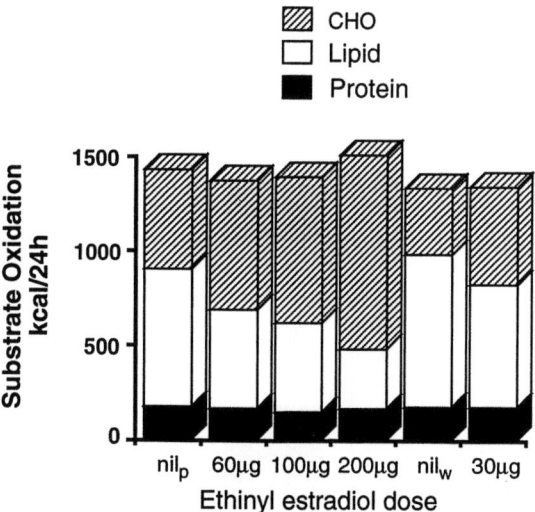

FIGURE 6.3. Rates of carbohydate (CHO), fat (lipid), and protein oxidation after an overnight fast in a girl with hypogonadism studied sequentially before (nil_p) or during oral treatment with daily doses of 60, 100, and 200 μg ethinyl estradiol, and again after stopping (nil_w) and recommencing daily ethinyl estradiol treatment at a dose of 30 μg. Reproduced with permision from O'Sullivan et al. (40).

lean body mass were observed between the two routes of estrogen therapy. When the effects of both routes of estrogen administration were compared, oral therapy led to a significant increase in fat mass of 1.2 ± 0.5 kg, equivalent to 4.9% ± 1.6% change in body fat (Fig. 6.4). This increase in fat mass arose from a significant increase occurring during 24 weeks of oral therapy with no significant change occurring during the transdermal estrogen phase (see Fig. 6.3). Oral estrogen therapy also induced a significant loss in lean body mass of 1.2 ± 0.4 kg ($p < 0.05$, equivalent to a 3.1% ± 0.8% change) compared to that observed during transdermal estrogen therapy. This difference was accounted for by a significant decrease in lean body mass of 0.8 ± 0.3 kg with oral therapy and a small but nonsignificant increase in lean body mass with transdermal estrogen therapy (Fig. 6.4). Thus, when compared to the transdermal route, oral estrogen therapy was accompanied by a significant decrease in lean body mass and a significant increase in whole body fat mass.

Mechanism

The mechanism(s) responsible for the changes in body composition are not known but may arise from long-term biological consequences of the suppression of lipid oxidation and IGF-I levels by oral but not transdermal estrogen administration. We speculate that the increase in body fat observed during oral estrogen therapy

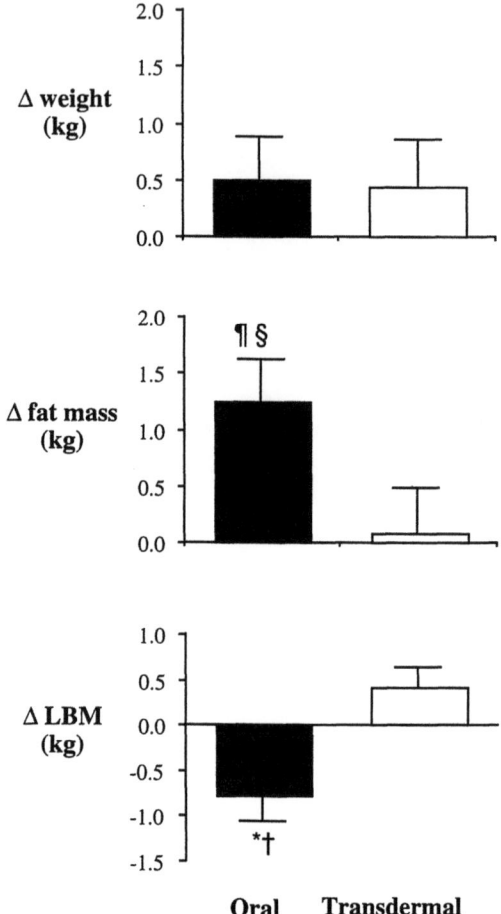

FIGURE 6.4. Change (mean ± SE) in body weight (*top panel*), fat mass (*middle panel*), and lean body mass (*LBM, lower panel*) after 24 weeks of oral and transdermal estrogen treatment. ¶, $p < 0.05$ oral vs. transdermal; *, $p < 0.01$ oral vs. transdermal; † $p < 0.02$ vs. before oral estrogen treatment; §, $p < 0.005$ vs. before oral estrogen treatment. Reproduced with permission from O'Sullivan et al. (39).

arises from chronic suppression of lipid oxidation. How lipid oxidation is reduced by estrogen is not known. One possible mechanism may involve the increase in circulating GH-binding protein, which could attenuate the stimulatory effects of endogenous GH on fat oxidation. The finding that insulin sensitivity was impaired during oral estrogen treatment argues against this possibility. Another mechanism may involve a direct effect of estrogen on the liver, the major site of fatty acid metabolism. In vitro studies have shown that pharmacological concentrations of estrogen reduce ketogenesis (a product of fatty acid oxidation) and increase fatty acid incorporation into triglycerides (41,42). These in vitro findings are in accordance with clinical observations that oral but not transdermal estrogen therapy stimulates hepatic triglyceride synthesis and increases triglyceride levels (43). Because intrahepatic fatty acid metabolism is partitioned between oxidative and nonoxidative (fatty acid incorporation into triglycerides) pathways, these findings suggest that estrogen may regulate the metabolic fate of intrahepatic free fatty acids by directing fatty acids away from oxidative into lipogenic pathways. Irrespective of the mechanism(s) involved, the impact of estrogen on hepatic fatty acid metabolism is likely to be the result of a first-pass pharmacological effect because estrogen-associated changes in triglycerides (43), GHBP, and IGF-I are seen only with the oral route of administration. The effects on fat oxidation and IGF-I induced by the oral route of administration are opposite to the effects on GH and strongly suggest that oral estrogen can act as a GH antagonist.

Summary

Hypogonadism in the menopause is characterized by reduced GH and IGF-I levels. The effects of estrogen on GH economy in the menopause are complex and are dependent on the route of delivery. Administration of oral estrogen, irrespective of type, reduces circulating IGF-I and increases GH and GHBP levels, while transdermal administration causes no change in GH or GHBP. The perturbation of the GH–IGF-I axis is accompanied by significant biological effects. When compared to the transdermal route, oral estrogen reduces lipid oxidation, increases fat mass, and reduces lean body mass. The route of estrogen therapy confers distinct and divergent effects on substrate oxidation and body composition. The suppression of lipid oxidation during oral estrogen therapy may increase fat mass while the fall in IGF-I may lead to a loss of lean body mass. Undesirable changes in body composition may be avoided in postmenopausal estrogen users by a nonoral route of administration. The route-dependent changes in body composition observed during estrogen replacement therapy may have important implications for postmenopausal health and estrogen use in general.

Acknowledgments. This work was supported in part by the National Health & Medical Research Council of Australia and by Ciba Geigy Australia.

References

1. Wennick JMB, Delemare van de Waal HA, Schoemaker R, Blaau G, van den Braken C, Schoemaker J. Growth hormone secretion patterns in relation to LH and estradiol secretion throughout normal female puberty. J Clin Endocrinol Metab 1990;124:129–35.
2. Faria ACS, Beckenstein LW, Booth RA, et al. Pulsatile growth hormone release in normal women during the menstrual cycle. Clin Endocrinol 1992;36:591–6.

3. Ho KY, Evans WS, Blizzard RM, et al. Effects of sex and age on the 24 hour secretory profile of GH secretion in man: importance of endogenous estradiol concentrations. J Clin Endocrinol Metab 1987;64:51–8.
4. Link K, Blizzard R, Evans W, Kaiser D, Parker M, Rogol A. The effect of androgens on the pulsatile release and the 24-hour mean concentration of growth hormone in peripubertal males. J Clin Endocrinol Metab 1986;62:159–64.
5. Liu L, Merriam GR, Sherins RJ. Chronic sex steroid exposure increases mean plasma growth hormone concentration and pulse amplitude in men with isolated hypogonadotropic hypogonadism. J Clin Endocrinol Metab 1987;64:58.
6. Weissberger AJ, Ho KKY. Activation of the somatotropic axis by testosterone in adult males: evidence for the role of aromatization. J Clin Endocrinol Metab 1993;76:1407–12.
7. Ross JL, Meyerson Long L, Loriaux DL, Cutler GB. Growth hormone secretory dynamics in Turner syndrome. J Pediatr 1985;106:202–6.
8. Mauras N, Rogol AD, Veldhuis JD. Specific, time-dependent actions of low-dose ethinyl estradiol administration on the episodic release of growth hormone, follicular-stimulating hormone and luteinising hormone in prepubertal girls with Turner's syndrome. J Clin Endocrinol Metab 1989;69:1053–8.
9. Ross JL, Cassorla FG, Skerda MC, Valk IM, Loriaux DL, Cutler GB. A preliminary study of the effect of estrogen dose on growth in Turner's syndrome. N Engl J Med 1983;309:1104–6.
10. Copeland KC. Effects of acute high dose and chronic low dose estrogen on plasma somatomedin C and growth in patients with Turners syndrome. J Clin Endocrinol Metab 1988;66:1278–82.
11. Wiedemann E, Schwartz E, Frantz A. Acute and chronic estrogen effects upon serum somatomedin activity, growth hormone, and prolactin in man. J Clin Endocrinol Metab 1976;42:942–52.
12. Clemmons DR, Underwood LE, Ridgway EG, Kliman B, Kjellberg RN, Van Wyk JJ. Estradiol treatment of acromegaly: reduction of immunoreactive somatomedin C and improvement of metabolic status. Am J Med 1980;69:571–5.
13. Weissberger AJ, Ho KY, Lazarus L. Contrasting effects of oral and transdermal routes of estrogen replacement therapy on 24-hour growth hormone (GH) secretion, insulin-like growth factor 1 and GH binding protein in post-menopausal women. J Clin Endocrinol Metab 1991;72:374–81.
14. Baumann G, Stolar MW, Amburn K, Barsano CP, DeVries BC. A specific growth hormone-binding protein in human plasma: initial characterisation. J Clin Endocrinol Metab 1986;62:134–41.
15. Herington AC, Ymer S, Stevenson J. Identification and characterisation of specific binding proteins for growth hormone in normal human sera. J Clin Invest 1986;77:1817–23.
16. Leung DW, Spencer SA, Cachaines G, et al. Growth hormone receptor and serum binding protein: purification, cloning and expression. Nature (Lond) 1987;330:537–43.
17. Baumann G, Amburn KD, Buchanan TA. The effect of circulating growth hormone-binding protein on metabolic clearance, distribution and degradation of human growth hormone. J Clin Endocrinol Metab 1987;64:657–60.
18. D'Ercole AJ, Stiles AD, Underwood LE. Tissue concentrations of somatomedin C: further evidence for multiple sites of synthesis and paracrine or autocrine mechanisms of action. Proc Natl Acad Sci USA 1984;81:935–9.

19. de Lignieres B, Basdevant A, Thomas G. Biological effects of 17-β-estradiol in postmenopausal women: oral versus percutaneous administration. J Clin Endocrinol Metab 1986;62:536–41.
20. Chetkowski RJ, Meldrum DR, Steingold KA, et al. Biological effects of transdermal estradiol. N Engl J Med 1986;314:1615–20.
21. Ho KY, Valiontis E, Waters MJ, Rajkovic IA. Regulation of growth hormone binding protein in man: comparison of gel chromatography and immunoprecipitation methods. J Clin Endocrinol Metab 1993;76:302–8.
22. Rajkovic IA, Valiontis E, Ho KKY. Direct quantitation of growth hormone binding protein in human serum by a ligand immunofunctional assay: comparison with immunoprecipitation and chromatography methods. J Clin Endocrinol Metab 1994;78: 772–7.
23. Duursma S, Bijlsma J, Van Paassen H, van Buul-Offers S, Skottner-Lundin A. Changes in serum somatomedin and growth hormone concentrations after 3 weeks oestrogen substitution in post-menopausal women; a pilot study. Acta Endocrinol 1984;106:527–31.
24. Dawson-Hughes B, Stern D, Goldman J, Reichlin S. Regulation of growth hormone and somatomedin-C secretion in postmenopausal women: effect of physiological estrogen replacement. J Clin Endocrinol Metab 1986;63:424–32.
25. Murphy LJ, Freisen HG. Differential effects of estrogen and growth hormone on uterine and hepatic insulin-like growth factor-1 expression in the ovariectomised hypophysectomised rat. Endocrinology 1988;122:325–32.
26. Lim L, Spencer SA, McKay P, Waters MJ. Regulation of growth hormone (GH) bioactivity by a recombinant human GH-binding protein. Endocrinology 1990;127:1287–91.
27. Baruch Y, Amit TPH, Enat R, Youdim MBH, Hochberg Z. Decreased serum growth hormone-binding protein in patients with liver cirrhosis. J Clin Endocrinol Metab 1991;73:777–80.
28. Maschak CA, Lobo R, Dozono TR, et al. Comparison of pharmacokinetic properties of various estrogen formulations. Am J Obstet Gynecol 1982;144:511–8.
29. Goebelsmann U, Maschak CA, Mishell DR Jr. Comparison of hepatic impact of oral and vaginal administration of ethinyl estradiol. Am J Obstet Gynecol 1985;151: 868–77.
30. Kelly JJ, Rajkovic IA, O'Sullivan AJ, Sernia C, Ho KKY. Effects of different oestrogen formulations on insulin-like growth factor-1, growth hormone and growth hormone binding protein in post-menopausal women. Clin Endocrinol 1993;39:561–7.
31. Mandel FP, Geola FL, Lu J, et al. Biological effects of various doses of ethinyl estradiol. Am J Obstet Gynecol 1982;59:673–9.
32. Ho KKY, Weissberger AJ. Impact of short-term estrogen administration on growth hormone secretion and action: distinct route-dependent effects on connective and bone tissue metabolism. J Bone Miner Res 1992;7:821–7.
33. Rizza R, Mandarino L, Gerich J. Dose-response characteristics for effects of insulin on production and utilization of glucose in man. Am J Physiol 1981;240:E630–9.
34. Ho KKY, Jenkins AB, Furler SM, Borkman M. Chisholm DJ. Impact of octreotide, a long-acting somatostatin analogue, on glucose tolerance and insulin sensitivity in acromegaly. Clin Endocrinol 1992;36:271–9.
35. O'Sullivan AJ, Ho KKY. A comparison of the effects of oral and transdermal estrogen replacement on insulin sensitivity in postmenopausal women. J Clin Endocrinol Metab 1995;80:1783–8.

36. Moller N, Jorgensen JOL, Alberti KGMM, Flyvbjerg A, Schmitz O. Short-term effects of growth hormone on fuel oxidation and regional substrate metabolism in normal man. J Clin Endocrinol Metab 1990;70:1179–86.
37. Jorgensen JOL, Theusen L, Ingemann-Hansen T, Pedersen SA, Jorgensen J, Christiansen JS. Beneficial effects of growth hormone treatment in GH-deficient adults. Lancet 1989;1:1221–5.
38. Salomon F, Cuneo RC, Hesp R, Sonksen PH. The effects of treatment with recombinant human growth hormone on body composition and metabolism in adults with growth hormone deficiency. N Engl J Med 1989;321:1797–803.
39. O'Sullivan AJ, Crampton L, Freund J, Ho KKY. Route of estrogen replacement confers divergent effects of energy metabolism and body composition in postmenopausal women. J Clin Invest 1998;102:1035–40.
40. O'Sullivan AJ, Hoffman DM, Ho KKY. Estrogen, lipid oxidation and body fat. N Engl J Med 1995;333:660–70.
41. Weinstein I, Soler-Argilaga C, Werner HV, Heimberg M. Effects of ethynyloestradiol on the metabolism of [$1\text{-}^{14}C$] oleate by perfused livers and hepatocytes from female rats. Biochem J 1979;180:265–71.
42. Ockner RK, Lysenko N, Manning JA, Monroe SE, Burnett DA. Sex steroid modulation of fatty acid utilization and fatty acid binding protein concentration in rat liver. J Clin Invest 1980;65:1013–23.
43. Walsh BW, Schiff I, Rosner B, Greenberg L, Ravnikar V, Sacks FM. Effects of postmenopausal estrogen replacement on the concentrations and metabolism of plasma lipoproteins. N Engl J Med 1991;325:1196–204.

7

Menstrual Cycle Interaction with the Growth Hormone Axis

PER OVESEN, NINA VAHL, JENS SANDAHL CHRISTIANSEN,
JOHANNES D. VELDHUIS, AND JENS OTTO LUNDE JORGENSEN

The periodicity of the menstrual cycle results from interactions among the hypothalamus, pituitary, ovaries, and genital tract. Each normal cycle culminates in menstrual bleeding, the first day of which is defined as the beginning of a menstrual cycle. The human menstrual cycle is divided into three phases: a follicular, or proliferative, phase; an ovulatory phase; and a luteal, or secretory, phase. The initiation of follicular growth begins during the late luteal phase of the preceding menstrual cycle, during which serum levels of progesterone (P) and estrogen (E_2) decline, because of termination of the corpus luteum lifespan, and levels of FSH (follicle-stimulating hormone) rise. The rise in FSH initiates follicular development, which continues after the onset of menstrual bleeding. However, FSH then declines as a result of negative feedback from the increasing E_2 secreted by the granulosa cells of the growing follicle. During the follicular phase, E_2 levels rise in parallel to the growth of the follicle and number of granulosa cells. LH levels increase at the midfollicular phase as a result of the positive feedback caused by increased E_2 release. Just before ovulation, E_2 secretion increases dramatically, which initiates the LH surge. LH promotes the process of luteinization of the granulosa cells and consequently enhances P secretion, which in turn facilitates the midcycle surge. After ovulation the remaining follicle undergoes striking changes resulting in the formation of the corpus luteum, which secretes a large amount of P and, to a lesser extent, E_2. Accordingly, healthy young women exhibit different gonadal hormone milieus during the normal menstrual cycle, and such variations provide a model to investigate the impact of endogenously secreted estrogen and progesterone on GH secretion.

Studies of GH Secretion During the Menstrual Cycle

In young normal women, the release of GH during the menstrual cycle has been reported to be variably elevated during the periovulatory (PO) and midluteal phases compared to the early follicular (EF) phase of the menstrual cycle (1–4). In one recent study (5), blood was obtained at 10-min intervals for 24 h and GH pulses were detected and characterized by cluster analysis. Thirty-eight healthy women were studied on one or two occasions during the menstrual cycle. Integrated serum GH concentrations (IGHC) were higher during the late follicular phase (identical with the PO phase, characterized by high estrogen and low progesterone levels) compared to the early follicular phase (characterized by low levels of estrogen and progesterone), while GH concentrations during the midluteal phase (characterized by high levels of estrogen and progesterone) were intermediate. There were no differences in pulse frequency, but GH pulse amplitude was increased during the late follicular phase compared to the other menstrual phases. Multiple linear regression analysis revealed that estradiol correlated positively and progesterone negatively with GH pulse amplitude. The authors concluded that late follicular phase concentrations of estradiol enhance circulating GH levels via an amplitude rather than a frequency-modulating effect on GH release. Progesterone may blunt this estrogen effect, thus resulting in more modest midluteal phase concentrations of GH.

Previous studies have not scrutinized whether this difference in GH status is caused by increased secretion or reduced clearance of pituitary GH. Furthermore, it has been unclear whether the PO phase is accompanied by changes in circulating IGF-I. We have investigated the 24-h GH release patterns in the EF versus the PO menstrual phase in the same women to minimize interindividual variation (6). Ten young healthy women with regular menses and comparable anthropometric measures were studied by deconvolution analysis of GH profiles obtained by blood sampling every 20 min for 24 h followed by an L-arginine stimulation test. Deconvolution analysis allows estimation of GH secretion rate and plasma half-life, in addition to amplitude and number of GH pulses. A high-sensitivity immunofluorometric GH assay was used, and the levels of IGF-I and IGF-binding protein (IGF-BP-3) were measured. There were no differences in the basal GH secretion rate or half-life during the two phases, whereas the pulsatile GH production rate was significantly elevated during the PO compared to the EF phase (Fig. 7.1). The number of GH secretory bursts identified during the 24-h sampling period was significantly increased by approximately 30% during the PO compared to the EF phase. There was no difference in GH pulse mass or amplitude between the two phases. Furthermore, the mean 24-h serum GH concentration was significantly increased in the PO versus the EF phase. Estradiol correlated positively with several features of GH status in the PO cycle phase, indicating that E_2 may be a significant determinant of GH secretion. Serum GH increased similarly after arginine infusion in both menstrual phases. Serum IGF-I levels were increased in the PO phase compared to the EF phase, whereas serum IGF-BP-3, IGF-II, and GHBP were similar during the two phases.

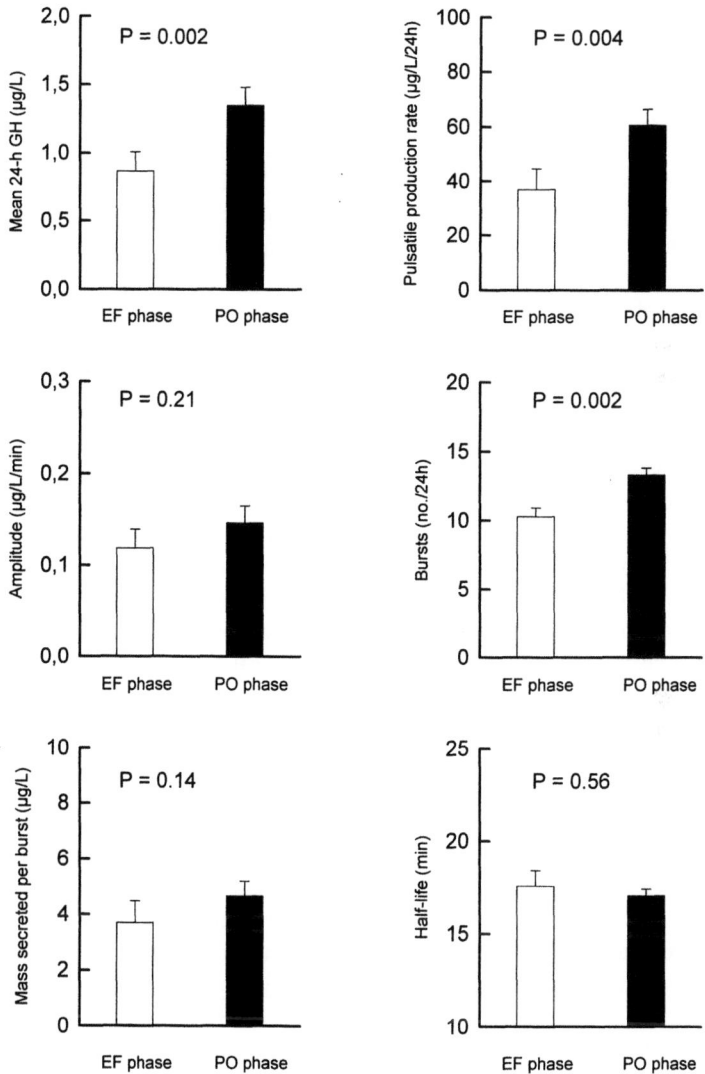

FIGURE 7.1. Quantitative characteristics of endogenous GH secretion and half-lives during the early follicular (EF) and periovulatory (PO) phases derived from deconvolution analysis. The data are mean ± SEM. *p* values reflect paired (within-subject) comparisons in *n* = 10 women. Data are derived from Ovesen et al. (6).

The conclusion of the study was that the elevated GH levels during the PO phase of the menstrual cycle are caused by increased GH production rate and burst frequency. Because there was a significant combined increase in plasma IGF-I and 24-h GH concentrations during the PO phase, it was suggested that endogenous estradiol acts to increase the hypothalamopituitary central drive of the GH–IGF-I axis or reduce IGF-I negative feedback actions on the hypothalamopituitary unit.

In an earlier study in normal women, 24-h sampling in the follicular and luteal phase of the menstrual cycle revealed no difference in the integrated concentrations of GH during the two phases (7). Similar observations have been made in studies based on daily blood samples throughout the menstrual cycle in which there were no changes in serum GH concentrations (1,8,9), while others have found a periovulatory rise in serum GH (2,4). The last five studies were all based on single GH measurements, which only poorly reflect GH secretion. Concerning stimulated GH levels, conflicting results have been observed; two studies reported increased GH response to arginine (3) and exercise (1), whereas other studies have failed to show a difference in GH responses throughout the menstrual cycle (6,10,11).

Our observation of an increased IGF-I level in the PO phase, which suggests a central stimulation of GH release, contrasts with studies showing unaltered IGF-I levels (12–14). The combined increase in serum IGF-I and serum GH observed during the PO phase in our analysis resembles that which occurs in pubertal girls (15), in whom high IGF-I concentrations are considered to result from sex hormone-induced increase in GH secretion. Both exogenous and endogenous estrogens influence circulating IGF-I. In postmenopausal women treated with oral estrogens, the rise in GH levels are accompanied by decreased rather than increased IGF-I levels (16,17). This result had led to the suggestion that oral estrogen administration predominantly acts to suppress hepatic IGF-I production, which subsequently feeds back to stimulate pituitary GH release. Occasional studies employing transdermal estrogen administration in postmenopausal women have observed small increments in both GH and IGF-I (16). This possible route-dependent difference in estrogen action is traditionally explained as a pharmacological first-pass effect of orally administered estrogen to inhibit hepatic IGF-I production and release. As these results were obtained in postmenopausal women receiving exogenous estrogens, they may not apply to premenopausal, regularly cycling women. Endogenous estradiol appears to correlate positively with IGF-I (18,19), which is in accord with our study, but it is not fully established whether this effect is mediated through the concomitant GH release.

Body composition is known to influence the activity of the GH axis. In a recent study of middle-aged individuals of both sexes, a strongly negative correlation between mean 24-h GH secretion (or peak GH secretion during an L-arginine stimulation test) and intraabdominal fat content assessed by quantitative computerized tomography was found (Fig 7.2) (20,21). Furthermore, this negative correlation between GH secretion and viceral fat mass accounted for the majority of the variation in activity of the GH axis, even after adjusting for E_2 concentrations and gender. The mechanism underlying the negative relationship between abdominal obesity and GH secretion is not understood, but the acute effects of E_2, which occur within hours or days, are not the result of changes in body composition.

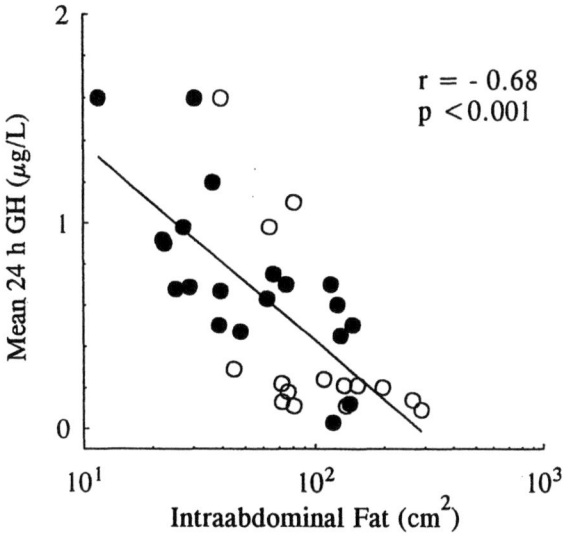

FIGURE 7.2. Log-linear relationship between viceral fat mass and secretory activity of the GH axis in a population of middle-aged women and men (age range, 27–59 years). Intraabdominal fat was quantified by computerized axial tomography. *Open circles*, males; *solid circles*, females. Data are derived from Vahl et al. (21).

Conclusion

In conclusion, the release of GH varies during the menstrual cycle, with elevated levels during the periovulatory phase caused by increased GH production rate and attendant higher burst frequency. The increased GH secretion was accompanied by an IGF-I elevation and positively correlated to serum E_2. Furthermore, body composition is a significant marker of GH status in both men and women, with abdominal obesity being a negative determinant of GH levels.

References

1. Hansen AP, Weeke J. Fasting serum growth hormone levels and growth hormone responses to exercise during normal menstrual cycles and cycles of oral contraceptives. Scand J Clin Lab Invest 1974;34:199–205.
2. Frantz AG, Rabkin MT. Effects of estrogen and sex difference on secretion of human growth hormone. J Clin Endocrinol Metab 1965;25:1470–80.
3. Merimee TJ, Fineberg SE, Tyson JE. Fluctuations of human growth hormone secretion during menstrual cycle: response to arginine. Metabolism 1969;18:606–8.
4. Genazzani AR, Lemarchand Beraud T, Aubert ML, Felber JP. Pattern of plasma ACTH, hGH, and cortisol during menstrual cycle. J Clin Endocrinol Metab 1975;41:431–7.
5. Faria AC, Bekenstein LW, Booth RAJ, Vaccaro VA, Asplin CM, Veldhuis JD, Thorner MO, Evans WS. Pulsatile growth hormone release in normal women during the menstrual cycle. Clin Endocrinol (Oxf) 1992;36:591–6.

6. Ovesen P, Vahl N, Fisker S, Veldhuis JD, Christiansen JS, Jorgensen JOL. Increased pulsatile but not basal growth hormone secretion rates and plasma insulin-like growth factor I levels during the periovulatory interval in normal women. J Clin Endocrinol Metab 1998;83(5):1662–7.
7. Zadik Z, Chalew SA, McCarter RJJ, Meistas M, Kowarski AA. The influence of age on the 24-hour integrated concentration of growth hormone in normal individuals. J Clin Endocrinol Metab 1985;60:513–6.
8. Holst N, Jenssen TG, Burhol PG, Haug E, Forsdahl F. Plasma gastrointestinal hormones during spontaneous and induced menstrual cycles. J Clin Endocrinol Metab 1989;68:1160–6.
9. Stone BA, Marrs RP. Growth hormone in serum of women during the menstrual cycle and during controlled ovarian hyperstimulation. Fertil Steril 1991;56:52–8.
10. Evans WS, Borges JL, Vance ML, Kaiser DL, Rogol AD, Furlanetto R, Rivier J, Vale W, Thorner MO. Effects of human pancreatic growth hormone-releasing factor-40 on serum growth hormone, prolactin, luteinizing hormone, follicle-stimulating hormone, and somatomedin-C concentrations in normal women throughout the menstrual cycle. J Clin Endocrinol Metab 1984;59:1006–10.
11. Gelato MC, Pescovitz OH, Cassorla F, Loriaux DL, Merriam GR. Dose-response relationships for the effects of growth hormone-releasing factor-(1-44)-NH2 in young adult men and women. J Clin Endocrinol Metab 1984;59:197–201.
12. Wang HS, Lee JD, Soong YK. Serum levels of insulin-like growth factor I and insulin-like growth factor-binding protein-1 and -3 in women with regular menstrual cycles. Fertil Steril 1995;63:1204–9.
13. Van Dessel HJHMT, Chandrasekher Y, Yap OWS, Lee PDK, Hintz RL, Faessen GHJ, Braat DDM, Fauser BCJM, Giudice LC. Serum and follicular fluid levels of insulin-like growth factor I (IGF-I), IGF-II, and IGF-binding protein-1 and -3 during the normal menstrual cycle. J Clin Endocrinol Metab 1996;81:1224–31.
14. Klein NA, Battaglia DE, Miller PB, Soules MR. Circulating levels of growth hormone, insulin-like growth factor-I and growth hormone binding protein in normal women of advanced reproductive age. Clin Endocrinol (Oxf) 1996;44:285–92.
15. Albertsson Wikland K, Rosberg S, Karlberg J, Groth T. Analysis of 24-hour growth hormone profiles in healthy boys and girls of normal stature: relation to puberty. J Clin Endocrinol Metab 1994;78:1195–201.
16. Weissberger AJ, Ho KK, Lazarus L. Contrasting effects of oral and transdermal routes of estrogen replacement therapy on 24-hour growth hormone (GH) secretion, insulin-like growth factor I, and GH-binding protein in postmenopausal women. J Clin Endocrinol Metab 1991;72:374–81.
17. Friend EF, Hartman ML, Pezzoli SS, Clasey JL, Thorner MO. Both oral and transdermal estrogen increase growth hormone release in postmenopausal women— a clinical research center study. J Clin Endocrinol Metab 1996;81:2250–6.
18. Ho KY, Evans WS, Blizzard RM, Veldhuis JD, Merriam GR, Samojlik E, Furlanetto R, Rogol AD, Kaiser DL, Thorner MO. Effects of sex and age on the 24-hour profile of growth hormone secretion in man: importance of endogenous estradiol concentrations. J Clin Endocrinol Metab 1987;64:51–8.
19. Massa G, Igout A, Rombauts L, Frankenne F, Vanderschueren Lodeweyckx M. Effect of oestrogen status on serum levels of growth hormone-binding protein and insulin-like growth factor-I in non-pregnant and pregnant women. Clin Endocrinol (Oxf) 1993;39:569–75.

20. Vahl N, Jorgensen JOL, Jurik AG, Christiansen JS. Abdominal adiposity and physical fitness are major determinants of the age-associated decline in stimulated GH secretion in healthy adults. J Clin Endocrinol Metab 1996;81:2209–15.
21. Vahl N, Jorgensen JOL, Skjaerbaeck C, Veldhuis JD, Orskov H, Christiansen JS. Abdominal adiposity rather than age and sex predicts the mass and patterned regularity of growth hormone secretion in mid-life healthy adults. Am J Physiol 1997;272:E1108–16.

8

Growth Hormone Axis Response to Superovulation

BRUCE R. CARR

There is growing evidence that both endogenous as well as exogenous steroids augment growth hormone (GH) synthesis and secretion. In men, as well as women, basal and stimulated GH levels correlate with circulating estrogen levels under a variety of circumstances. To further support this hypothesis, there are a number of specific clinical examples. First, a significant increase in GH secretion occurs during normal and precocious pubertal development (1–3). Second, in girls with Turner syndrome, pulsatile GH activity is augmented during estrogen treatment (4). Third, postmenopausal women treated with exogenous estrogens have increased basal and growth hormone-releasing hormone-stimulated GH secretion (5). Finally, basal plasma GH levels rise in men treated with estrogens and fall to basal levels after cessation of therapy (6). In contrast, gonadotropin-releasing hormone (GnRH) agonists typically suppress estrogen production and inhibit GH release in women with endometriosis or leiomyomas, as well as in children treated with GnRH agonists for precocious puberty (7,8).

Previous investigations have suggested an important role for GH in the stimulation of estrogen by the growing follicle and in augmenting ovulation in infertile women undergoing superovulation (9–11). Infertility patients treated with human menopausal gonadotropins (HMG) to induce superovulation provide an ideal population to study the influence of high endogenous estrogen levels on the GH axis. In this chapter, the effects of GH on estradiol production, in vitro as well as in vivo, reviewed. In addition, the impact of superovulation on the GH–IGF-I axis is discussed.

Effect of Growth Hormone on Estradiol Production: In Vitro Studies

The presence of GH and insulin-like growth factor-I (IGF-I) were determined in follicular fluid obtained from ovarian follicles of women undergoing superovulation with gonadotropins with or without GH therapy (12). The levels of GH in the follicular fluid in the GH-treated subjects were significantly greater than in the follicular fluid from patients treated with gonadotropins alone [8.5 ± 0.6 mU/l (mean ± SEM) versus 6.2 ± 0.8 mU/l, respectively]. Follicular fluid IGF-I levels also tended to be higher in patients treated with GH plus gonadotropins compared to the patients treated with gonadotropin alone, but the values did not reach statistical significance.

In another study, investigators determined the levels of IGF-I and IGF-II in follicular fluid obtained from follicles from women undergoing superovulation with or without the addition of GH treatment (13). The effect of superovulation on IGF-I levels in follicular fluid was greater in the GH-treated subjects (32 nmol/l) than in controls (28, $p = 0.05$). On the other hand, the effect of GH plus superovulation on IGF-II levels in follicular fluid was not different from that of superovulation alone.

Barreca et al. investigated both in vivo and in vitro effects of GH on estradiol secretion as well as IGF-I and IGF-II production by human granulosa cells (11). Women undergoing standard ovulation induction with human menopausal gonadotropin (HMG) alone or HMG plus GH (0.1 IU/kg per body weight per day) were investigated. Granulosa cells were obtained at the time of follicular aspiration for in vitro fertilization. The cells were first incubated with serum-free medium for the first 24 h; the medium was then removed and analyzed for estradiol and progesterone content. Estradiol and progesterone levels were greater in granulosa cells obtained from patients who were previously treated with GH plus gonadotropins versus gonadotropins alone ($p < 0.001$).

In other studies, granulosa cells were obtained from patients treated with gonadotropins alone (11). These cells were placed in a culture and exposed to various doses of GH from 0 to 200 µg/l in the presence or absence of antagonists to the IGF receptor. In these experiments estradiol levels increased as a function of dose of GH in the culture media, reaching a peak at 100 µg/l GH. In cells treated with GH plus the antagonist to the IGF-I receptor, there was no stimulation of estradiol secretion compared to control cells. In addition, granulosa cells obtained from women treated with gonadotropins alone were placed in a culture for 24 h. The media was removed and replaced with serum-free media containing increasing concentrations of IGF-I or IGF-II with or without GH (50 µg/l). After 24 h in the presence or absence of GH, estradiol secretion increased as a function of the IGF-I concentration in the culture media. The concentrations of estradiol was greater in cells treated with GH plus IGF-I. Likewise, in cells treated with IGF-II alone, the concentration of estradiol secretion increased modestly, but the levels of estradiol were greatest in cells treated with GH plus IGF-II.

In summary, the results of these in vivo and in vitro investigations suggest that GH directly (or indirectly via IGFs) stimulates granulosa cells to secrete estrogen and progesterone. In addition, it appears that both IGF-I and IGF-II secretion can be augmented in granulosa cells treated with GH in vitro and, to a lesser extent, in follicular fluid samples after in vivo treatment.

Effect of Growth Hormone Plus Superovulation on Follicular Response: In Vivo Studies

Stone and Marrs compared data in women with low and high basal GH levels during superovulation (14). More oocytes were recovered from patients with higher GH levels. In addition, a number of investigators have studied the addition of GH treatment in women who exhibited a poor response to gonadotropins alone (9,10,15,16). These investigations demonstrated that patients treated with GH and gonadotropins can have an improved follicular response and increased pregnancy rates than patients treated with gonadotropins alone. Owen et al. reported a controlled, randomized, and double-blind prospective trial of the effect of gonadotropins with and without GH (16). Thirteen patients received GH at 24 IU intramuscularly on alternate days for a maximum of 2 weeks, and 12 patients received placebo beginning on the first day of human menopausal gonadotropin treatment. Women who received the GH required significantly fewer ampoules of HMG compared to the group receiving placebo. GH plus gonadotropin treatment did not significantly affect the number of developing follicles or the number of oocytes collected. However, more oocytes were fertilized and a greater number of pregnancies were achieved in patients receiving GH than in those receiving placebo.

The effect of GH-releasing hormone (GHRH) combined with gonadotropin for superovulation was studied in a group of women undergoing superovulation for in vitro fertilization (17). All patients received standard doses of GnRH agonist to downregulate endogenous gonadotropin secretion followed by treatments with either exogenous gonadotropins (a standard dose) or GHRH (5 µg/kg per body weight per day) on day 1 to the last day of gonadotropin administration. The administration of GHRH resulted in a shortening of the stimulatory cycle by approximately 2 days and a significant reduction in the total number of gonadotropin ampoules utilized per patient. The number of follicles recruited and the number of oocytes retrieved in both groups were the same.

Effect of Superovulation with Human Menopausal Gonadotropins with and Without GnRH Analogs on Growth Hormone Levels in Women

In our investigation, three groups of women were recruited (18). Group one (I) consisted of 8 healthy regularly cycling women who were used as controls during a natural cycle. Group two (II) consisted of women undergoing superovulation combined with intrauterine insemination for infertility ($n = 30$). Group three (III) consisted of women undergoing treatment for in vitro fertilization (IVF) that included a standard treatment with GnRH agonists plus HMG ($n = 30$). The characteristics of these patients are presented in Table 8.1. The mean age and body weight between the groups were not statistically

TABLE 8.1. Patient characteristics and peak serum GH and estradiol concentrations.

Protocol	Number of patients	Age[a]	Weight[a] (kg)	Total ampoules of pergonal[a]	Peak estradiol (nmol/l)[a]	Peak GH (nmol/l)[a]
Group I (natural cycle)	8	28 ± 2.8	54 ± 5.2	—	1.19 ± 0.21[c,d]	2.54 ± 1.15[g]
Group II (hMG)	30	33 ± 1.0	60 ± 3.1	23.8 ± 2.9	5.44 ± 0.62[b,d]	8.70 ± 1.58[f]
Group III (hMG + GnRH agonist)	30	35 ± 1.0	60 ± 1.2	27.5 ± 0.7	8.73 ± 0.91[b,e]	7.54 ± 1.12[f]

[a] Mean ± SEM.
[b]>[c] $p = 0.01$.
[d]>[e] $p = 0.01$.
[f]>[g] $p = 0.01$.
Reproduced with permission from Wilson et al. (18).

significantly different. The total number of ampoules of HMG given to group II, 23.8 ± 2.94 (mean ± SEM), was not statistically different from that of patients in group III (27.5 ± 0.7). Peak estradiol levels were significantly greater in group III (8.73 ± 0.91) than in group II (5.44 ± 0.62) or group I (1.19 ± 0.2). Also, peak estradiol levels were significantly greater in group II than in group I. Peak GH levels were similar in group II and III (8.70 ± 1.58 and 7.54 ± 1.2, respectively) but significantly greater than in group I (2.54 ± 1.5).

The length of a follicular phase and peak estradiol levels of all groups were variable. The data were normalized by assigning the peak estradiol value for each patient on day 10, regardless of the length of the stimulation cycle. An increase in the estradiol levels as a function of time is illustrated in Figure 8.1. Estradiol levels increased modestly in group I throughout the menstrual cycle. As expected, the greatest increase in estradiol levels was seen in group II and III. Follicular-phase basal-to-maximal peak levels of estradiol in superovulated and nonstimulated women was also investigated. Each group demonstrated a significant increase in peak estradiol levels compared to basal levels (Fig. 8.2A).

Daily GH levels in superovulated and nonstimulated women during the follicular phase of the menstrual cycle were also determined. Daily GH levels varied throughout the early part of the menstrual cycle until approximately day 9, at which time the levels of estradiol rose in all groups. At day 9, the increase in GH levels was only modest in group I subjects, but in groups II and III, GH increased markedly just before the injection of human chorionic gonadotropin hCG (Fig. 8.3). Basal-to-peak GH increments were also investigated (Fig. 8.2B). GH levels increased from basal to peak in all three groups, but the increase in GH levels in group I was not statistically significant. In contrast, basal-to-peak GH increments in group II and III were statistically significant.

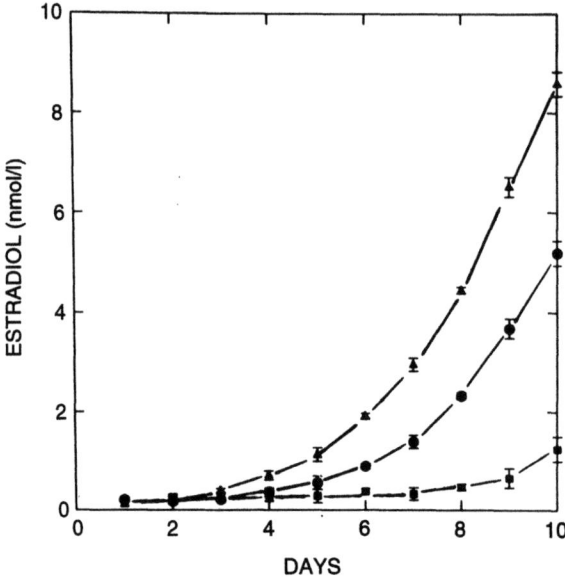

FIGURE 8.1. Follicular phase serum estradiol levels in unstimulated and superovulated women. Daily estradiol levels in superovulated and unstimulated women were measured during the follicular phase of the menstrual cycle. Because the length of the follicular phase was different in each of the women studied, the data were centered by adjusting the peak day of estradiol to day 10. Each data point represents the mean (± SEM) estradiol levels, expressed in nanomoles per liter. *Squares*, group I (unstimulated) patients; *circles*, group II (human menopausal gonadotropin, HMG); *triangles,* group III (HMG-leuprolide acetate). Reproduced with permission from Wilson et al. (18).

Several investigators have reported an increase in estradiol levels associated with increased GH release. We have also demonstrated that the inhibition of estradiol secretion in reproductive-age women by treatment with GnRH agonists results in diminished GH levels or GHRH-stimulated GH release (8). As stated previously, during the normal cycle, estradiol and GH levels increased modestly but the increase in GH levels did not reach statistical significance. However, in superovulated women the increase in estradiol and GH levels was marked. The dramatic GH response in HMG-stimulated cycles may reflect the higher levels of estradiol achieved. As a function of time, the gradual increase in estradiol levels precedes the sharp rise in GH levels, suggesting a possible threshold effect for estradiol for pituitary GH secretion. It appears that the inhibitory effects of GnRH agonists on GH secretion can be overcome by treating with gonadotropins. In fact, patients treated with GnRH agonists plus HMG (group III) exhibited higher levels of estradiol and similar levels in GH patients compared to group II. This study suggests estradiol is a primary regulator of GH release in women. Possibly in normal cycles, and definitely in superovulated cycles, increased GH levels also may directly or indirectly increase the effects of IGF on the ovarian follicle to further augment follicular growth and estrogen secretion. The mechanisms by which estradiol increases GH release are not clearly understood.

8. GH Axis Response to Superovulation   79

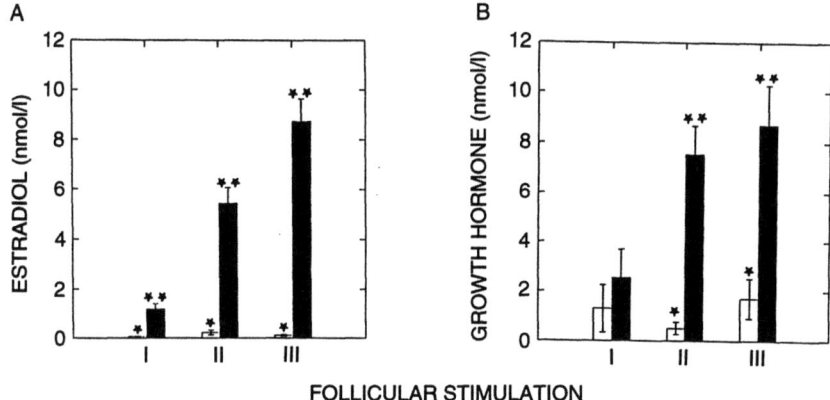

FIGURE 8.2A,B. Follicular phase basal and maximal peak serum levels of (*A*) estradiol and (*B*) GH in superovulated and unstimulated women. Each *bar* represents the mean ± SEM. *Open bars*, serum estradiol and GH levels on day 1 of follicular stimulation with HMG or day 3 of the unstimulated normal cycle; *solid bars*, peak estradiol and GH response in the same women. Group I represents normal unstimulated control women; group II women were treated with HMG alone; and group III women were treated with GnRH agonist plus HMG. In paired *t* tests, ★ ★ is significantly different from ★ ($p = 0.005$). Reproduced with permission from Wilson et al. (18).

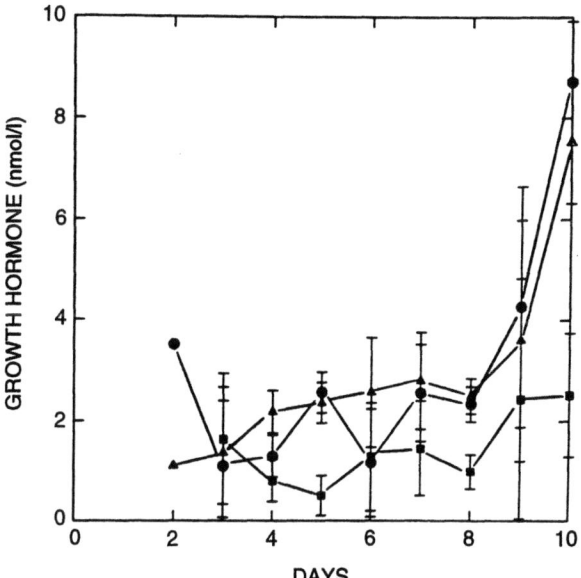

FIGURE 8.3. Daily serum GH levels in superovulated and unstimulated women during the follicular phase of the menstrual cycle. Because the length of the follicular phase was different in each woman studied, the data were normalized by adjusting the peak day of GH secretion to day 10. Each data point represents the mean (± SEM) GH levels, expressed in nanomoles per liter. *Squares*, group I (unstimulated); *circles*, group II (HMG alone); *triangles*, group III (HMG plus leuprolide acetate). Reproduced with permission from Wilson et al. (18).

In summary, the results of our investigation as well as those of others suggest that estrogen levels are correlated with an increased release of GH. We have demonstrated that superovulation with HMG is associated with an increase in GH levels compared with values in nonstimulated normal ovulatory cycles.

References

1. Mauras N, Blizzard RM, Link K, Johnson ML, Rogol AD, Veldhuis JD. Augmentation of growth hormone secretion during puberty: evidence for a pulse amplitude-modulated phenomenon. J Clin Endocrinol Metab 1987;64:596–601.
2. Mansfield MJ, Rudin CR, Criler JF, et al. Changes in growth and serum growth hormone and plasma somatomedin-C levels during suppression of gonadal sex steroid secretion in girls with central precocious puberty. J Clin Endocrinol Metab 1988;66:3–9.
3. Miller JD, Tannenbaum GS, Colle E, Guyda HJ. Daytime pulsatile growth hormone secretion during childhood and adolescence. J Clin Endocrinol Metab 1982;55:989–91.
4. Mauras N, Rogol AD, Veldhuis JD. Specific, time-dependent actions of low-dose ethinyl estradiol administration on the episodic release of growth hormone, follicle-stimulating hormone, and luteinizing hormone in prepubertal girls with Turner's syndrome. J Clin Endocrinol Metab 1989;69:1053–8.
5. Dawson-Hughes B, Stern D, Goldman J, Reichlin S. Regulation of growth hormone and somatomedin C secretion in postmenopausal women: effect of physiological estrogen replacement. J Clin Endocrinol Metab 1986;63:424–32.
6. Wiedemann E, Schwartz E, Frantz AG. Acute and chronic estrogen effects upon serum somatomedin activity, growth hormone, and prolactin in man. J Clin Endocrinol Metab 1976;42:942–52.
7. Harris DA, Van Vliet G, Egli CA, et al. Somatomedin-C in normal puberty and in true precocious puberty before and after treatment with a potent luteinizing hormone-releasing hormone agonist. J Clin Endocrinol Metab 1985;61:152–9.
8. Word RA, Odom MJ, Byrd W, Carr BR. The effect of gonadotropin-releasing agonists on growth hormone secretion in adult premenopausal women. Fertil Steril 1990;54:73–8.
9. Homburg R. Ovulation induction in gonadotrophin-resistant women. Baillière's Clin Obstet Gynecol 1993;7:349–61.
10. Homburg R, West C, Torresani T, Jacobs HS. A comparative study of single-dose growth hormone therapy as an adjuvant to gonadotrophin treatment for ovulation induction. Clin Endocrinol 1990;32:781–5.
11. Barreca A, Artini PG, Del Monte P, Ponzani P, Pasquini P, Cariola G, Volpe A, Genazzani AR, Giordano G, Minuto F. *In vivo* and *in vitro* effect of growth hormone on estradiol secretion by human granulosa cells. J Clin Endocrinol Metab 1993;77:61–7.
12. Barreca A, Minuto F, Volpe A, Ceccheli E, Cella F, Del Monte P, Artini P, Giordano G. Insulin-like growth factor-I (IGF-I) and IGF-I binding protein in the follicular fluids of growth hormone treated patients. Clin Endocrinol 1990;32:497–505.
13. Owen EJ, Torresani T, West C, Mason BA, Jacobs HS. Serum and follicular fluid insulin-like growth factors I and II during growth hormone co-treatment for in vitro fertilization and embryo transfer. Clin Endocrinol 1991;35:327–34.

14. Stone BA, Marrs RP. Ovarian response to menopausal gonadotropins in groups of patients with differing basal growth hormone levels. Fertil Steril 1992;58:32–6.
15. Homburg R, West C, Torresani T, Jacobs HS. Cotreatment with human growth hormone and gonadotropins for induction of ovulation: a controlled clinical trial. Fertil Steril 1990;53:254–60.
16. Owen EJ, Shoham Z, Mason BA, Ostergaard H, Jacobs HS. Cotreatment with growth hormone, following pituitary suppression, for ovarian stimulation in in vitro fertilization: a randomized, double-blind, placebo-controlled trial. Fertil Steril 1991;56:1104–10.
17. Volpe A, Coukos G, Barreca A, Giordano G, Artini PG, Genazzani AR. Clinical use of growth hormone-releasing factor for induction of superovulation. Hum Reprod 1991;6:1228–32.
18. Wilson EE, Word RA, Byrd W, Carr BR. Effect of superovulation with human menopausal gonadotropins on growth hormone levels in women. J Clin Endocrinol Metab 1991;73:511–5.

9

Androgenic Modulation of the Growth Hormone–IGF Axis and Its Impact on Metabolic Outcomes

David A. Fryburg, Arthur L. Weltman, Linda A. Jahn,
Judy Y. Weltman, Eugene Samojlik, Raymond L. Hintz,
and Johannes D. Veldhuis

The maintenance of lean body mass, particularly skeletal muscle mass, has much relevance to human health because skeletal muscle serves two important functions. First, it is a major protein reservoir and provides amino acids for the synthesis of protein in more critical tissues during periods of caloric depletion and catabolic illness (1,2). Second, the loss of skeletal muscle that occurs with age or chronic illness likely contributes to the risk of falls, fractures, and subsequent deterioration of functional capacity and health. Comprehension of the physiological factors that regulate skeletal muscle (and whole-body) protein mass are therefore relevant to human health.

The protein anabolic actions of growth hormone (GH) and testosterone have been recognized for many years (3–5). Experimental or clinical withdrawal of either hormone yields significant declines in corporal protein (lean body) mass, whereas replacement or excessive use of these hormones increases protein mass (6,7). Of particular interest has been the effects of these hormones on skeletal muscle. Studies in both humans and animals suggest that both testosterone and GH increase skeletal muscle protein mass via increases in muscle protein synthesis (8,9). The effect of GH appears relatively rapidly and at the level of skeletal muscle, as previous studies have demonstrated that local, intraarterial infusions of GH for 6 h stimulates human muscle protein synthesis without affecting muscle protein degradation (8). This effect of GH, which has also been observed for IGF-I (10), stands in contrast to that of insulin, which promotes net anabolism by decreasing protein degradation (11).

One of the complicating facets of studying the effects of testosterone on human protein metabolism is the confounding effect of testosterone on GH secretion. Other investigators have observed that testosterone administration may increase GH secretion, an effect that may be caused by aromatization of testosterone to estradiol which, in turn, may act to stimulate GH secretion (12–14). Indirect stimulation of GH secretion may therefore prohibit the identification of the effects of testosterone, per se, as distinct from those of GH. The overall goal of the present study was to quantify GH secretion under short-term conditions of testosterone manipulation in healthy young men as well as to examine the impact of these manipulations on two metabolic outcomes.

Methods

Hormonal Manipulation

To determine the effects of varied testosterone on GH secretion and metabolic outcomes, a double-blind, randomized, three-way crossover trial was undertaken. Five healthy young (age, 21–26 years) subjects within 20% of ideal body weight were studied three times under hypogonadal, eugonadal, and hyperandrogenic conditions. All studies were performed after 5 weeks of different injection paradigms. To induce the hypogonadal state, a single injection of the long-acting gonadotropin-releasing hormone (GnRH) agonist leuprolide (7.5 mg i.m.) was administered on week 1 with saline injections for the next 4 weeks. For the eugonadal state, subjects remained in their own physiological eugonadal state by using saline injections for 5 weeks. For the hyperandrogenic state, saline was injected weekly for the first 2 weeks, followed by testosterone enanthate (3 mg kg^{-1} $week^{-1}$ i.m.) for the last 3 weeks. All subjects were counseled to maintain entry-level exercise and dietary routines.

General Clincial Research Center (GCRC) Admission Phase

All subjects were readmitted on the fifth week and received their last injection then. During the inpatient phase, all subjects consumed a weight-maintenance, defined-content diet that was the same during each of the three inpatient stays. The same inpatient protocol was performed under each of the three conditions listed earlier. Generally, the sequence of testing included exercise-stimulated GH secretion (day 1); 24-h-nonstimulated GH secretion (day 2); and GHRH-stimulated GH secretion (day 3). Each subject was admitted to the GCRC (University of Virginia, Charlottesville) the afternoon of day 0, before beginning the inpatient sampling paradigm.

On day 1, after an overnight fast and bed rest, blood samples were withdrawn every 5 min for GH concentrations for 1 h (0700–0800) before the subject exercised from 0800 to 0830 on a cycle ergometer at an exercise intensity associated with a predetermined blood lactate concentration of 2.5 mM, which has been shown to stimulate GH secretion. During all three admissions, each subject exerted the same power output. Samples for GH secretion were withdrawn during the exercise bout (every 5 min) as well as for 3 h after completion of the exercise (every 10 min).

The following morning at 0600, indirect calorimetry was performed over 30 minutes (Delta Trac; Sensor Medics, Anaheim, CA) after the subject awoke and lay in bed in a temperature-stable environment for approximately 1 h. At 0800, quantification of 24-h GH secretion was initiated. Samples for GH were collected every 10 min through 0750 on day 3. After the last GH sample at 0750 on day 3, each subject received an i.v. bolus of growth hormone releasing factor (GRF) at a dose of 1 µg/kg (Serono). Sampling for GH continued every 10 min for the next 4 h. After the post-GRF sampling was completed, the IVs were removed and the subject discharged.

Samples were also collected day 2 for total and free testosterone and estradiol, total IGF-I, and IGF-binding protein (IGF-BP3). Urine for nitrogen and creatinine were collected in two 24-h aliquots from days 1 through 3. There was an 8-week washout period between each of these three treatment protocols.

Assays

All samples were processed together in batch by subject for all three treatment states. GH was assayed with a commercially available immunoradiometric assay (Nichols, San Capistrano, CA) that has a lower limit of detection of 0.2 µg/l. Total plasma IGF-I concentrations were measured by radioimmunoassay after acid chromatographic separation (27). IGF-BP-3 was determined using a two-site, non-competitive IRMA (Diagnostic Systems, Webster, TX). Total and free serum testosterone and estradiol concentrations and serum DHT (dihydrotestosterone) concentrations were measured using previously described methods (28,29). Urinary nitrogen was determined by the Kjeldahl method.

Hormone Secretion Analysis

A multiparameter deconvolution technique was utilized to estimate specific measures of GH secretion from the exercise, GHRH, and the 24-h serum GH concentration profiles (30). The daily pulsatile GH production rate was computed as the product of the mean GH secretory burst mass and frequency.

Statistical Analysis

The randomization code was broken after all the subjects samples had been processed and deconvolved. Statistical comparisons for all five subjects for the leuprolide, saline, and testosterone-treated states were then made via ANOVA for repeated measures and post hoc comparisons using Duncan's multiple range test. Interactions between gonadal steroid modulation and GH secretion on basal metabolic rate and urinary nitrogen excretion were tested with one-way ANCOVA. Statistical analysis was conducted with True Epistat statistical software (Epistat Services, Richardson, TX).

Results

Hormone Responses to the Experimental Manipulations

Table 9.1 summarizes the time course of total serum testosterone levels by treatment condition, as determined by single, morning testosterone samples. At week 1 before any manipulation, subjects started each arm of the study with approximately the same total serum testosterone. For the leuprolide-treated subjects, total testosterone levels rose slightly at week 2 and then declined by week 3, approaching nearly nondetectable levels by week 4. During treatment with saline alone, testosterone concentrations remained fairly stable. Testosterone treatment, initiated after the week 3 sample was withdrawn, caused the serum testosterone to rise to approximately 1500 ng/dl and remained at that level for the remainder of the study. As shown in Table 9.1, the duration of the hypogonadal period approximated the duration of the high-testosterone period.

Table 9.2 summarizes and contrasts the resulting serum concentrations of total and free testosterone and estradiol at week 5. The gonadal steroid manipulations and stanozolol treatment elicited significant alterations in total testosterone concentrations comparable to those presented in Table 9.1. Parallel changes were observed in serum total estradiol as well as free testosterone, and free estradiol concentrations.

Finally, despite fairly parallel alterations in total and free hormone levels, the relationship between total testosterone and estradiol and free testosterone and estradiol diverged during this study. That is, across treatments the total testosterone/total estradiol ratio increased from 24 ± 1 (leuprolide) to 180 ± 40 (saline; $p < 0.001$), but did not change further with testosterone treatment, 189 ± 19. In contrast, the free testosterone/free estradiol ratio increased from 8 ± 1 (leuprolide) to 130 ± 15 (saline) and further to 235 ± 19 (ANOVA: $p < 0.0001$; leuprolide versus saline, $p < 0.0001$; saline versus testosterone, $p < 0.001$).

Metabolic Responses to Gonadal Steroid Manipulation

Leuprolide suppression of gonadal steroid levels resulted in the most marked urinary excretion of nitrogen to 8.7 ± 1.5 g N/g Cr, which diminished in a

TABLE 9.1. Time course for the changes in serum testosterone by treatment.

Treatment/time	Wk 1	Wk 2	Wk 3	Wk 4	Wk 5
Leuprolide	908 ± 77	1190 ± 145^a	457 ± 106^b	121 ± 35^b	54 ± 17^b
Saline	937 ± 63	883 ± 165	896 ± 113	1051 ± 102	1061 ± 138
Testosterone	989 ± 184	1075 ± 97	1000 ± 92	1429 ± 186^b	1515 ± 185^b

All data expressed as mean ± SEM. Serum testosterone concentration, ng/dl.
[a] $p < 0.05$ vs. Wk 1.
[b] $p < 0.01$ vs. Wk 1.

TABLE 9.2. Serum concentrations of total and free testosterone and estradiol at week 5.

Treatment/ [hormone]	Total testosterone	Free testosterone	Total estradiol	Free estradiol
Leuprolide	33 ± 20	0.4 ± 0.3	76 ± 32	0.4 ± 0.3
Saline	796 ± 328[a]	10.5 ± 1.4[a]	309 ± 37[a]	0.9 ± 0.3[a]
Testosterone	1609 ± 314	44.0 ± 10.7	668 ± 84	1.9 ± 0.5

All data expressed as mean ± SEM. Total and free serum testosterone concentrations, ng/dl. Total and free serum estradiol, pg/ml.
[a] $p < 0.01$ vs. leuprolide or testosterone.

stepwise manner to 6.4 ± 1.6 (saline) and 5.4 ± 0.7 (testosterone). These changes were statistically different from one another [$p < 0.005$ by ANOVA; leuprolide versus saline ($p < 0.01$); saline versus testosterone ($p < 0.05$); leuprolide versus testosterone ($p < 0.001$)].

Modulation of gonadal steroid levels, similar to urinary nitrogen loss, also affected oxygen consumption and calculated basal metabolic rate. Oxygen consumption increased by 11% from 236 ± 29 (leuprolide) to 262 ± 25 (saline) and to 272 ± 23 (testosterone) ml O_2 per minute ($p < 0.01$ by ANOVA; leuprolide versus saline or testosterone, $p < 0.02$; saline versus testosterone, $p = NS$). Similar inferences were also made for basal metabolic rate (BMR), which increased from 1628 ± 90 (leuprolide) to 1818 ± 78 (saline) to 1898 ± 104 (testosterone) and 1817 ± 130 (stanozolol) kcal/24 h ($p < 0.01$ by ANOVA; $p < 0.02$ leuprolide versus saline or testosterone; saline versus testosterone, $p = NS$). This increase was still present when energy expenditure values were normalized for fat free mass (data not shown). In addition to affecting basal oxygen consumption, modulation of testosterone levels also altered oxygen consumption at the end of exercise from 2.05 ± 0.83 (leuprolide) to 2.32 ± 0.83 (saline) to 2.47 ± 0.99 (testosterone) l/min ($p < 0.05$ by ANOVA; $p < 0.05$ testosterone versus leuprolide; $p < 0.08$ saline versus testosterone).

GH Secretory Dynamics

In response to altering testosterone levels, the mean serum GH concentration rose from 1.55 ± 0.28 (leuprolide) to 1.92 ± 0.38 (saline) to 2.67 ± 0.38 µg/l (testosterone) (ANOVA:$p < 0.01$; leuprolide or saline versus testosterone, $p < 0.03$). Similarly, the area under the 24-h serum GH concentration curve rose from 2234 ± 811 (leuprolide) to 2780 ± 1146 (saline) to 4578 ± 2629 (testosterone) µg l^{-1}· min^{-1} (ANOVA: $p = 0.05$; leuprolide or saline versus testosterone $p < 0.05$; leuprolide versus saline, $p = NS$).

Similar conclusions were reached using deconvolution analysis; that is, short-term reduction in serum gonadal steroids did not substantially alter GH production rates, whereas short-term testosterone treatment elevated GH production by 22% over saline and 62% over leuprolide (leuprolide, 62 ± 26; saline, 88 ± 43; testosterone, 107 ± 39 μg l^{-1}·day^{-1} (ANOVA: $p < 0.05$; leuprolide versus testosterone, $p < 0.05$; saline versus testosterone, or leuprolide, $p = NS$). The increment in mean integrated serum GH concentrations or estimated production rates from leuprolide to testosterone was not caused by either a change in the number of GH bursts per 24 h or the estimated half-life of GH (data not shown).

Stimulated GH Secretion

Table 9.3 summarizes the measures of exercise and GHRH-stimulated GH secretion during the four different treatment states. All subjects achieved the same exercise intensity during each test (data not shown). Modulation of the gonadal steroid environment did not alter exercise-stimulated GH secretion. Similarly, modulation of gonadal steroids did not alter GHRH-stimulated GH release.

Serum IGF-I Concentrations

As observed for 24-h GH secretion, serum total IGF-I concentrations were also altered by short-term variations in the gonadal steroids. Between leuprolide and saline treatment, IGF-I concentrations did not change (314 ± 24 to 348 ± 16 μg/l), but increased significantly with testosterone treatment (421 ± 29 μg/l; $p < 0.01$). IGF-BP-3 concentrations did not change among leuprolide (3295 ± 164 ng/ml) versus saline (3287 ± 124 ng/ml) versus testosterone (3293 ± 198 ng/ml) treatment conditions.

TABLE 9.3. Effect of gonadal steroid manipulation on exercise and GHRH-stimulated GH secretion.

	Leuprolide	Saline	Testosterone
Exercise			
Mean [GH]	2.1 ± 1.9	1.2 ± 0.6	3.0 ± 1.2
GH Area	1159 ± 1003	652 ± 310	1602 ± 677
GH Prod	28 ± 25	22 ± 12	48 ± 22
GHRH			
Mean [GH]	7.5 ± 2.9	4.5 ± 3.8	8.7 ± 1.9
GH Area	1867 ± 717	1189 ± 909	2037 ± 384
GH Prod	52 ± 22	33 ± 19	68 ± 18

Data expressed as mean ± SEM.
Variables: (a) mean GH (in mg/l); (b) GH area (mg/l × min); (c) GH production (mass, mg/l released after secretagogue).

One of the goals of this experiment was to determine the relative contributions of GH and gonadal steroids. One statistical approach is to examine outcome measures using one-way analysis of covariance. In this analysis, GH production is the covariate and testosterone levels by tertiles are the interactive factors. When organized by tertile of testosterone (i.e., low, normal, or high), accounting for the variability caused by GH production yields a significant relation between urinary nitrogen and testosterone ($p < 0.02$). A similar ANCOVA result with urinary nitrogen excretion as the outcome variable has the same relationship as observed between testosterone and BMR ($p = 0.01$).

Figures 9.1 and 9.2 graphically depict the ANCOVA statistical results. In both of these figures, the x and z axes segregate data by tertile of testosterone and 24-h pulsatile GH secretion. The first tertile for either indicates the lowest third of measured values and the third tertile, the highest. Thus, there are a total of nine cells in each graph. The value displayed as a column in each cell (y axis) reflects the mean of urinary nitrogen/creatinine excretion or BMR

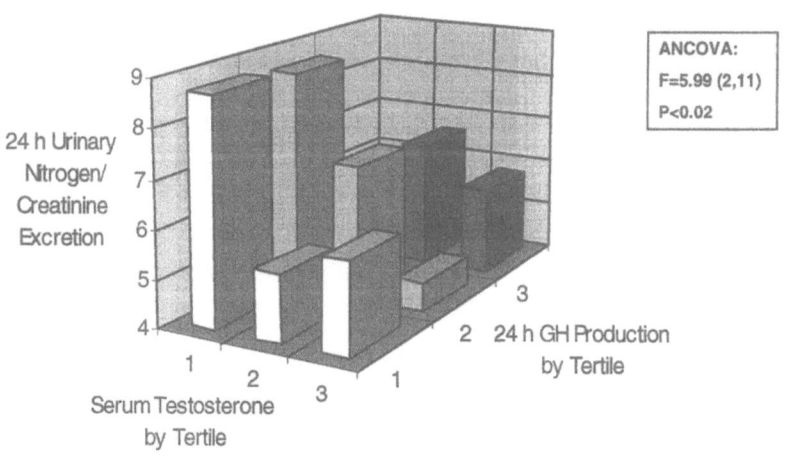

FIGURE 9.1. Triaxial display of urinary nitrogen excretion (normalized for creatinine) segregated by tertiles of serum testosterone concentrations and 24-h pulsatile GH production rate. Tertiles run from low (=1) to high (=3) hormone levels. Movement from low to high serum testosterone tertiles discloses significant changes in urinary nitrogen excretion, whereas a similar pattern is not observed between the low and high GH production tertiles (by ANCOVA). Reproduced with permission from Fryburg DA, Weltman A, Jahn LA, Weltman JY, Samojlik E, Hintz RL, Veldhuis JD. Short-term modulation of the androgen milieu alters pulsatile, but not exercise- or growth hormone (GH)-releasing hormone-stimulated GH secretion in healthy men: impact of gonadal steroid and GH secretory changes on metabolic outcomes. J Clin Endocrinol Metab 1997;82(11):3710–3719. © The Endocrine Society.

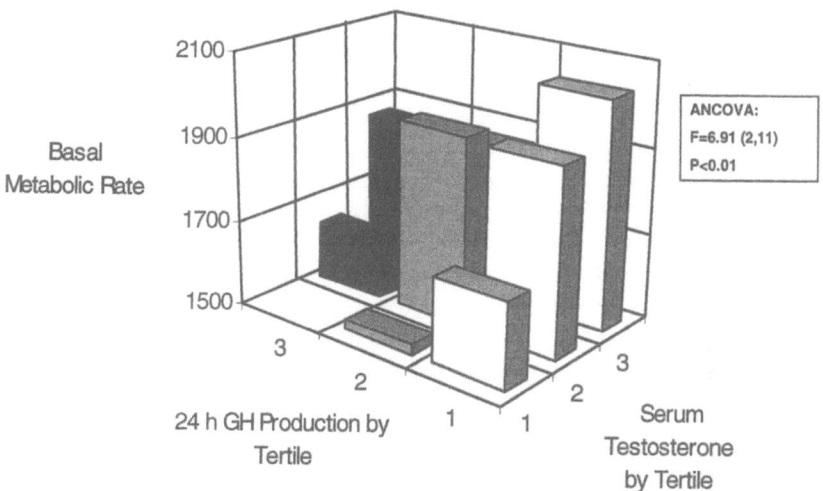

FIGURE 9.2. Triaxial display of basal metabolic rate segregated by tertiles of serum testosterone concentrations and 24-h pulsatile GH production. Tertiles run from low (=1) to high (=3) hormone levels. Movement from low to high serum testosterone tertiles discloses significant changes in basal metabolic rate, whereas a similar pattern is not observed between the low and high GH production tertiles (by ANCOVA) (with permission of publisher). Reproduced with permission from Fryburg DA, Weltman A, Jahn LA, Weltman JY, Samojlik E, Hintz RL, Veldhuis JD. Short-term modulation of the androgen milieu alters pulsatile, but not exercise- or growth hormone (GH)-releasing hormone-stimulated GH secretion in healthy men: impact of gonadal steroid and GH secretory changes on metabolic outcomes. Reproduced with permission from J Clin Endocrinol Metab 1997;82(11):3710–3719. © The Endocrine Society.

measurements of subject data sorted by these tertiles; this value is displayed on the y axis. Note that there is one cell (testosterone tertile 1, GH production tertile 3) into which no subject's data segregated.

Figure 9.1 displays the corresponding ANCOVA for urinary nitrogen excretion. Viewing the data from the lowest to highest testosterone tertile within the lowest or middle GH secretion tertiles suggests that increasing testosterone, even when GH secretion is low, substantially decreases urinary nitrogen loss. Conversely, viewing the data from the lowest to highest GH tertile does not yield the same visual conclusion as that for testosterone. Figure 9.2 depicts a similar finding, that is, serum testosterone and not GH production is the key statistical determinant of the increase in basal metabolic rate.

Discussion

The results of this experiment demonstrated that, in a small cohort of healthy young men: (1) between the short-term hypogonadal and eugonadal states, there is little change in either GH secretion or serum IGF-I concentrations; (2) between the eugonadal and the hyperandrogenic states, GH secretion and IGF-I concentrations increase significantly; (3) the metabolic effects (including urinary nitrogen excretion and basal metabolic rate) of this variation in testosterone are mostly observed, especially between the hypogonadal and eugonadal states in contrast to the hyperandrogenic state; and (4) in this model, testosterone rather than GH accounts for the variability in either urinary nitrogen excretion or basal metabolic rate.

In other studies, elucidating the distinct actions of these two hormones has required the use of hypophysectomized animals in which the singular and combined effects of gonadal steroids and GH could be evaluated. Although some early observations suggested that androgens do not exert GH-independent effects on growth or the accretion of muscle mass (except gender-specific muscles such as rat levator ani) (15), subsequent studies in rats, lambs, and humans supported the notion that androgens either directly promote growth or augment GH-stimulated growth. Klindt et al. (16) observed that in hypophysectomized rats testosterone alone is a weak but demonstrable anabolic agent. By conducting a dose–response study of GH in hypophysectomized animals with or without testosterone supplementation, these investigators demonstrated that the provision of a fixed dose of testosterone significantly augmented GH-stimulated weight gain, particularly at low (but not maximal) doses of GH (16).

Studies in other species, including lambs and humans, support the premise that gonadal steroids can act to promote growth independently of GH. Young et al. (17) demonstrated that hypophysectomized and castrated prepubertal lambs grew in response to testosterone alone. In children with precocious puberty and GH deficiency, Attie et al. (18) observed that premature endogenous production of gonadal steroids per se augmented linear growth without GH. Moreover, gonadal steroids could be additive with GH in promoting growth (18).

The data from the present study are consistent with many of these earlier observations and reemphasize the need to modify the precept that gonadal steroids are *passive* actors in the growth process. Instead of using models in which pituitary function has been experimentally or pathologically eliminated, we investigated this issue by examining some simple metabolic endpoints in the context of controlling for GH secretion. Despite the limitations of urinary nitrogen collections (19), the decline in urinary nitrogen excretion discloses that net protein anabolism has been stimulated. The parallel response in basal metabolic rate—measured independently of urinary nitrogen excretion—provides additional support to the interactive nature of gonadal steroids and physiological amounts of GH at the level of tissue metabolism.

The results of the present study strongly suggest that gonadal steroids are potent modulators of the observed metabolic response and are more highly correlated with these responses than GH secretion. This is not to imply that GH is not necessary for these changes. Rather, based upon a large body of previous experimental work, it is likely that GH is *necessary* for the full anabolic expression of androgens. Thus, the present studies reorient perspective on the relative contributions by androgens and GH to this outcome.

The concept that androgens alter the anabolic response to GH was also suggested by Malhotra et al., who studied the effects of oxandrolone on 24-h GH secretion and linear growth in boys at varying stages of puberty (20). They found that oxandrolone did not affect GH secretion, and that in a multivariate regression model only serum testosterone level and treatment with oxandrolone (not GH secretion) were correlated with the observed growth. Thus, for both growing adolescents as well as fully developed (adult) men, manipulation of androgens can influence metabolic outcome without substantive alterations in GH secretion.

In summary, short-term variation in the gonadal steroid environment alters the 24-h, but not stimulated, GH secretion. In contrast to the relative lack of effect on GH secretion, these large changes in gonadal steroids substantially alter urinary nitrogen balance, basal metabolic rate, and exercise-induced oxygen consumption. Such responses were statistically independent of the observed variation in GH secretion. In the context of previously published data, these observations suggest that gonadal steroids modulate target tissue responses in the presence of physiological amounts of GH.

Acknowledgments. The authors thank the nurses and staff of the General Clinical Research Center of the University of Virginia for their excellent care of the subjects, G. Bauer and K. Kern of the GCRC Core Laboratory for conducting the hormone assays, and Dr. Robert Abbott for statistical advice. This work was supported by USPHS grants AR01881, DK38578, RR00847 to the University of Virginia GCRC, and research grants from the Muscular Dystrophy Association and Eli Lilly, Inc.

References

1. Kinney JM, Elwyn DH. Protein metabolism and injury. Annu Rev Nutr 1983;3: 433–66.
2. Fryburg DA, Barrett EJ. Insulin, GH, and IGF-I regulation of protein metabolism. Diabetes Rev 1994;2:209–38.
3. Evans HM, Simpson ME, Li CH. The gigantism produced in normal rats by injection of the pituitary growth hormone. I. Body growth and organ changes. Growth 1948;12:15–32.
4. Kostyo JL, Nutting DF. Growth hormone and protein metabolism. In: Greep RO, Astwood EB, eds. Handbook of physiology, section 7, vol 4, part 2. Washington, D.C.: American Physiological Society, 1974:187–210.
5. Kochakian CD. Testosterone and testosterone acetate and the protein energy metabolism of castrate dogs. Endocrinology 1937;21:75–755.
6. Salomon F, Cuneo R, Hesp R, Sonksen P. The effects of treatment with recombinant human growth hormone on body composition and metabolism in adults with growth hormone deficiency. N Engl J Med 1989;321:1797–803.

7. Bhasin S, Storer TW, Berman N, Yarasheski KE, Clevenger B, et al. Testosterone replacement increases fat-free mass and muscle size in hypogonadal men. J Clin Endocrinol Metab 1997;82(2):407–13.
8. Fryburg DA, Gelfand RA, Barrett EJ. Growth hormone acutely stimulates forearm muscle protein synthesis in normal humans. Am J Physiol 1991;260:E499–504.
9. Griggs RC, William K, Jozefowicz RF, Herr BE, Forbes G, Halliday D. Effect of testosterone on muscle mass and muscle protein synthesis. J Appl Physiol 1989;66: 498–503.
10. Fryburg DA. Insulin-like growth factor-I exerts growth hormone- and insulin-like actions on human muscle protein metabolism. Am J Physiol 1994;267:E331–6.
11. Louard RJ, Fryburg DA, Gelfand RA, Barrett EJ. Insulin sensitivity of protein and glucose metabolism in human forearm skeletal muscle. J Clin Invest 1992;90:2348–54.
12. Illig R, Prader A. Effect of testosterone on growth hormone secretion in patients with anorchia and delayed puberty. J Clin Endocrinol Metab 1970;30:615–8.
13. Dawson-Hughes B, Stern D, Goldman J, Reichlin S. Regulation of growth hormone and somatomedin-C secretion in postmenopausal women: effect of physiological estrogen replacement. J Clin Endocrinol Metab 1986;63(2):424–32.
14. Ho KY, Evans WS, Blizzard RM, Veldhuis JD, Merriam GR, Samojlik E, Furlanetto EM, Rogol AD, Kaiser DL, Thorner MO. Effects of sex and age on the 24-hour profile of growth hormone secretion in man: importance of endogenous estradiol concentrations. J Clin Endocrinol Metab, 1987;64:51–8.
15. Scow RO, Hagan SN. Effect of testosterone propionate and growth hormone on growth and chemical composition of muscle and other tissues in hypophysectomized male rats. Endocrinology 1965;77:852–8.
16. Klindt J, Ford JJ, MacDonald GJ. Synergism of testosterone propionate with growth hormone in promoting growth of hypophysectomized rats: effect of sexual differentiation. J Endocrinol 1990;127:249–56.
17. Young JR, Mesiano S, Hintz R, Caddy DJ, Ralph MM, Browne CA, Thorburn GD. Growth hormone and testosterone can independently stimulate the growth of hypophysectomized prepubertal lambs without any alteration in circulating concentrations of insulin-like growth factors. J Endocrinol 1989;121:563–70.
18. Attie KM, Ramirez-Nelson R, Conte FA, Kaplan SL, Grumbach MM. The pubertal growth spurt in eight patients with true precocious puberty and growth hormone deficiency: evidence for a direct role of sex steroids. J Clin Endocrinol Metab 1990;71 (4):975–83.
19. Kopple JD. Uses and limitations of the balance technique. J Parenter Enteral Nutr 1987;11(5):79S–85.
20. Malhotra A, Poon E, Tse W-Y, Pringle PJ, Hindmarsh PC, Brook CDG. The effects of oxandrolone on the growth hormone and gonadal axes in boys with constitutional delay of growth and puberty. Clin Endocrinol 1993;38:393–8.

10

Proposed Mechanisms of Sex-Steroid–Hormone Neuromodulation of the Human GH–IGF-I Axis

JOHANNES D. VELDHUIS, WILLIAM S. EVANS, NIKHITA SHAH,
E. SHANNON STORY, MEGAN J. BRAY, AND STACEY M. ANDERSON

Sex-steroid hormones are integral to pubertal development in boys and girls, and sustain sexual function and body composition throughout the human lifetime (1–14). A corollary tenet of human physiology in both the male and female is that sex steroids interact with the growth-promoting GH–IGF-I axis at multiple loci of control (15–17). Although this presentation focuses on the hypothalamopituitary actions of estrogen and androgen in modulating output of the GH–IGF-I axis, other important sites of sex-steroid–GH interplay also operate in health and disease, such as at the levels of the gonad, the IGF-I-binding proteins, GH-receptor expression, and target tissues such as fat, muscle, and bone (18). Extensive experimental data in the rodent further establish sex-steroid–GH axis interactions throughout the pre- and postpubertal lifetimes (19,20). The present symposial update primarily highlights issues explored recently in human-based investigations of sex hormone–GH axis interactions. We also identify some of the exciting and as yet unaddressed challenges within this clinical neuroendocrine theme.

Clinical Paradigms of Prominent Sex-Steroid–GH Axis Neuroendocrine Interactions

Puberty

Clinical studies document several paradigms of remarkable gonadal sex-steroid interactions on GH–IGF-I neuroregulation. Foremost is natural puberty, at which time there is a logarithmic amplification of sex-steroid hormone concentrations with an attendant 1.5- to 10 fold augmentation of daily GH (growth hormone) secretion rates in healthy girls and boys (14,16,17,20–22). This physiological context of joint enhancement of estrogen/androgen and

GH production stimulates remarkable growth in bone mass (thereby conditioning the adult's maximal bone mass endowment), linear height, muscle (lean) mass, sexual development, and a reciprocal decline in visceral fat content (9,16,21,23–28). The dominant corresponding neuroendocrine reactions within the GH–IGF-I axis impact three dynamic modes of neuroregulatory adaptation, as illustrated in Figure 10.1: (i) increased pulsatile GH secretion with enhancement by 1.5- to 10 fold of GH secretory pulse mass (and amplitude) (6,16,19,21,29–31); (ii) a heightening of the 24-h rhythmicity of GH

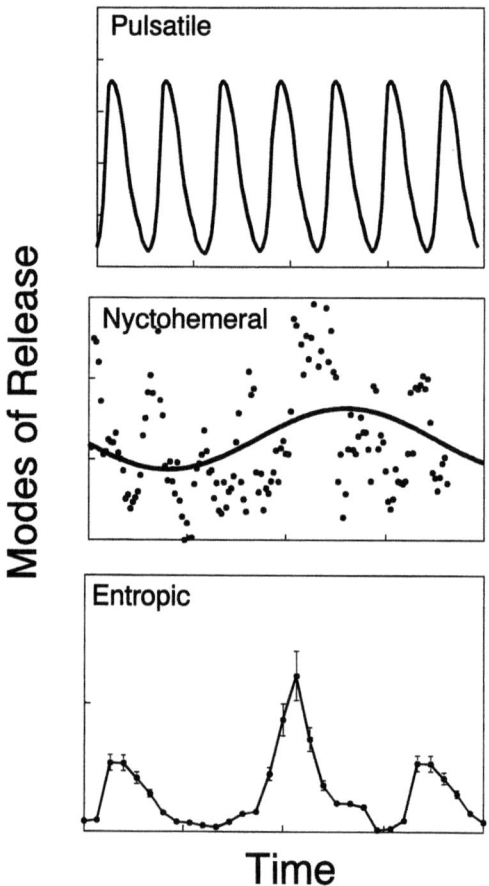

FIGURE 10.1. Schematized illustration of three recognized dynamic endpoints of sex-steroid actions on the GH axis in the human: *top,* pulsatile GH secretion; *middle,* nyctohemeral rhythms; and *bottom,* the entropic (orderliness) mode of GH release. All three principal regulatory mechanisms are amplified in puberty and by aromatizable androgen (e.g., testosterone) or estradiol treatment.

TABLE 10.1. Primary plausible dynamic neuroregulatory adaptations of the GH axis to sex-hormone stimulation.

---

1. Pulsatile mode of GH secretion
   (a) Frequency$^+$
   (b) Amplitude (or mass)*
   (c) Basal secretion rate^{NS}
   (d) Half-life$^+$

2. Nyctohemeral GH release*
   (a) Circadian rhythm^{NS}
   (b) Sleep-activity cycle^{NS}

3. Entropic pattern*
   (orderliness of release process)

$^+$Refuted in studies to date
*Documented in the clinical literature (see Discussion).
^{NS}Not well studied.

release, which is likely endowed by both true circadian and sleep–activity variations over the day and night (20,21,31–33); and (iii) a marked reduction in the moment-to-moment patterned orderliness of the GH release process, which can be quantified by the approximate entropy statistic (8,26,34–36). The neuroendocrine bases for these three prominent modulations of GH secretion are summarized in Table 10.1 and are examined further next.

Nonpubertal Contexts

Other clinical paradigms also underscore the pivotal role of estrogen and androgen in directing the neuroregulation of GH secretion in the human. Such experimental examples include reversible estrogen deficiency states, such as anorexia nervosa (37); gonadal failure in men and women before versus during sex hormone replacement (4,13,16,20,25,26,34,38–58); healthy aging in men, in whom declining serum testosterone concentrations correlate with falling IGF-I and GH production rates (59–63); gonadotropin-releasing hormone (GnRH) agonist-treated children with precocious puberty (64,65) or adults treated experimentally or therapeutically with GnRH agonists (11,27,28,66,67); ovulation induction (68,69) or the spontaneous menstrual cycle (68,70–73); premenopausal ovariectomy followed by estrogen replacement (45); children with constitutionally delayed puberty or isolated GnRH deficiency administered graded amounts of testosterone (26,34,43,54,58,74); GnRH-deficient adults administered GnRH by pulsatile infusion or given testosterone parenterally (43,54,58); and acute androgen deprivation via steroidogenic enzyme inhibitors (75). In general, these clinical studies show parallel effects of testosterone and estradiol repletion, but with two notable

TABLE 10.2. Differences and similarities between testosterone (T) and estrogen (E_2) actions on the GH axis in the (postpubertal) human.

Actions	T	E_2
GH secretory burst mass	↑	↑
Entropy (disorderliness)	↑	↑
Nyctohemeral rhythm	↑	↑
IGF-I concentration*	↑	↓ or no change[+]
Linear growth/epiphyseal fusion[#]	↑	↑
Whole-body protein anabolism*	↑	No
Muscle IGF-I expression*	↑	?
Bone mass	↑	↑

*Gender differences are evident clinically (see text).
[+]Adult.
[#]Pubertal.
?Uncertain status.

exceptions: (i) testosterone but not estradiol replacement increases plasma IGF-I concentrations in the (postpubertal) individual; and (ii) testosterone but not estradiol elevates whole-body anabolism or muscle IGF-I gene expression (15,20,27,28,38): (see Table 10.2).

Role of Specific Sex-Steroid Receptors

The mechanisms by which sex steroids govern the GH axis likely include mediation via corresponding estrogen or androgen receptors. Extensive clinical evidence indicates that certain potent tissue-anabolic actions of testosterone and nonaromatizable androgens can be exerted not only without aromatization [e.g., oxandrolone or stanozolol effects (27,38,76)], but also without necessarily increasing GH release or serum IGF-I levels measurably (15,27,28,38,58,77). In contrast, many other tissue actions of testosterone are exerted following its aromatization in the brain, pituitary gland, fat, or muscle to estrogen. For example, in the human, testosterone stimulation of GH secretion is mediated predominantly, if not exclusively, after aromatization to estrogen (11,20,38,53,58,76,78). Thus, nonaromatizable androgens (e.g., oxandrolone, 5α-dihydrotestosterone, stanozolol) all fail to stimulate GH secretion; conversely, the antiestrogen, tamoxifen, blocks testosterone enhancement of GH release in hypogonadal men (11,53). This fundamental mechanistic principle is illustrated in Figure 10.2. In contrast, such a distinction is not evident in the rat, in which species testosterone and nonaromatizable androgens typically control the GH axis similarly (20). Hence, clinical investigations are especially pertinent to understanding androgenic control of the human GH axis, which in this context shows regulatory differences from that of the rodent.

SEX-STEROID MODULATION OF GH AXIS

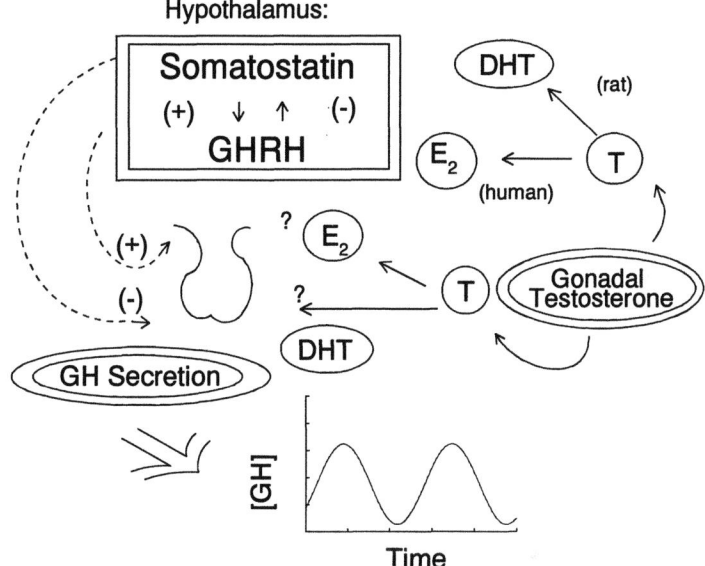

FIGURE 10.2. Concept of the primary actions of testosterone on the GH axis; namely, acting either after aromatization via the estrogen receptor or via the androgen receptor (before or after its 5-alpha reduction). Adapted with permission from (20).

Regulation of GHRH–Somatostatin Actions

Albeit not fully excluded, few if any sex-steroid actions are documented directly on isolated normal human somatotrope cells in vitro, or directly on the human anterior pituitary gland somatotrope population in vivo. Most clinical observations (e.g., of augmented GH pulse mass in response to estrogen therapy) can be accounted for by changes in either hypothalamic growth hormone-releasing hormone (GHRH) or somatostatin regulation, or possibly via other intrapituitary paracrine factors (20) (Table 10.3). A putative, but controversial, exception may be the pituitary feedback actions of IGF-I on somatotrope cells (79–82). To our knowledge, this feedback pathway has not been demonstrated to serve as a target of sex hormone actions in the human (see following). Analogously, possible sex-steroid hormone modulation of somatostatin actions directly on somatotropes in the human remains largely undefined, whereas estrogen does appear to enhance maximal GHRH and

TABLE 10.3. Presumptive mechanisms of amplified GH pulse mass: heightened net GHRH/somatostatin input to responsive somatotropes.

Proposed mechanism	Possible example
1. ↑ GHRH release	? Estrogen/testosterone, GHRPs
2. ↓ SS release	Fasting, sleep, hypoglycemia, ?? sex steroids
3. Concurrent (1) and (2)	? Puberty in boys/girls
4. Sensitization to GHRH	Coordinated SS withdrawal preceding GHRH pulse ("rebound" GH release) (?estrogen/testosterone effect)
5. ↑ GHRP-receptor driven GH secretion	i.v. GHRP-2 infusion (estrogen may sensitize)

dose-responsive GHRP-2 (nonclassical GH-releasing peptide) actions in some (but not all) recent studies in women (6,11,17,19,83,84). In young men, testosterone is not known to definitively regulate GHRH (or GHRP) dose-responsive stimulation of GH secretion, although in several studies the effect of a maximal GHRH stimulus was influenced by short-term testosterone treatment or deprivation (via castration or GnRH analog treatment) in the male rat or young men (6,11,85), but not in women (86).

Putative Hypothalamic Actions of Sex-Steroids on GHRH or Somatostatin Interactive Neuroregulation of the Human GH Axis

Clinical and animal studies have unveiled conspicuous effects of specific relevant sex-steroid hormones on hypothalamic neuroregulation of GH release; e.g., via neuropeptides, such as GHRH, somatostatin, neuropeptide Y(NPY), galanin, or via neurotransmitters, such as dopamine, serotonin, or acetylcholine (6,11,17,19,20,87–89). Although brain GHRH and somatostatin expression and release are not directly measurable in the human during life, several important inferences have been achievable in clinical investigations, as reviewed recently (4–6,8,11,14,16,17,19,20,26,34).

A consensus among human studies to date is that puberty or testosterone or estrogen treatment amplifies serum GH concentrations via increased peak amplitude (or area), which results from augmented GH secretory burst mass (16,20,21,26,31,34,38,90,91). Given responsive somatotrope cells, the mass of GH secreted per burst is controlled jointly by the effective hypothalamic GHRH stimulus (its magnitude, and possibly its timing or waveform of release) and the prior and prevailing somatostatin concentration; thus, augmented GH secretory burst mass induced by estrogen strongly implicates an altered GHRH–somatostatin balance in favor of net effective GHRH action (20). In principle, heightened effective GHRH–somatostatin drive would arise if (a) GHRH release or action alone increased; (b) somatostatin actions alone decreased; (c) both (a) and (b) occurred; or (d) cosecretagogues of GHRH or modulators of somatostatin participated in the altered response (20) (see Table 10.3).

In addition to the countervailing amounts of GHRH and somatostatin acting at any given moment, prior somatostatin exposure influences subsequent GHRH action. Specifically, abrupt withdrawal of a somatostatin infusion facilitates both rebound and GHRH-stimulated GH release in the rat and in children and adults (92–100). In the sheep, pig, and rat, multiple and varying patterns of GHRH–somatostatin release into hypothalamopituitary portal blood can precede or accompany a spontaneous GH pulse (20). How and whether sex steroids modulate such relative timing of GHRH and somatostatin release is not known in either animals or the human. Direct, frequently sampled, portal-venous measurements of GHRH and somatostatin in various sex-steroid milieus have not been performed in experimental animals to discriminate among the foregoing mechanistic issues. Thus, any (or all) of the preceding considerations should remain pertinent in humans pending further studies.

The frequency of pulsatile GH secretion does not vary across puberty (21) or in response to sex hormone treatment of adults or children (16,34,84,90,91). This suggests, but does not prove, that an isolated decline in somatostatin tone at pubertal awakening is unlikely. Indeed, clinical conditions of putatively reduced somatostatin release, such as sleep and fasting, stimulate a detectable rise in GH pulse frequency (59,98,101–103). Conversely, somatostatin infusions suppress GH pulse frequency (and amplitude) in young men (104), supporting the a priori notion that somatostatin controls GH pulse frequency in the human. In contrast, elevated GH pulse frequency is not observed either at puberty or in response to gonadal-steroid manipulations in the human. Thus, although concurrent (partial) somatostatin withdrawal cannot yet be excluded, enhanced (somatostatin-independent) GHRH actions or increased GHRH release are plausible hypotheses to explicate puberty and central actions of estrogen and aromatizable androgens in the human. Moreover, a modulating impact of sex steroids on nonGnRH cosecretagogues of GH (e.g., NPY, galanin, putative endogenous GHRP-like agonists, etc.) remains tenable.

Actions of Estrogen or Testosterone on the GHRP Effector Pathway

GH-releasing peptides, GHRPs, are nonclassical, non-GHRH, oligopeptide derivatives of metenkephalin originally identified in the laboratory of C.Y. Bowers (105,106), and more recently produced as nonpeptidyl (orally active) agonistic analogs (e.g., MK-0677) (107,108). These agents potently release GH in the human, rat, sheep, guinea pig, pig, dog, and hamster (20). Receptors for the GHRP ligand family were cloned recently in the rat, human, and pig, and are expressed in the hypothalamus and pituitary gland, as well as in other tissues (109). An endogenous natural ligand has been postulated, but not yet identified definitively (110).

GHRPs act to release GH by one or more plausible and likely complementary mechanisms (Fig. 10.3). For example, GHRPs facilitate basal and GHRH-stimulated GH release in vivo and in vitro, are more potent in the hypothalamopituitary-intact animal or patient, release GHRH itself (e.g., in the sheep), and overcome in part the inhibitory actions of somatostatin and the autonegative feedback effects of GH (20). This litany illustrates the complex and likely multivalent actions of GHRPs.

The most potent clinically available GHRP is GHRP-2. We have evaluated GHRP-2 actions in healthy ovariprival postmenopausal women studied before versus during estradiol replacement (84,111–114). In a paired analysis, pretreatment with oral estradiol for 7–12 days increased significantly the steepness (rate of ascent, or maximal slope change) of the GHRP-2 dose–response curve in 10 older women (GHRP-2 doses of 0, 0.03, 0.1, 0.3, and 1.0 µg/kg i.v. bolus). Thus, estrogen enhances the sensitivity of GH release to escalating doses of GHRP-2, indicating that this sex steroid augments GHRP-2 potency. Maximal GH secretory responses were achieved during estrogen deprivation at a GHRP-2 dose of 1 µg/kg, but not during estrogen treatment;

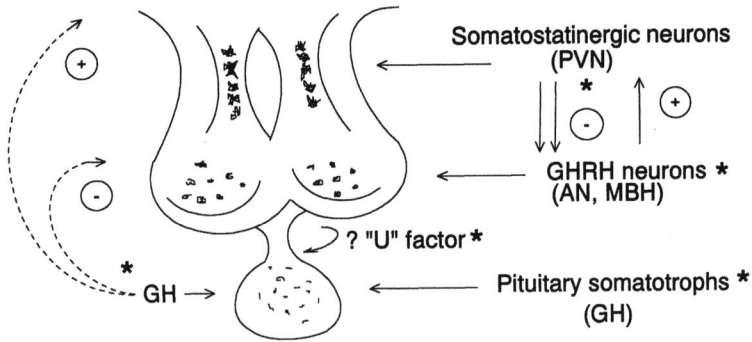

(a) direct pituitary actions (human)

(b) stimulate GHRH release (sheep)

(c) oppose SS's inhibition of GHRH neurons (rat)

(d) attenuate GH autonegative feedback (human, rat)

(e) "U" factor?

FIGURE 10.3. Plausible mechanistic sites of single or joint actions of GHRP agonists to release GH. Multiple complementary loci of GHRP action are likely. Adapted with permission from (20).

thus, no definitive inference is possible yet on whether the efficacy (maximal action) of GHRP-2 is estrogen dependent (Fig. 10.4). The observed increase in GHRP-2 potency in response to estrogen replacement could reflect heightened endogenous GHRH release under estrogen drive because GHRH typically synergizes with GHRP (see preceeding). Other complementary mechanisms cannot be excluded, such as reduced somatostatin inhibition or attenuated GH–IGF-I negative feedback in the face of estrogen replacement.

Enhanced maximal GHRP-2-stimulated GH secretion, if confirmed by studies using higher doses of GHRP secretagogues, would indicate conversely that estrogen can recruit other (non-GHRP-dependent) response pathways that promote GH secretion and can alter one or more rate-limiting steps in the effector sequence mediating GHRP actions, e.g., increased GHRP receptor number, etc. Indeed, in a recent analysis of short-term GHRP-2–estrogen interactions in postmenopausal women, we observed a significant joint (GHRP-2–estradiol) impact on the orderliness of GH release and on the nyctohemeral rhythmicity of GH release; namely, although a constant GHRP-2 infusion increased the mesor (mean) and amplitude of the daily GH cosine rhythm and shifted the timing of the GH maximum from the night to daytime, cotreatment with estradiol augmented the amplitude effect and restored the phase shift (Fig. 10.5) (84).

To our knowledge, comparable clinical data are not yet available concerning the possible impact of testosterone on dose-responsive GHRP actions in men. However, in the intact rat, both estrogen and testosterone increase GHRP stimulation of GH release (115). Sexual dimorphism of GHRP-receptor expression is also evident in the male and the female rat as the male animal exhibits fewer GHRP receptors in the ventromedial (but not arcuate) nucleus (116). Clinical data show that single-dose (near-maximal) GHRP actions are increased by two- to threefold in puberty (versus prepuberty) (117,118), which further suggests sex-steroidal modulation of GH axis responsiveness to GHRP agonists.

Although one recent study reported no evident dependence of GHRP effects on stage or the menstrual cycle (119), only a single (near-maximal) dose of GHRP was evaluated, and the truly preovulatory window of maximal (endogenous) pulsatile GH secretion was not necessarily appraised specifically. Thus, dose-responsive studies of GHRP actions (without versus with concurrent maximal GHRH stimulation, or pharmacologically achieved somatostatin withdrawal) in the early follicular, mid- to late follicular, preovulatory, and luteal phases of the human menstrual cycle will be important. Investigations should also be undertaken in estrogen-withdrawn versus estrogen-replete postmenopausal women because age and estrogen may exert separable and potentially confounding effects on the GH–IGF-I axis (20,120,121). For example, in the ovariectomized (macaque) primate, GH secretion declines in early adulthood, and diminished GH release (but not reduced IGF-I production) in older animals can be rescued by estrogen replenishment. Analogously, in women, the majority of (oral) estrogen replacement studies document stimu-

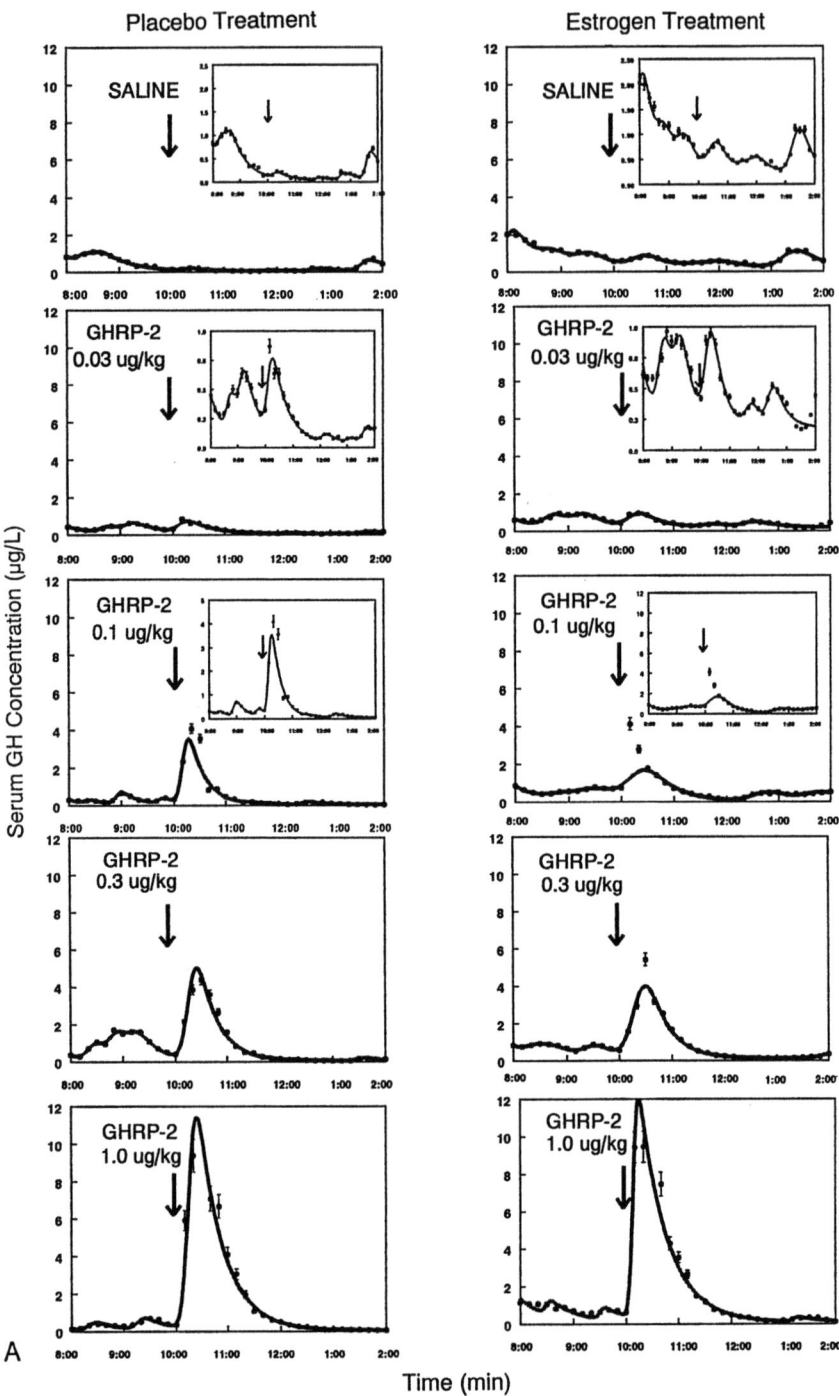

FIGURE 10.4A. Impact of estrogen on GHRP-2 dose-responsive stimulation of serum GH concentrations (*A*) in one postmenopausal woman. GHRP-2 was infused by i.v. bolus injection in randomly ordered doses of 0, 0.03, 0.1, 0.3, and 1.0 µg/kg before versus during oral estradiol replacement. Serum GH concentrations were measured in blood collected every 10 min (*B*). Adapted with permission from Shah et al. (84).

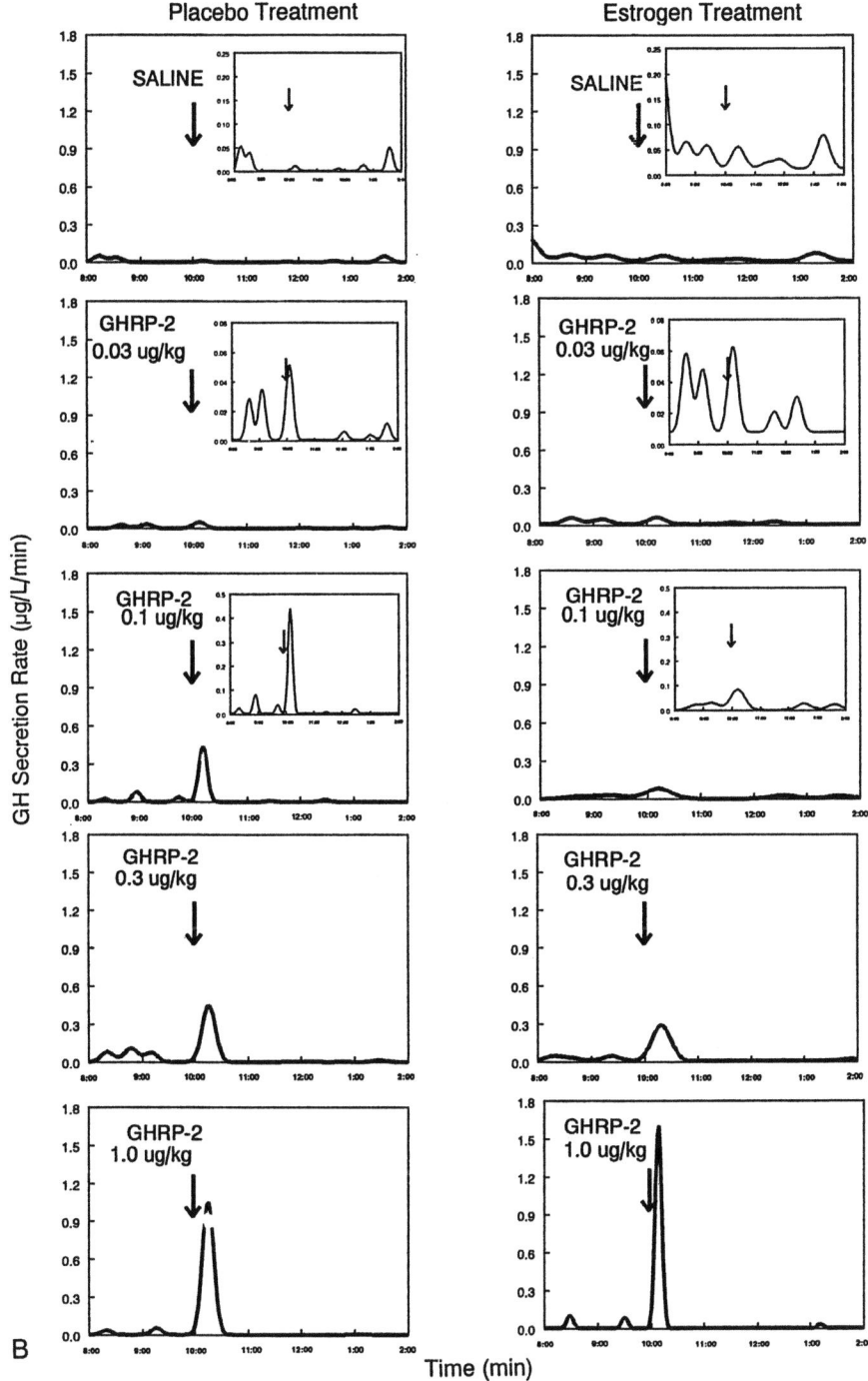

FIGURE 10.4B. (*continued*)

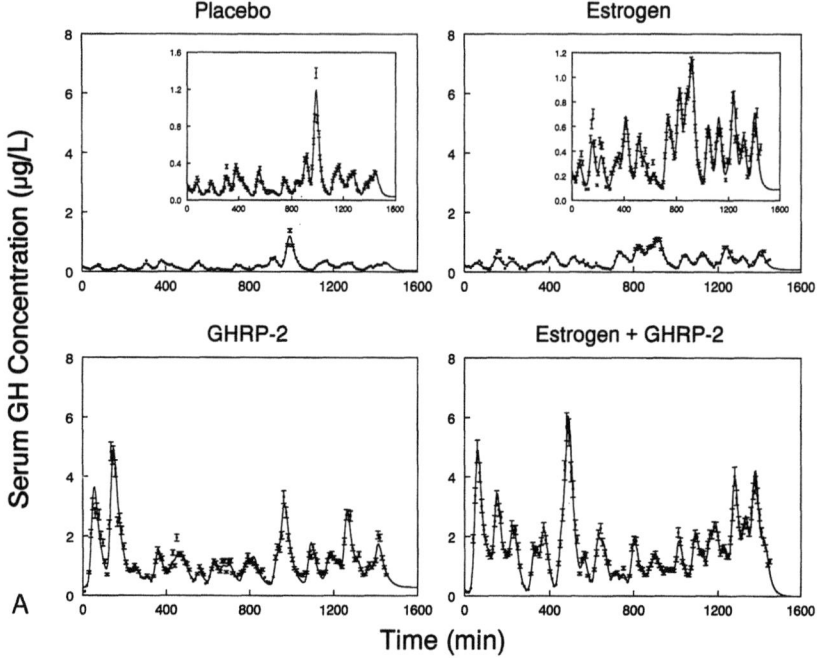

FIGURE 10.5A. Illustrative profiles of 24-h serum GH (*A*) concentrations in one postmenopausal woman treated with placebo (control), estrogen (1 mg micronized estradiol twice daily orally for 10 days), continuous GHRP-2 infusion (1 μg/kg/h × 24 h), or both active agents. Estrogen and GHRP-2 treatment each amplified pulsatile GH secretion, but the GHRP-2 effect was severalfold greater in magnitude. Estrogen and GHRP-2 interacted principally at the level of nyctohemeral (24-h) rhythmicity of GH release (*B*). Adapted with permission from Shah et al. (84).

lation of GH secretion (20,39,45,49,122–124), but a concomitant lowering of (or no change in) serum IGF-I concentrations (44,46,125). Thus, although estrogen stimulates GH secretion in young and older women, postpubertal age may impair the ability to stimulate IGF-I production in both the human and monkey. A further complexity of estrogen interaction with the GH–IGF-I axis arises because estrogen also may impair tissue actions of GH (e.g., hepatic IGF-I production) in the rabbit and human (15,49,126–136).

Concept of Sex-Steroid Actions on the GH Axis as a Feed-Forward and Feedback Network

The critical physiological thesis that the GH–IGF-I axis is an interconnected ensemble that embodies multiple, physiologically relevant, and interactively

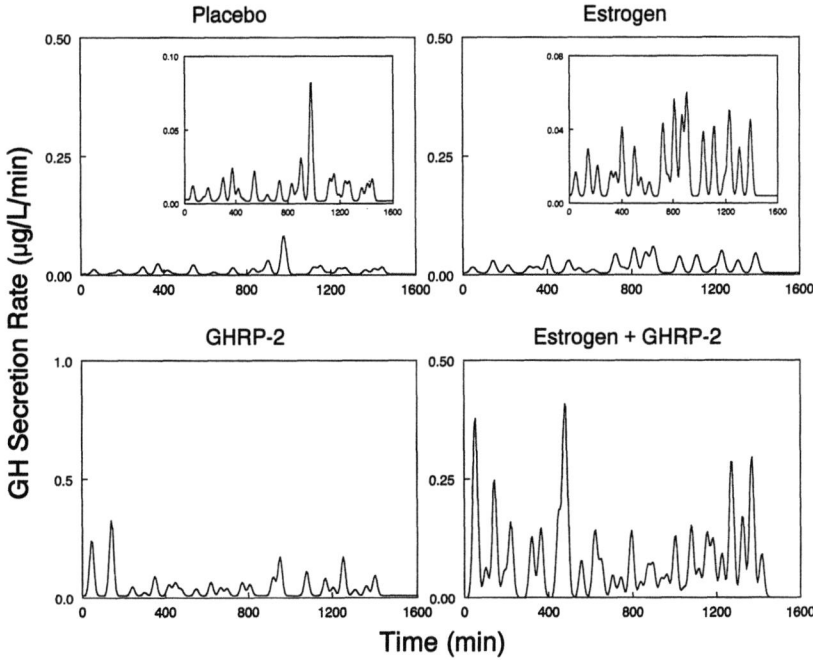

B

FIGURE 10.5B. (*continued*).

controlled regulatory loci is depicted schematically in Figure 10.6. In this network-based construction, somatostatin and GHRH neurons also maintain reciprocal connectivity, intrahypothalamically (137–140). Based on data in the rat and sheep, somatostatinergic neurons in the periventricular nucleus project to and inhibit GHRH-secreting neurons in the arcuate nucleus (20). Conversely, GHRH neurons stimulate somatostatin neurons, creating reciprocal coupling, and possibly thereby endowing cyclical GHRH–somatostatin rhythms driving pulsatile GH release (137,141–144). The latter notion assumes that GH secretory output by somatotrope cells in the anterior pituitary gland largely reflects the net effective ratio of GHRH–somatostatin inputs.

Both GH and IGF-I likely exert multisite feedback inhibition of the GH axis (20). Although the exact CNS loci of GH negative feedback signaling are imperfectly defined in the human, indirect evidence points to a mediatory role for hypothalamic somatostatin; e.g., as inferred from studies with L-arginine, pyridostigmine, or repeated GHRH challenges (20). How androgen or estrogen might modulate GH autofeedback actions clinically is not known, but GH autofeedback effects are sexually dimorphic in the rat and mouse.

POSSIBLE FEEDBACK MODULATION OF HUMAN GH AXIS BY SEX STEROIDS

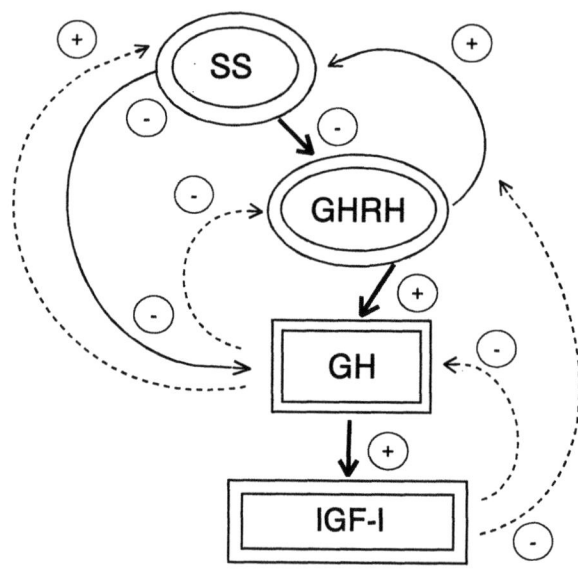

FIGURE 10.6. Schematized concept of the GH axis as a multicomponent, networked, feedforward, and feedback control sytem. Adapted with permission from (137).

Specifically, the male but not the female rodent is highly susceptible to GH negative-feedback inhibition of GHRH or GHRP actions (20,145–147). Although some of the IGF-I negative feedback in the human may be exerted directly on anterior pituitary somatotropes (see earlier), hypothalamic actions of human recombinant (hr) IGF-I (at least in combination with hrIGF-II) also are inferred from experiments using intracerebroventricular infusions of IGF-I in the rat (81), and were interpreted indirectly in one clinical study using peripheral IGF-I infusions in young men (80). How or whether puberty or sex steroids modulate such feedback effects is not yet established, but can be examined in indirect studies (36) and by IGF-I infusions (following).

In recent clinical investigations, we observed that pretreatment of healthy postmenopausal women ($n = 6$) with oral estrogen or placebo for 7–12 days did not modify the impact of i.v. human recombinant IGF-I infusions (5 µg $kg^{-1} h^{-1}$) on either the time-dependent rise in plasma total (or free) IGF-I concentrations or the minimum serum concentration of GH achieved during the infusion (Fig. 10.7). However, as serum GH concentrations were not suppressed maximally (as measured in an ultrasensitive GH chemiluminescence-based assay), we cannot yet rule out a possible interaction of estrogen with "high-dose" IGF-I negative feedback. Analogous preliminary data are not yet available for IGF-I infusions following placebo versus i.m. testosterone replacement

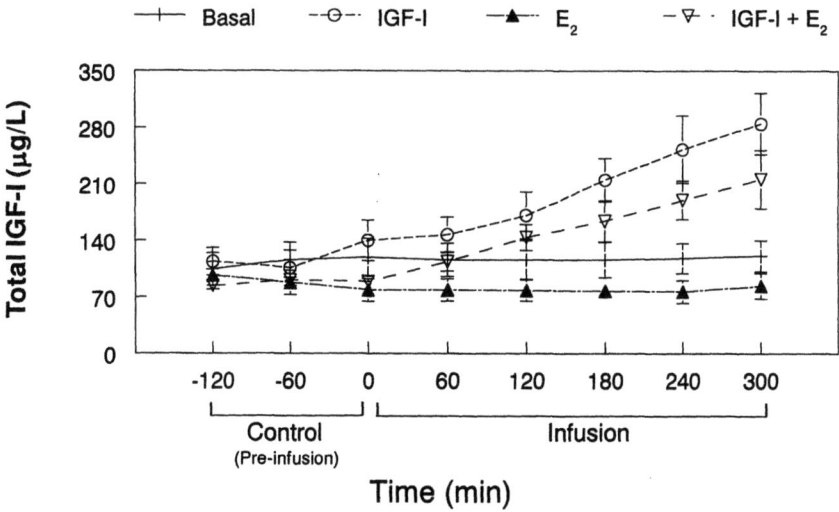

FIGURE 10.7. Apparent lack of impact of estrogen (versus placebo) replacement on rate of rise of hourly plasma (total) IGF-I concentrations during constant saline versus human recombinant IGF-I infusions (5 μg/kg/h i.v.) in postmenopausal women ($n = 6$). The slopes of the (± estrogen pretreated) IGF-I infusion curves are statistically identical, which indicates the absence of any major influence of estrogen on IGF-I distribution volume or (first-order) of nonsteady-state clearance. An effect of estrogen on the slower (18–30 h irreversible) removal of IGF-I is not excluded in this acute infusion experiment. Unpublished data (Storey S, Frystak J, Orskov H, and Veldhuis JD).

in older men. The foregoing observations speak against sex-steroidal control of IGF-I negative feedback of the GH axis in older women, but do not necessarily exclude such actions in men or in puberty or young adulthood.

Plausible intrahypothalamic sites of sex hormone regulation of the GHRH–somatostatin–GH–IGF-I axis are adumbrated in Figure 10.8. For example, in principle, estrogen (or aromatizable androgen) could increase GH secretion by antagonizing GHRH intrahypothalamic stimulation of somatostatin release or by facilitating GHRH escape from inhibition by neuronally released somatostatin. These important, albeit presumptive, brain mechanisms remain largely unexplored directly in either the human or experimental animal, but are suggested by inferential data in the rat (148). Moreover, exactly how sex steroids modulate the multiple specific CNS neurotransmitters that govern GHRH or somatostatin secretion singly and jointly remains essentially unknown clinically.

Other Considerations

Recent preliminary data from our laboratory indicate that estradiol does not alter direct dose-responsive somatostatin inhibition of pituitary GH secretion in (older)

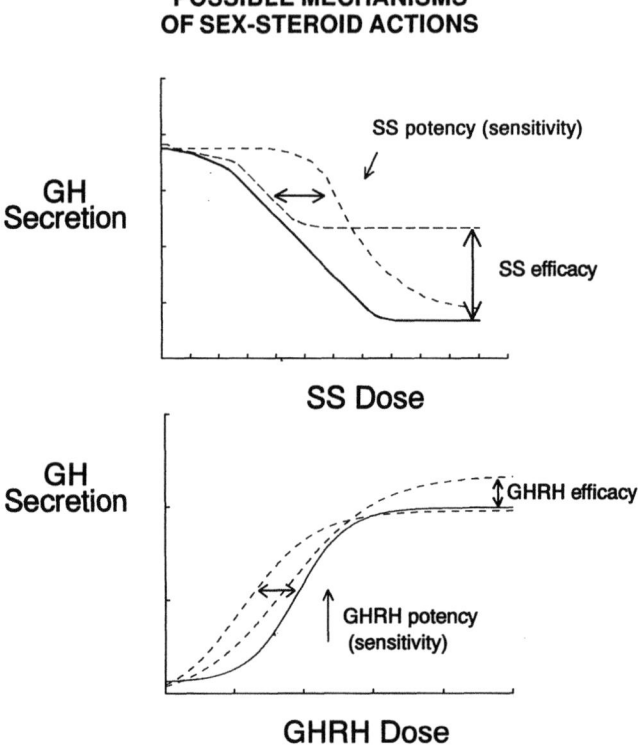

FIGURE 10.8. Plausible (multiple) sites of actions of sex steroids on specific regulatory loci within the GHRH–somatostatin–GH neuroendocrine axis. Both feed-forward (e.g., GHRH on GH) and feedback (e.g., GH on somatostatin) points of control are represented.

women. For example, continuous i.v. infusion of (randomly ordered) 0, 30, 100, or 300 $\mu g/m^2$ somatostatin hourly for each of 3 hours in six healthy postmenopausal women dose-dependently inhibited GH release, which occurred independently of, and equivalently for, prior oral estrogen versus placebo replacement for 7–12 days. The time course and the magnitude of GH suppression, were similar in the estrogen-replaced and placebo sessions (Fig. 10.9). Thus, under experimental conditions in which one can demonstrate estrogen amplification of pulsatile GH secretion, estrogen does not appear to modify somatostatin inhibitory actions on the pituitary gland. This inference should be distinguished from any possible effects of estrogen on *endogenous* hypothalamic somatostatin release per se. To our knowledge, comparable studies with respect to androgen action are not available.

Estrogen or androgens hypothetically could also increase the half-life of GH (slow its metabolic clearance rate, MCR), as suggested in one study, and thereby elevate blood GH concentrations (149). Studies of GH half-life, evaluated with

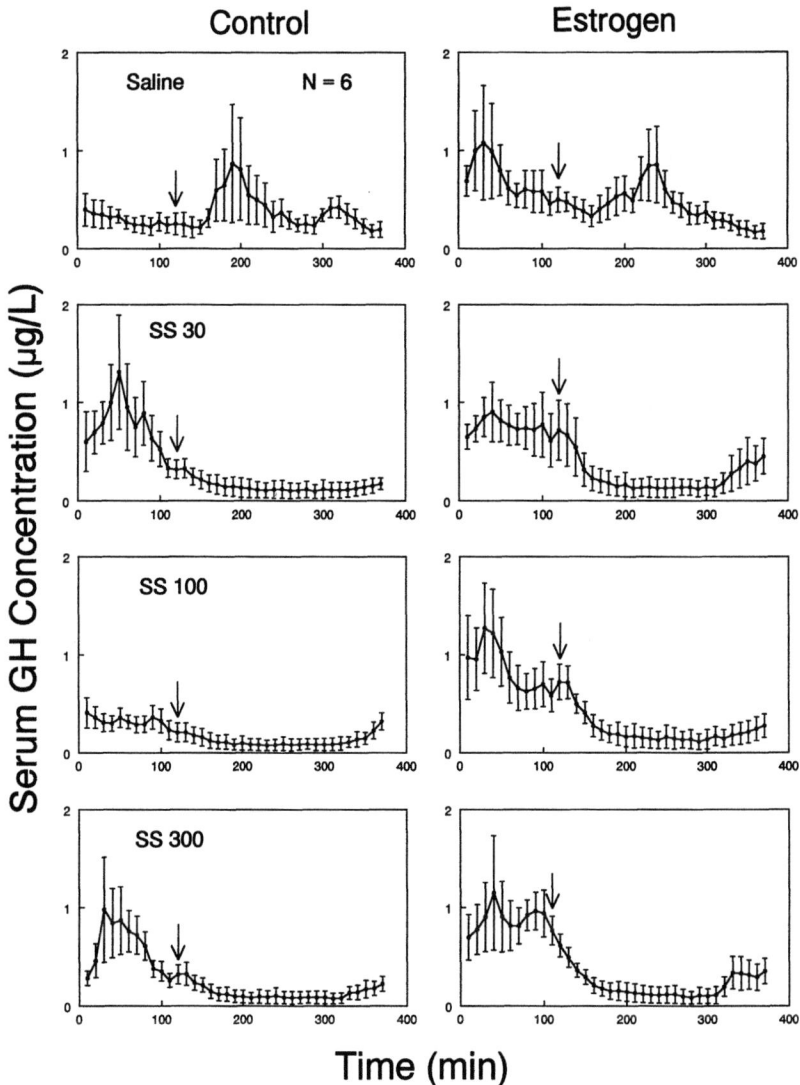

FIGURE 10.9. Somatostatin dose-dependent (0, 30, 100, and 300 μg/m^2/h i.v. infusion for 3 h) inhibition of pulsatile serum GH concentrations (mean ± SEM, $n = 6$ postmenopausal volunteers) during estrogen withdrawal (control) versus estrogen replacement (7–12 days of oral micronized estradiol-17β, 1 mg twice daily). Serum GH concentrations were measured by ultrasensitive chemiluminescence-based assay in blood sampled every 10 min for 6 h. *Arrows* denote the start of the somatostatin or placebo infusion (after a 1-h baseline). Unpublished data (Bray M, Shah N, Evans WS, and Veldhuis JD).

versus without testosterone or estrogen replacement, across puberty, in various age strata, over the normal human menstrual cycle, and in men compared to women make this conjecture largely untenable (21,26,27,29,73,84,150). For example, after i.v. bolus infusion of human recombinant GH in octreotide-suppressed adults, the mean (monocomponent) GH half-life is actually slightly but consistently higher in young men (9.3 min) than young women (8.2 min). Therefore, the 1.5- to 2 fold higher mean (24-h) serum GH concentration recognized in premenopausal women (compared to comparably aged men) is not caused by a corresponding 150%–200% decrease in GH removal rates in women. Moreover, recent GH infusion studies in six women at each of three stages of the menstrual cycle showed that GH half-life (as well as its MCR and distribution volume) are invariant of stage of the cycle. Thus, the doubling of serum GH concentrations in the late follicular phase of young women (70,71,73) reflects truly increased daily GH secretion rates, rather than reduced GH elimination, as also inferred recently by deconvolution analysis (73).

Unresolved Clinical Issues

Despite the foregoing significant clinical knowledge base, many important clinical mechanistic questions remain unaddressed or unresolved. Table 10.4 highlights several such queries.

Summary

Sex hormones control the neuroendocrine activity of the GH–IGF-I axis via modulating three dynamic facets of GH secretion: the pulsatile, 24-h (nyctohemeral) rhythmic, and entropic (moment-to-moment orderliness) modes of GH release. Aromatizable androgens and estrogen markedly and selectively stimulate GH pulse mass, augment the 24-h rhythmicity of GH release,

TABLE 10.4. Selected major unresolved clinical issues.

1. How do estrogen or aromatizable androgens modulate intrahypothalamic GHRH–somatostatin interactions?
2. Do sex steroids attenuate IGF-I or GH (auto-) negative feedback in the human?
3. Do gonadal steroids activate the putative (human) endogenous GHRP effector pathway?
4. How do sex hormone effects on the GH–IGF-I axis vary with age (e.g., at puberty versus at older ages)?
5. When should GH treatment (or GH secretagogues) and sex-steroid replacement be combined in older men and women?

and increase the randomness (entropy) of the GH release process. Sex steroids likely achieve these modulatory actions by governing GHRH or somatostatin release or effects singly and jointly, so as to favor net effective GHRH excess. In addition, estrogen sensitizes GHRP-2 stimulation of GH release (in older women), thus permitting the speculation that sex steroids might also impact the endogenous (putative) GHRP-effector pathway. Indeed, continuous GHRP-2 infusion alone in gonadoprival (older) women mimics all three principal dynamic actions of sex steroids on the GH axis. Although the extent and nature of any regulatory effects of sex steroids on (intrahypothalamic) reciprocal GHRH–somatostatin interactions in the human are unknown, the rise in quantifiable entropy of the GH release process at puberty (34), and during estrogen or testosterone (but not 5α-dihydrotestosterone) treatment (36,148), strongly points to a sex-steroid impact on the GHRH–somatostatin–GH–IGF-I network (feedback–feed-forward interactions). The exact character of such proposed network changes will require further study. Lastly, clinical investigations are needed to identify the specific CNS neurotransmitter pathways and effector molecules that mediate estrogen and aromatizable androgen drive of GHRH–somatostatin (and possibly GHRP) actions on the GH–IGF-I axis.

Acknowledgments. We thank Patsy Craig for her skillful preparation of the manuscript; Paula P. Azimi for the data analysis, management, and graphics; Ginger Bauler for performance of various assays; and Sandra Jackson and the expert nursing staff at the University of Virginia General Clinical Research Center (GCRC) for conducting the research protocols. This work was supported in part by NIH Grant MO1 RR00847 (to the GCRC of the University of Virginia Health Sciences Center), the National Science Foundation Center for Biological Timing (Grant DIR89-20162), the NIH U-54 Specialized Cooperative Centers' Program in Reproductive Research (HD-28934), and NIH NCRR Clinical Research Scholar's Award (N.S.), a NIH NCRR Clinical Associate Physician Award (S.M.A), and RO1 # NIA AG14799 (J.D.V.).

References

1. Merimee TJ, Fineberg SE. Studies of the sex-based variation of human growth hormone secretion. J Clin Endocrinol Metab 1971;33:896–902.
2. Veldhuis JD. The hypothalamic pituitary-testicular axis. In: Yen SSC, Jaffe RB, eds. Reproductive endocrinology. Philadelphia: Saunders, 1991:409–59.
3. Veldhuis JD, Evans WS. Endocrine testing of gonadal function: testes and ovary. In: Kovacs K, Asa SL, eds. Biochemical tests in functional endocrine pathology. Boston: Blackwell, 1990:34.
4. Veldhuis JD, Iranmanesh A, Rogol AD, Urban RJ. Regulatory actions of testosterone on pulsatile growth hormone secretion in the human: studies using deconvolution analysis. In: Adashi EY, Thorner MO, eds. Somatotropic axis and the reproductive process in health and disease. New York: Springer-Verlag, 1995:40–57.

5. Rogol AD, Martha PM Jr, Johnson ML, Veldhuis JD, Blizzard RM. Growth hormone secretory dynamics during puberty. In: Adashi EY, Thorner MO, eds. The somatotrophic axis and the reproductive process in health and disease. New York: Springer-Verlag, 1996:69–82.
6. Veldhuis JD. Neuroendocrine control of pulsatile growth-hormone release in the human: relationship with gender. Growth Horm IGF Res 1998;8:49–59.
7. Evans WS, Booth RA, Ho KKY, Faria ACS, Asplin CM, Veldhuis JD, et al. Growth hormone economy in normally cycling women. In: Adashi EY, Thorner MO, eds. The somatotropic axis and the reproductive process in health and disease. New York: Springer-Verlag, 1996:107–23.
8. Veldhuis JD. New modalities for understanding dynamic regulation of the somatotropic (GH) axis: explication of gender differences in GH neuroregulation in the human. J Pediatr Endocrinol Metab 1996;9:237–53.
9. Aynsley-Green A, Zachmann M, Prader A. Interrelation of the therapeutic effects of growth hormone and testosterone on growth in hypopituitarism. J Pediatr 1976;89:992–9.
10. Delemarre-Van De Wall HA, Wennink JM, Odink RJ. Gonadotrophin and growth hormone secretion through puberty. Acta Paediatr Scand 1991;372:26–31.
11. Deversa J, Lois N, Arce V, Diaz MJ, Lima L, Tresguerres JA. The role of sexual steroids in the modulation of growth hormone (GH) secretion in humans. J Steroid Biochem Mol Biol 1991;40:165–73.
12. Merimee TJ, Quinn S, Russell B, Riley W. The growth hormone-insulin-like growth factor I axis: studies in man during growth. In: Raizada MK, Leoith D, eds. Molecular biology and physiology of insulin and insulin-like growth factors. New York: Plenum Press, 1991:86–96.
13. Zachmann M. Interrelations between growth hormones and sex hormones: physiology and therapeutic consequences. Horm Res 1992;1:1–8.
14. Veldhuis JD. Neuroendocrine mechanisms mediating awakening of the gonadotropic axis in puberty. Pediatr Nephrol 1996;10:304–17.
15. Mauras N, Doi SQ, Shapiro JR. Recombinant human insulin-like growth factor-1, recombinant human growth hormone, and sex steroids: effects on markers of bone turnover in humans. J Clin Endocrinol Metab 1996;81:2222–6.
16. Mauras N, Rogol AD, Haymond MW, Veldhuis JD. Sex steroids, growth hormone, IGF-I: neuroendocrine and metabolic regulation in puberty. Horm Res 1996;45:74–80.
17. Veldhuis JD. The neuroendocrine regulation and implications of pulsatile GH secretion: gender effects. Endocrinologist 1995;5:198–213.
18. Iranmanesh A, Veldhuis JD. Clinical pathophysiology of the somatotropic (GH) axis in adults. In: Veldhuis JD, ed. Endocrinology and metabolism clinics of North America. Philadelphia: Saunders, 1992:783–816.
19. Veldhuis JD. Gender differences in secretory activity of the human somatotropic (growth hormone) axis. Eur J Endocrinol 1996;134:287–95.
20. Giustina A, Veldhuis JD. Pathophysiology of the neuroregulation of growth hormone secretion in experimental animals and the human. Endocr Rev 1998;19:717–97.
21. Martha PM Jr, Goorman KM, Blizzard RM, Rogol AD, Veldhuis JD. Endogenous growth hormone secretion and clearance rates in normal boys as determined by deconvolution analysis: relationship to age, pubertal status and body mass. J Clin Endocrinol Metab 1992;74:336–44.

22. Wennink JM, Delemarre-van de Waal HA, Schoemaker R, Blaauw G, van den Brakern C, Schoemaker J. Growth hormone secretion patterns in relation to LH and estradiol secretion throughout normal female puberty. Acta Endocrinol (Copenh) 1991;124:129–35.
23. Attie KM, Ramirez NR, Conte FA, Kaplan SL, Grumbach MM. The pubertal growth spurt in eight patients with true precocious puberty and growth hormone deficiency: evidence for a direct role of sex steroids. J Clin Endocrinol Metab 1990;71:975–83.
24. Roemmich JN, Clark PA, Mai V, Berr SS, Weltman A, Veldhuis JD, et al. Alterations in growth and body composition during puberty: III. Influence of maturation, gender, body composition, fat distribution, aerobic fitness, and energy expenditure on nocturnal growth hormone release. J Clin Endocrinol Metab 1998;83:1440–7.
25. Bhasin S, Storer TW, Berman N, Yarasheski KE, Clevenger B, Phillips J, et al. Testosterone replacement increases fat-free mass and muscle size in hypogonadal men. J Clin Endocrinol Metab 1997;82:407–13.
26. Giustina A, Scalvini T, Tassi C, Desenzani P, Poiesi C, Wehenberg W, et al. Maturation of the regulation of growth hormone secretion in young males with hypogonadotropic hypogonadism pharmacologically exposed to progressive increments in serum testosterone. J Clin Endocrinol Metab 1997;82:1210–9.
27. Fryburg DA, Weltman A, Jahn LA, Weltman JY, Samolijik E, Veldhuis JD. Short-term modulation of the androgen milieu alters pulsatile but not exercise or GHRH-stimulated GH secretion in healthy men. J Clin Endocrinol Metab 1997;82:3710–9.
28. Mauras N, Hayes V, Welch S, Rini A, Helgeson K, Dokler M, et al. Testosterone deficiency in young men: marked alterations in whole body protein kinetics, strength and adiposity. J Clin Endocrinol Metab 1998;83:1886–92.
29. Van den Berg G, Veldhuis JD, Frolich M, Roelfsema F. An amplitude-specific divergence in the pulsatile mode of GH secretion underlies the gender difference in mean GH concentrations in men and premenopausal women. J Clin Endocrinol Metab 1996;81:2460–6.
30. Mauras N, Blizzard RM, Link K, Johnson ML, Rogol AD, Veldhuis JD. Augmentation of growth hormone secretion during puberty: evidence for a pulse amplitude-modulated phenomenon. J Clin Endocrinol Metab 1987;64:596–601.
31. Martha PM, Rogol AD, Veldhuis JD, Kerrigan JR, Goodman DW, Blizzard RM. Alterations in the pulsatile properties of circulating growth hormone concentrations during puberty in boys. J Clin Endocrinol Metab 1989;69:563–70.
32. Van Cauter E, Kerkhofs M, Caufriez A, Van Onderbergen A, Thorner MO, Copinschi G. A quantitative estimation of growth hormone secretion in normal man: reproducibility and relation to sleep and time of day. J Clin Endocrinol Metab 1992;74:1441–50.
33. Golstein J, Van Cauter E, Desir D, Noel J, Spire JP, Refetoff S, et al. Effects of "jet lag" on hormonal patterns. IV. Time shifts increase growth hormone release. J Clin Endocrinol Metab 1983;56:433–40.
34. Veldhuis JD, Metzger DL, Martha PM Jr, Mauras N, Kerrigan JR, Keenan B, et al. Estrogen and testosterone, but not a non-aromatizable androgen, direct network integration of the hypothalamo-somatotrope (growth hormone)-insulin-like growth factor I axis in the human: evidence from pubertal pathophysiology and sex-steroid hormone replacement. J Clin Endocrinol Metab 1997;82:3414–20.

35. Veldhuis JD, Pincus SM. Orderliness of hormone release patterns: a complementary measure to conventional pulsatile and circadian analyses. Eur J Endocrinol 1998;138:358–62.
36. Pincus SM, Gevers E, Robinson ICAF, Van den Berg G, Roelfsema F, Hartman ML, et al. Females secrete growth hormone with more process irregularity than males in both human and rat. Am J Physiol 1996;270:E107–15.
37. Genazzani AD, Petraglia F, Volpogni C, Gastaldi M, Pianazzi F, Montanini V, et al. Modulatory role of estrogens and progestins on growth hormone episodic release in women with hypothalamic amenorrhea. Fertil Steril 1993;60:465–70.
38. Ulloa-Aguirre A, Blizzard RM, Garcia-Rubi E, Rogol AD, Link K, Christie CM, et al. Testosterone and oxandrolone, a non-aromatizable androgen, specifically amplify the mass and rate of growth hormone (GH) secreted per burst without altering GH secretory burst duration or frequency or the GH half-life. J Clin Endocrinol Metab 1990;71:846–54.
39. Dawson-Hughes B, Stern D, Goldman J, Reichlin S. Regulation of growth hormone and somatomedin-C secretion in postmenopausal women: effect of physiological estrogen replacement therapy. J Clin Endocrinol Metab 1986;63:424–32.
40. Link K, Blizzard RM, Evans WS, Kaiser DL, Parker MW, Rogol AD. The effect of androgens on the pulsatile release and the twenty-four hour mean concentrations of growth hormone in peripubertal males. J Clin Endocrinol Metab 1986;62:159–65.
41. Frantz AG, Rabkin MT. Effects of estrogen and sex difference on secretion of human growth hormone. J Clin Endocrinol Metab 1965;25:1470–80.
42. Jasper HC. Somatomedin response to testosterone stimulation in male pseudohermaphroditism, cryptorchidism, anorchia, or micropenis. J Clin Endocrinol Metab 1985;60:910.
43. Keenan BS, Richards GE, Ponder SW, Dallas JS, Nagamani M, Smith ER. Androgen-stimulated pubertal growth: the effects of testosterone and dihydrotestosterone on growth hormone and insulin-like growth factor-I in the treatment of short stature and delayed puberty. J Clin Endocrinol Metab 1993;76:996–1001.
44. Friend KE, Hartman ML, Pezzoli SS, Clasey JL, Thorner MO. Both oral and transdermal estrogen increase growth hormone release in postmenopausal women— a clinical research center study. J Clin Endocrinol Metab 1996;81:2250–6.
45. De Leo V, Lanzetta D, D'Antona D, Danero S. Growth hormone secretion in premenopausal women before and after ovariectomy: effect of hormone replacement therapy. Fertil Steril 1993;60:268–71.
46. Weissberger AJ, Ho KKY, Lazarus L. Contrasting effects of oral and transdermal routes of estrogen replacement therapy on 24-hour growth hormone (GH) secretion, insulin-like growth factor I, and GH-binding protein in postmenopausal women. J Clin Endocrinol Metab 1991;72:374–81.
47. Martin LG, Grossman LS, Connor TB, Levitsky LL, Clark JW, Camitta FD. Effect of androgen on growth hormone secretion and growth in boys with short stature. Acta Endocrinol (Copenh) 1979;91:201–12.
48. Parker MW, Johansson AJ, Rogol AD, Kaiser DL, Blizzard RM. Effect of testosterone on somatomedin-C concentrations in prepubertal boys. J Clin Endocrinol Metab 1984;58:87.
49. Bellantoni MF, Harman SM, Cho DE, Blackman MR. Effects of progestin-opposed transdermal estrogen administration on growth hormone and insulin-like growth factor-I in postmenopausal women of different ages. J Clin Endocrinol Metab 1991;72:172–8.

50. Wiedemann E, Schwartz E, Frantz AG. Acute and chronic estrogen effects upon serum somatomedin activity, growth hormone, and prolactin in man. J Clin Endocrinol Metab 1976;42:942–52.
51. Thompson RG, Rodriguez A, Kowarski AA, Migeon CJ, Blizzard RM. Integrated concentrations of growth hormone correlated with plasma testosterone and bone age in preadolescent and adolescent males. J Clin Endocrinol Metab 1978;62:1341–4.
52. Hagenfeldt Y, Linde K, Sjoberg H-E, Zumkeller W, Arver S. Testosterone increases serum 1,25-dihydroxyvitamin D and insulin-like growth factor-I in hypogonadal men. Int J Androl 1992;15:93–102.
53. Weissberger AJ, Ho KKY. Activation of the somatotropic axis by testosterone in adult males: evidence for the role of aromatization. J Clin Endocrinol Metab 1993;1407:1412.
54. Guo C-Y, Jones TH, Eastell R. Treatment of isolated hypogonadotropic hypogonadism effect on bone mineral density and bone turnover. J Clin Endocrinol Metab 1997;82:658–65.
55. Chalew SA, Udoff LC, Hanukoglu A, Bistritzer T, Armour KM, Kowarski AA. The effect of testosterone therapy on spontaneous growth hormone seretion in boys with constitutional delay. Am J Dis Child 1989;142:1345–8.
56. Craft WH, Underwood LE. Effect of androgens on plasma somatomedin-C/insulin-like growth factor 1 responses to growth hormone. Clin Endocrinol 1984;20:549–54.
57. Drop SL, Sabbe-Claus L, Visser HK. The effect of puberty and short-term oral administration of testosterone undecoanoate on GH tests and sex-steroid related plasma compounds in GH deficient patients. Clin Endocrinol 1982;16:375–81.
58. Eakman GD, Dallas JS, Ponder SW, Keenan BS. The effect of testosterone and dihydrotestosterone on hypothalamic regulation of growth hormone secretion. J Clin Endocrinol Metab 1996;81:1217–23.
59. Veldhuis JD, Liem AY, South S, Weltman A, Weltman J, Clemmons DA, et al. Differential impact of age, sex-steroid hormones, and obesity on basal versus pulsatile growth hormone secretion in men as assessed in an ultrasensitive chemiluminescence assay. J Clin Endocrinol Metab 1995;80:3209–22.
60. Abbasi AA, Drinka PJ, Mattson DE, Rudman D. Low circulating levels of insulin-like growth factors and testosterone in chronically institutionalized elderly men. J Am Geriatr Soc 1993;41:975–82.
61. Nicklas BJ, Ryan AJ, Treuth MM, Harman SM, Blackman MR, Hurley BF, et al. Testosterone, growth hormone and IGF-I responses to acute and chronic resistive exercise in men aged 55–70 years. Int J Sports Med 1995;16:445–50.
62. Rudman D, Shetty KR. Unanswered questions concerning the treatment of hyposomatotropism and hypogonadism in elderly men. J Am Geriatr Soc 1994;42:522–7.
63. Tenover JS. Testosterone in the aging male. J Androl 1997;18:103–6.
64. Mansfield MJ, Rudlin CR, Crigler JF Jr, Karol KA, Crawford JD, Boepple PA, et al. Changes in growth and serum growth hormone and plasma somatomedin-C levels during suppression of gonadal sex steroid secretion in girls with central precocious puberty. J Clin Endocrinol Metab 1988;66:3–9.
65. Harris DA, Van Vliet G, Egli CA, Grumbach MM, Kaplan SL, Styne DM, et al. Somatomedin-C in normal puberty and in true precocious puberty before and after treatment with potent luteinizing hormone-releasing hormone agonist. J Clin Endocrinol Metab 1985;61:152.

66. Friedman AJ, Rein MS, Pandian MR, Barbieri RL. Fasting serum growth hormone and insulin-like growth factor-I and -II concentrations in women with leiomyomata uteri treated with leuprolide acetate or placebo. Fertil Steril 1990;53:250–3.
67. Word RA, Odom MJ, Byrd W, Carr BR. The effect of gonadotropin-releasing hormone agonists on growth hormone secretion in adult premenopausal women. Fertil Steril 1990;54:73–8.
68. Blumenfeld Z, Barkey RJ, Youdim MB, Brandes JM, Amit T. Growth hormone (GH)-binding protein regulation by estrogen, progesterone, and gonadotropins in human: the effect of ovulation induction with menopausal gonadotropins, GH, and gestation. J Clin Endocrinol Metab 1992;75:1242–9.
69. Wilson EE, Word RA, Byrd W, Carr BR. Effect of superovulation with human menopausal gonadotropins on growth hormone levels in women. J Clin Endocrinol Metab 1991;73:511–55.
70. Faria ACS, Bekenstein LW, Booth RA Jr, Vaccaro VA, Asplin CM, Veldhuis JD, et al. Pulsatile growth hormone release in normal women during the menstrual cycle. Clin Endocrinol 1992;36:591–6.
71. Yen SSC, Vela P, Rankin J, Littell AS. Hormonal relationships during the menstrual cycle. JAMA 1970;211:1513–7.
72. Jimena P, Castilla JA, Peran F, Molina R, Ramirez JP, Acebal M, et al. Insulin and insulin-like growth factor I in follicular fluid after induction of ovulation in women undergoing *in vitro* fertilization. J Reprod Fertil 1992;96:641–7.
73. Ovesen P, Vahl N, Fisker S, Veldhuis JD, Christiansen JS, Jorgensen JOL. Increased pulsatile, but not basal, growth hormone secretion rates and plasma insulin-like growth factor I levels during the preovulatory interval in normal women. J Clin Endocrinol Metab 1998;83:1662–7.
74. Liu L, Merriam GR, Sherins RJ. Chronic sex steroid exposure increases mean plasma growth hormone concentration and pulse amplitude in men with isolated hypogonadotropic hypogonadism. J Clin Endocrinol Metab 1987;64:651–6.
75. Schnorr J, Santen J, Veldhuis JD. Role of endogenous aromatase in testosterone's short-term feedback regulation of 24-hour LH, FSH, and GH release in young men. Presented at the Society for Gynecologic Investigation Annual Meeting, Atlanta, Georgia, March 11, 1999 (Abstract).
76. Malhotra A, Poon E, Tse WY, Pringle PJ, Hindmarsh PC, Brook CG. The effects of oxandrolone on the growth hormone and gonadal axes in boys with constitutional delay of growth and puberty. Clin Endocrinol 1993;38:393–8.
77. Bowers CY, Momany FA, Reynolds A, Hong A. On the in vitro and in vivo activity of a new synthetic hexapeptide that acts on the pituitary to specifically release growth hormone. Endocrinology 1984;114:1537–45.
78. Postel-Vinay MC, Tar A, Hocquette JF, Clot JP, Fontoura M, Brauner R, et al. Human plasma growth hormone (GH)-binding proteins are regulated by GH and testosterone. J Clin Endocrinol Metab 1991;73:197–202.
79. Chapman IM, Hartman ML, Pezzoli SS, Harrell FE Jr, Hintz RL, Alberti KGMM, et al. Effect of aging on the sensitivity of growth hormone secretion to insulin-like growth factor-I negative feedback. J Clin Endocrinol Metab 1997;82:2996–3004.
80. Bermann M, Jaffe CA, Tsai W, DeMott-Friberg R, Barkan AL. Negative feedback regulation of pulsatile growth hormone secretion by insulin-like growth factor I: involvement of hypothalamic somatostatin. J Clin Invest 1994;94:138–45.

81. Minami S, Suzuki N, Sugihara H, Tamura H, Emoto N, Wakabayashi I. Microinjection of rat GH but not human IGF-I into a defined area of the hypothalamus inhibits endogenous GH secretion in rats. J Endocrinol 1997;153:283–90.
82. Hartman ML, Clayton PE, Johnson ML, Celniker A, Perlman AJ, Alberti KK, et al. A low dose euglycemic infusion of recombinant human insulin-like growth factor I rapidly suppresses fasting-enhanced pulsatile growth hormone secretion in humans. J Clin Invest 1993;91:2453–62.
83. Shah N, Anderson SM, Azimi P, Evans WS, Bowers CY, Veldhuis JD. Estradiol augments basal GH pulse mass and amplifies GH-secretory responsiveness to low doses of GHRP-2. Presented at the Fifth International Pituitary Congress, Naples, FL, June 28–30, 1998.
84. Shah N, Anderson SM, Azimi P, Evans WS, Bowers CY, Veldhuis JD. Growth hormone releasing peptide-2 (GHRP-2) induced growth hormone (GH) release is influenced by oral estrogen replacement in postmenopausal women. Presented at the Growth Hormone Research Society Conference, San Francisco, CA, September 3–6, 1998.
85. Ohlsson L, Isaksson O, Jansson JO. Endogenous testosterone enhances growth hormone (GH)-releasing factor-induced GH secretion in vitro. J Endocrinol 1987;113:249–53.
86. Lima L, Arce V, Lois N, Fraga C, Lechuga MJ, Tresquerres JA, et al. Growth hormone (GH) responsiveness to GHRH in normal adults is not affected by short-term gonadal blockade. Acta Endocrinol (Copenh) 1989;120:31–6.
87. Giustina A, Licini M, Bussi AR, Grelli A, Pizzocolo G, Schettino M, et al. Effects of sex and age on the growth hormone response to galanin in healthy human subjects. J Clin Endocrinol Metab 1993;76:1369–72.
88. O'Keane V, Dinan TG. Sex steroid priming effects on growth hormone response to pyridostigmine throughout the menstrual cycle. J Clin Endocrinol Metab 1992;75:11–4.
89. Schober E, Frisch H, Waldhauser F, Bieglmayr C. Influence of estrogen administration on growth hormone response to GHRH and L-Dopa in patients with Turner's syndrome. Acta Endocrinol (Copenh) 1989;120:442–6.
90. Mauras N, Rogol AD, Veldhuis JD. Increased hGH production rate after low-dose estrogen therapy in prepubertal girls with Turner's syndrome. Pediatr Res 1990;28(6):626–30.
91. Mauras N, Rogol AD, Veldhuis JD. Specific, time-dependent actions of low-dose estradiol administration on the episodic release of GH, FSH and LH in prepubertal girls with Turner's syndrome. J Clin Endocrinol Metab 1989;69:1053–8.
92. degli Uberti EC, Ambrosio MR, Cella SG, Margutti AR, Trasforini G, Rigamonti AE, et al. Defective hypothalamic growth hormone (GH)-releasing hormone activity may contribute to declining GH secretion with age in man. J Clin Endocrinol Metab 1997;82:2885–8.
93. Tannenbaum GS, Painson JC, Lengyel AM, Brazeau P. Paradoxical enhancement of pituitary growth hormone (GH) responsiveness to GH-releasing factor in the face of high somatostatin tone. Endocrinology 1989;124:1380–8.
94. Spoudeas HA, Matthews DR, Brook CGD, Hindmarsh PC. The effect of changing somatostatin tone on the pituitary growth hormone and thyroid-stimulating hormone responses to their respective releasing factor stimuli. J Clin Endocrinol Metab 1992;75:453–8.

95. Stachura ME, Tyler JM, Farmer PK. Combined effects of human growth hormone (GH)-releasing factor-44 (GRF) and somatostatin (SRIF) on post-SRIF rebound release of GH and prolactin: a model for GRF-SRIF modulation of secretion. Endocrinology 1988;123:1476–82.
96. Steiger A, Guldner J, Hemmeter U, Rothe B, Wiedemann K, Holsboer F. Effects of growth hormone-releasing hormone and somatostatin on sleep EEG and nocturnal hormone secretion in male controls. Neuroendocrinology 1992;56:566–73.
97. Dickerman Z, Guyda H, Tannenbaum GS. Pretreatment with somatostatin analog SMS 201-995 potentiates growth hormone responsiveness to growth hormone-releasing factor in short children. J Clin Endocrinol Metab 1993;77:652–7.
98. Thorner MO, Vance ML, Hartman ML, Holl RW, Evans WS, Veldhuis JD, et al. Physiological role of somatostatin on growth hormone regulation in humans. Metabolism 1990;39:40–2.
99. Sugihara H, Minami S, Wakabayashi I. Post-somatostatin rebound secretion of growth hormone is dependent on growth hormone-releasing factor in unrestrained female rats. J Endocrinol 1989;122:583–91.
100. Tzanela M, Guyda H, Van Vliet G, Tannenbaum GS. Somatostatin pretreatment enhances growth hormone (GH) responsiveness to GH-releasing hormone: a potential new diagnostic approach to GH deficiency. J Clin Endocrinol Metab 1996;81:2487–94.
101. Hartman ML, Veldhuis JD, Johnson ML, Lee MM, Alberti KG, Samojlik E, et al. Augmented growth hormone (GH) secretory burst frequency and amplitude mediate enhanced GH secretion during a two-day fast in normal men. J Clin Endocrinol Metab 1992;74:757–65.
102. Ho KY, Veldhuis JD, Johnson ML, Furlanetto R, Evans WS, Alberti KG, et al. Fasting enhances growth hormone secretion and amplifies the complex rhythms of growth hormone secretion in man. J Clin Invest 1988;81:968–75.
103. Hartman ML, Pezzoli SS, Hellmann PJ, Suratt PM, Thorner MO. Pulsatile growth hormone secretion in older persons is enhanced by fasting without relationship to sleep stages. J Clin Endocrinol Metab 1996;81:2694–701.
104. Calabresi E, Ishikawa E, Bartolini L, Delitala G, Fanciulli G, Oliva O, et al. Somatostatin infusion suppresses GH secretory burst number and mass in normal men: a dual mechanism of inhibition. Am J Physiol 1996;270:E975–9.
105. Bowers CY, Reynolds GA, Durham D, Barrera CM, Pezzoli SS, Thorner MO. Growth hormone (GH)-releasing peptide stimulates GH release in normal men and acts synergistically with GH-releasing hormone. J Clin Endocrinol Metab 1990;70:975–82.
106. Bowers CY, Veeraragavan K, Sethumadhavan K. Atypical growth hormone releasing peptides. In: Bercu BB, Walker RF, eds. Growth hormone. II. Basic and clinical aspects. New York: Springer-Verlag, 1994:203–22.
107. Svensson J, Lonn L, Jansson J-O, Murphy G, Wyss D, Krupa D, et al. Two-month treatment of obese subjects with the oral growth hormone (GH) secretagogue MK-677 increases GH secretion, fat-free mass, and energy expenditure. J Clin Endocrinol Metab 1998;83:362–9.
108. Chapman IM, Bach MA, Cauter EV, Farmer M, Krupa D, Taylor AM, et al. Stimulation of the growth hormone (GH)-insulin-like growth factor I axis by daily oral administration of a GH secretagogue (MK-0677) in healthy elderly subjects. J Clin Endocrinol Metab 1996;81:4249–57.

109. Howard AD, Feighner SD, Cully DF, Arena JP, Liberator PA, Rosenblum CI, et al. A receptor in pituitary and hypothalamus that functions in growth hormone release. Science 1996;273:974–7.
110. Leong DA, Pomes A, Gaylinn BD, Thorner MO, Hellmann PH, Veldhuis JD. A novel hypothalamic hormone measured in hypophyseal portal plasma drives rapid bursts of GH secretion. Presented at the 80th Annual Meeting of the Endocrine Society, New Orleans, LA, June 24–27, 1998.
111. Pihoker C, Middleton R, Reynolds GA, Bowers CY, Badger TM. Diagnostic studies with intravenous and intranasal growth hormone-releasing peptide-2 in children of short stature. J Clin Endocrinol Metab 1995;80:2987–92.
112. Pihoker C, Kearns GL, French D, Bowers CY. Pharmacokinetics and pharmacodynamics of growth hormone-releasing peptide-2: a phase I study in children. J Clin Endocrinol Metab 1998;83:1168–72.
113. Van den Berghe G, de Zegher F, Bowers CY, Wouters P, Muller P, Soetens F, et al. Pituitary responsiveness to growth hormone (GH) releasing hormone, GH-releasing peptide-2 and thyrotropin releasing hormone in critical illness. Clin Endocrinol 1996;45:341–51.
114. Tiulpakov AN, Brook CG, Pringle PJ, Peterkova VA, Volevodz NN, Bowers CY. GH responses to intravenous bolus infusions of GH releasing hormone and GH releasing peptide 2 separately and in combination in adult volunteers. Clin Endocrinol 1995;43:347–50.
115. Mallo F, Alvarez CV, Benitez L, Burguera B, Coya R, Casanueva FF, et al. Regulation of His-dTrp-Ala-Trp-dPhe-Lys-NH2 (GHRP-6)-induced GH secretion in the rat. Neuroendocrinology 1993;57:247–56.
116. Bennett PA, Thomas GB, Howard AD, Feighner SD, van den Ploeg LH, Smith RG, et al. Hypothalamic growth hormone secretagogue-receptor (GHS-R) expression is regulated by growth hormone in the rat. Endocrinology 1997;138:4552–7.
117. Arvat E, Ramunni J, Bellone J, Di Vito L, Baffoni C, Broglio F, et al. The GH, prolactin, ACTH and cortisol responses to Hexarelin, a synthetic hexapeptide, undergo different age-related variations. Eur J Endocrinol 1997;1237:635–42.
118. Ghigo E, Arvat E, Muccioli G, Camanni F. Growth hormone releasing peptides. Eur J Endocrinol 1997;136:445–60.
119. Penalva A, Pombo M, Carballo A, Barreiro J, Casanueva FF, Dieguez C. Influence of sex, age and adrenergic pathways on the growth hormone response to GHRP-6. Clin Endocrinol 1993;38:87–91.
120. Kerrigan JR, Rogol AD. The impact of gonadal steroid hormone action on growth hormone secretion during childhood and adolescence. Endocr Rev 1992;13:281–98.
121. Lai Z, Roos P, Zhai O, Olsson Y, Fholenhag K, Larsson C, et al. Age-related reduction of human growth hormone-binding sites in the human brain. Brain Res 1993;621:260–6.
122. Cicognani A, Cacciari E, Tacconi M, Pascucci MG, Tonioli S, Pirazzoli P, et al. Effect of gonadectomy on growth hormone, IGF-I and sex steroids in children with complete and incomplete androgen insensitivity. Acta Endocrinol (Copenh) 1989;121:777–83.
123. Campagnoli C, Biglia N, Altare F, Lanza MG, Lesca L, Cantamessa C, et al. Differential effects of oral conjugated estrogens and transdermal estradiol on insulin-like growth factor I, growth hormone and sex hormone binding globulin serum levels. Gynecol Endocrinol 1993;7:251–8.

124. Kelly JJ, Rajkovic IA, O'Sullivan AJ, Sernia C, Ho KKY. Effects of different oral estrogen formulations on insulin-like growth factor-1, growth hormone and growth hormone binding protein in post-menopausal women. Clin Endocrinol 1993;39: 561–7.
125. Goodman-Gruen D, Barrett-Connor E. Effect of replacement estrogen on insulin-like growth factor-I in postmenopausal women: the Rancho Bernardo study. J Clin Endocrinol Metab 1996;81:4268–71.
126. Domene HM, Marin G, Sztein J, Yum YM, Baron J, Cassorla FG. Estradiol inhibits growth hormone receptor gene expression in rabbit liver. Mol Cell Endocrinol 1994;103:81–7.
127. Lieberman SA, Mitchell AM, Marcus R, Hintz RL, Hoffman AR. The insulin-like growth factor I generation test: resistance to growth hormone with aging and estrogen replacement therapy. Horm Metab Res 1994;26:229–33.
128. Holloway L, Butterfield G, Hintz R, Gesundheit N, Marcus R. Effects of recombinant human growth hormone on metabolic indices, body composition, and bone turnover in healthy elderly women. J Clin Endocrinol Metab 1994;79:470–9.
129. Bjorntorp P. Growth hormone, insulin-like growth factor-I and lipid metabolism: interactions with sex steroids. Horm Res 1996;46:188–91.
130. Thompson JL, Butterfield GE, Gylfadottir UK, Yesavage J, Marcus R, Hintz RL, et al. Effects of human growth hormone, insulin-like growth factor I, and diet and exercise on body composition of obese postmenopausal women. J Clin Endocrinol Metab 1998;83:1477–84.
131. Baum HBA, Biller BMK, Finkelstein JS, Cannistraro KB, Oppenheim DS, Schoenfeld DA, et al. Effects of physiologic growth hormone therapy on bone density and body composition in patients with adult-onset growth hormone deficiency. Ann Intern Med 1996;125:883–90.
132. Burman P, Johansson AG, Siegbahn A, Vessby B, Karlsson FA. Growth hormone (GH)-deficient men are more responsive to GH replacement therapy than women. J Clin Endocrinol Metab 1997;82:550–5.
133. Massa G, Bouillon R, Vanderschueren-Lodeweyckx M. Serum growth hormone (GH)-binding protein and insulin-like growth factor-I levels in Turner's syndrome before and during treatment with recombinant human GH and ethinyl estradiol. J Clin Endocrinol Metab 1992;75:1298–302.
134. Lucidi P, Lauteri M, Laureti S, Celleno R, Santoni S, Volpi E, et al. A dose-response study of growth hormone (GH) replacement on whole body protein and lipid kinetics in GH-deficient adult. J Clin Endocrinol Metab 1998;83:353–7.
135. Johannsson G, Bjarnason R, Bramnert M, Carlsson LM, Degerblad M, Manhem P, et al. The individual responsiveness to growth hormone (GH) treatment in GH-deficient adults is dependent on the level of GH-binding protein, body mass index, age, and gender. J Clin Endocrinol Metab 1996;81:1575–81.
136. Janssen YJH, Frolich M, Deurenberg P, Roelfsema F. Serum leptin levels during recombinant human GH therapy in adults with GH deficiency. Eur J Endocrinol 1997;137:650–4.
137. Straume M, Chen L, Johnson ML, Veldhuis JD. Systems-level analysis of physiological regulation interactions controlling complex secretory dynamics of growth hormone axis: a connectionist network model. Methods Neurosci 1995;28:270–310.
138. Straume M, Veldhuis JD, Johnson ML. Realistic emulation of highly irregular temporal patterns of hormone release: a computer-based pulse simulator. Methods Neurosci 1995;28:220–43.

139. Tannenbaum GS. Multiple levels of cross-talk between somatostatin (SRIF) and growth hormone (GH)-releasing factor in genesis of pulsatile GH secretion. Clin Pediatr Endocrinol 1994;3:97–110.
140. Epelbaum J. Intrahypothalamic neurohormonal interactions in the control of growth hormone secretion. Ciba Found Symp 1992;168:54–64.
141. Tannenbaum GS, Farhadi-Jou F, Beaudet A. Ultradian oscillation in somatostatin binding in the arcuate nucleus of adult male rats. Endocrinology 1993;133:1029–34.
142. Wehrenberg WB, Ling N, Bohlen P, Esch F, Brazeau P, Guillemin R. Physiological roles of somatocrinin and somatostatin in the regulation of growth hormone secretion. Biochem Biophys Res Commun 1982;109:562–7.
143. Ge F, Tsagarakis S, Rees LH, Besser GM, Gorman A. Relationship between growth hormone-releasing hormone and somatostatin in the rat: effects of age and sex on content and in vitro release from hypothalamic explants. J Endocrinol 1989;123:53–8.
144. Turner JP, Tannenbaum GS. In vivo evidence of a positive role for somatostatin to optimize pulsatile growth hormone secretion. Am J Physiol 1995;269:E683–90.
145. Mathews LS, Enberg B, Norstedt G. Regulation of rat growth hormone receptor gene expression. J Biol Chem 1989;264:9905–10.
146. Maiter DM, Gabriel SM, Koenig JI, Russell WE, Martin JB. Sexual differentiation of growth hormone feedback effects on hypothalamic growth hormone-releasing hormone and somatostatin. Neuroendocrinology 1990;51:174–80.
147. Wells T, Flavell DM, Wells SE, Carmignac DF, Robinson IC. Effects of growth hormone secretagogues in the transgenic growth-retarded (Tgr) rat. Endocrinology 1997;138:580–7.
148. Gevers E, Pincus SM, Robinson ICAF, Veldhuis JD. Differential orderliness of the GH release process in castrate male and female rats. Am J Physiol: Regul Integr Comp Physiol 1998;274:R437–44.
149. Veldhuis JD, Lassiter AB, Johnson ML. Operating behavior of dual or multiple endocrine pulse generators. Am J Physiol 1990;259:E351–61.
150. Shah N, Evans WS, Veldhuis JD. Mode of GH entry into the bloodstream, rather than gender or sex-steroid hormones, determines GH half-life in the human. Presented at the 79th Endocrine Society Annual Meeting, Minneapolis, MN, June 11–14, 1997.

11

The Growth Hormone Axis in Precocious Puberty

PAUL A. BOEPPLE

Children with central precocious puberty (CPP) recapitulate normal sexual maturation, albeit at an early age. The as yet poorly understood sexual dimorphism of pubertal disorders, with females predominating among patients with CPP and males dramatically outnumbering females among those presenting with constitutional delay of puberty, means that one cannot rely exclusively on any single clinical model to achieve comprehensive insights into pubertal maturation. However, children with idiopathic CPP who combine normal neuroendocrine function with normal skeletal responsiveness are an important model for the study of pubertal physiology. The goals for therapy of children with CPP include the ability to induce suppression of pubertal gonadal sex steroid secretion in a safe, selective, and reversible manner and thus diminish the differences between CPP patients and their peers regarding early sexual maturation and long-term growth, including final adult height. Since the early 1980s, these goals have been met largely by the ability of gonadotropin-releasing hormone (GnRH) agonist-analog (GnRHa) to induce pituitary gonadotroph desensitization (1). Thus, patients with idiopathic CPP evaluated before and during therapy with GnRHa provide a unique perspective on the modulation of the GH axis by endogenous sex-steroids addition and withdrawal.

The GH Axis in CPP During Active Puberty

Characterization of the growth hormone (GH) axis in untreated children with CPP has been largely consistent with the now well documented impacts of sex steroids in other human models. However, extensive normative data regarding the neurosecretory profiles of GH in healthy prepubertal children are difficult to come by. Indeed, there has only been one report in which the 24-h patterns of GH secretion in patients with CPP were compared directly with those of age-matched prepubertal children (Fig. 11.1) (2). As had been the case with several previous cross-sectional studies of GH secretion in healthy prepubertal children and pubertal adolescents, the wide variations in parameters of normal GH secretion make it difficult to discern statistically significant differences between groups, especially if the number of subjects studied is relatively small (3,4). With these limitations in mind, it appears that 24-h integrated serum GH concentrations are increased in children with CPP when they are studied during active puberty and that these increases stem predominantly from an augmentation of GH pulse amplitude, at least in girls (Fig. 11.1) (2). Interestingly, in the small number of boys studied, the increases in integrated serum GH concentration appeared to stem from an in increase in the frequency of GH pulses, whereas the amplitude of GH pulses in patients and controls were comparable (Fig. 11.1) (2).

Several groups have reported that plasma levels of somatomedin-C or insulin-like growth factor-I (IGF-I) are increased compared to values in prepubertal age-matched controls, but are appropriate for children matched for pubertal development (5–10). These consistent findings provide additional support for the conclusion that GH secretion is effectually increased in patients with CPP, a conclusion which is further bolstered by studies of IGF-I in

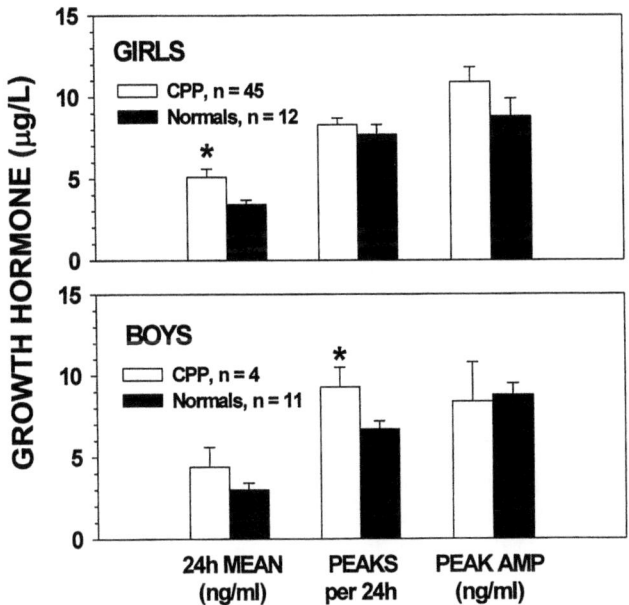

FIGURE 11.1. Parameters of spontaneous 24-h growth hormone (GH) secretion in children with central precocious puberty (CPP) and prepubertal age- and sex-matched controls (*, significantly increased vs. controls). Reprinted with permission from (2).

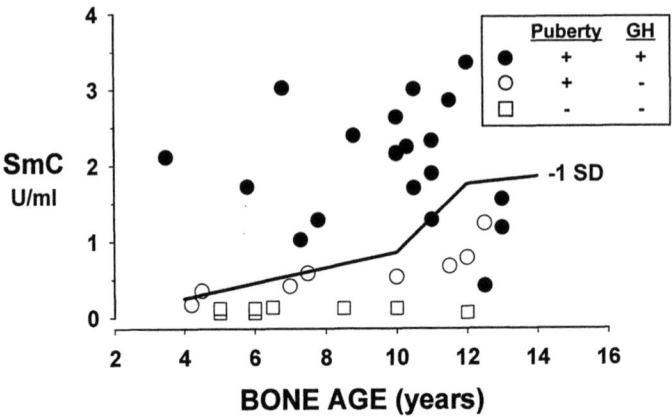

FIGURE 11.2. IGF-I (somatomedin-C, SmC) in patients with central precocious puberty (CPP) and intact GH secretion (*closed circles*), CPP and deficient GH secretion (*open circles*), and prepubertal GH deficient patients (*open squares*) plotted relative to bone age. The *jagged line* indicates −1 SD for normal controls. Reprinted with permission from Rappaport et al. (8).

CPP patients with and without intact GH axes (Fig. 11.2) (8–10). Children with CPP and GH deficiency do not have the exuberant rise in IGF-I exhibited by children with intact GH secretion when they are exposed to pubertal levels of gonadal steroids. These observations support the idea that it is a sex steroid-induced augmentation of GH secretion which is largely responsible for the rise in IGF-I both in CPP and in normal puberty.

The GH Axis in CPP During GnRHa Administration

The goal of GnRH agonist therapy is to induce consistent pituitary desensitization and thus a prepubertal gonadal sex-steroid milieu. One might expect the effects of appropriate GnRHa administration on the GH axis to be uniform and compatible with a resumption of prepubertal physiology. However, reports have varied considerably regarding the changes in GH and IGF-I associated with GnRHa therapy.

As anticipated, most but not all studies have reported a decrease in GH secretion during GnRHa treatment, largely related to a decrease in GH pulse amplitude (Fig 11.3) (7,11–13). In at least one of the studies that failed to document a decline in GH secretion during GnRHa administration, some patients were included who had neurogenic CPP, thus raising questions about the normality of the neuroendocrine regulation of GH (12). In addition, the regimen of GnRHa employed may not have been so effective in inducing complete pituitary gonadotropin desensitization, thus not achieving a clear model of full gonadal sex-steroid suppression (14).

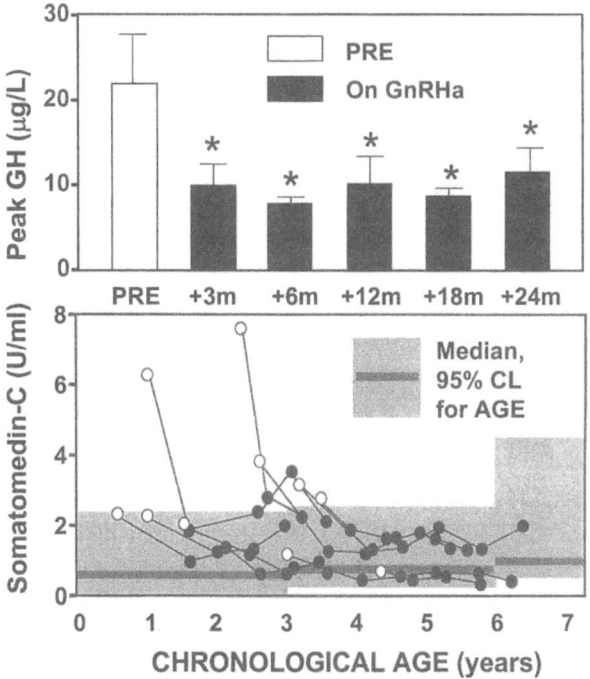

FIGURE 11.3. *Upper panel*: serial peak nighttime GH levels (mean ± SEM) in 10 girls with central precocious puberty (CPP) before and during the first 24 months (m) of GnRH agonist (GnRHa) administration (*, significantly decreased vs. PRE). Lower panel: serial IGF-I (somatomedin-C, SmC) levels plotted relative to chronological age in the same 10 girls with CPP before (*open circles*) and during (*closed circles*) GnRH agonist administration. Reprinted with permission from Mansfield et al. (7).

Changes in plasma IGF-I levels in association with GnRHa administration in children with CPP have also been variable. Several studies have reported the expected decrease in IGF-I as gonadal sex steroids returned to prepubertal levels and GH secretion declined (5–7), while others found no significant change from baseline (11,12). In fact, in one study IGF-I levels remained unchanged even though 24-h integrated GH concentrations were shown to decrease in the same patients (11). Invariably, serum IGF-I levels in children with CPP treated with GnRHa have remained higher than those in age-matched, prepubertal controls (see Fig. 11.3) (5–7). These observations, along with the fact that there are age-related increases in plasma IGF-I in normal childhood which occur despite unchanging GH secretion and gonadal sex steroid secretion, support the idea that the control of IGF-I generation is complex and goes beyond the influence of GH and sex steroids alone.

Serum levels of growth hormone-binding protein (GHBP), the circulating extracellular domain of the GH receptor, have been hypothesized to reflect the state of tissue GH responsiveness. Nonetheless, various interpretations have been proposed with important differences among them. On the one hand, increased serum GHBP may correlate with increased numbers of cell-surface GH receptors (and thus reflect increased responsiveness to circulating GH), while on the other hand increased serum GHBP results in more plasma GH in the bound state, perhaps limiting its biological activity (15). In addition, the regulation of the circulating levels of GHBP is complex, with GH administration in GH-deficient subjects resulting in increased levels (16) while a negative correlation is evident between integrated GH concentration and GHBP levels in normal adolescent boys studied longitudinally through puberty (15). Sex-steroid administration in hypogonadal subjects has, for the most part, been associated with a decrease in GHBP levels, but most studies comparing prepubertal children and pubertal adolescents have not found significant differences in serum GHBP (15,16). Finally, GHBP has been reported to correlate positively with bone mineral index (BMI) and serum leptin concentrations (15,17,18). With such a complex set of interrelated variables influencing the circulating levels of GHBP, it is perhaps no surprise that data stemming from the CPP model are not entirely consistent. Although some studies have reported an increase in GHBP in patients with CPP when gonadal suppression is induced (16,19,20), other investigations have found no difference in GHBP levels in association with GnRHa administration (14).

Pituitary–Gonadal Suppression During GnRHa Administration

Desensitization of pituitary gonadotropin secretion by potent, long-acting agonists of GnRH is not an "all-or-none" phenomenon. It is clear that some regimens of GnRHa administration may induce partial desensitization such that each dose of the agonist maintains its capacity to stimulate some degree of gonadotropin secretion and thus gonadal stimulation. This difference is particularly evident if weaker agonists are administered by daily injection (21,22). Thus, the degree to which treatment with GnRHa represents a "complete withdrawal" of gonadal sex steroids must be assessed carefully. The potency, formulation, dose, route, and frequency of administration of the GnRHa regimen all may affect the extent to which gonadotropin secretion is suppressed (22).

Given that serum levels of testosterone and estradiol fall considerably even when pituitary-gonadal suppression is incomplete, the query arises whether low residual levels of gonadal sex-steroid secretion that persist during some regimens of GnRHa administration are relevant to observations regarding changes in GH and IGF-I during therapy. A variety of lines of evidence support the notion that low levels of estrogen may represent a potent stimulus of linear growth and GH secretion. First is the observation that many normal girls measured serially at the peripubertal transition begin to exhibit accelerated rates of linear growth before any sign of breast development can be detected clinically (23). In accord with these observations in normal adolescents, a dose–response study of ethinyl estradiol replacement therapy in hypogonadal girls with Turner syndrome defined that the maximal short-term linear growth rates were achieved by small amounts of estrogen administration (24). Similarly low estrogen doses have also been shown to augment GH and IGF-I secretion in the same primary hypogonadal female model (25).

Studies employing a supersensitive estrogen bioassay have provided evidence that prepubertal levels of serum estradiol are far lower in prepubertal boys and girls than previously recognized (26). The transition into puberty and its associated increases in linear growth and GH secretion likely occurs as serum estradiol levels in the female rise within a concentration range far below the sensitivity of clinically available radioimmunoassays. A recent study employing the supersensitive estrogen assay found a relationship between doses of the potent GnRHa, deslorelin, and the degree of ovarian suppression in girls with CPP at levels that were well below the working range of estradiol RIAs (27). Estradiol concentrations remained at or just above the upper limits of the prepubertal range when serum estradiol by RIA was undetectable and pituitary gonadotropin secretion appeared appropriately suppressed. It is likely that the persistence of ovarian estradiol secretion at levels that exceed those of prepubertal girls during some GnRHa regimens in large part explains the inconsistent data generated in this model with respect to changes in the GH–IGF-I axis during therapy.

Growth Velocity Correlates in CPP Patients

The lowered rates of GH secretion during GnRHa therapy in CPP, combined with growth velocities that in some patients fall well below those of normal, prepubertal children, have prompted some investigators to administer GH in combination with GnRHa in this setting (28,29). However, levels of GH secretion in CPP patients on therapy correlate poorly with their growth velocities despite the fact that spontaneous and stimulated levels of GH secretion in a substantial number of CPP patients fall below the range defined in normal prepubertal children (30). Although there was no correlation of GH with growth velocity in one study (30), the same investigators found that integrated GH concentrations correlated negatively with BMI, as has also been described in normal puberty (31).

If not related to levels of GH secretion, what explains the low linear growth rates in some children with CPP during GnRHa therapy? Their growth velocities are determined in large part by their degree of skeletal maturation (32). The "intrinsic growth rate" in the absence of gonadal sex steroids correlates negatively with bone age across a wide range skeletal maturation. Thus, GH administration in combination with GnRHa in patients with CPP is not to be considered "standard of care," but must be considered carefully on a case-by-case basis.

Conclusions

The model of reversible, selective suppression of gonadal sex-steroid secretion induced by GnRHa administration in children with CPP has provided further support for the important role of sex steroids in the modulation of the GH axis. However, data stemming from this model have also underscored the realization that the regulation of GH and IGF-I secretion during childhood and adolescent growth is extraordinarily complex and goes beyond the influences exerted by maturation and suppression of the hypothalamic–pituitary–gonadal axis.

References

1. Boepple PA, Mansfield MJ, Wierman ME, Rudlin CR, Bode HH, Crigler JF, Crawford JD, Crowley WF. Use of a potent, long-acting agonist of gonadotropin-releasing hormone (GnRH) in the treatment of precocious puberty. Endocr Rev 1986;7:24–33.
2. Ross JL, Pescovitz OH, Barnes K, Loriaux DL, Cutler GB Jr. Growth hormone secretory dynamics in children with precocious puberty. J Pediatr 1987;110:369–72.
3. Thompsom RG, Rodriguez A, Kowarski A, Migeon CJ, Blizzard RM. Integrated concentrations of growth hormone correlated with plasma testosterone and bone age in preadolescent and adolescent males. J Clin Endocrinol Metab 1972;35:334–7
4. Plotnick LP, Thompson RG, Beitins I, Blizzard RM. Integrated concentrations of growth hormone correlated with stage of puberty and estrogen levels in girls. J Clin Endocrinol Metab 1974;38:436–9.
5. Harris DA, Van Vliet G, Egli CA, Grumbach MM, Kaplan SL, Styne DM, Vainsel M. Somatomedin-C in normal puberty and in true precocious puberty before and after treatment with a potent luteinizing hormone-releasing hormone agonist. J Clin Endocrinol Metab 1985;61:152–9.
6. Pescovitz OH, Rosenfeld RG, Hintz RL, Barnes K, Hench K, Comite F, Loriaux DL, Cutler GB Jr. Somatomedin-C in accelerated growth of children with precocious puberty. J Pediatr 1985;107:20–5.
7. Mansfield MJ, Rudlin CR, Crigler JF Jr, Karol KA, Crawford JD, Boepple PA, Crowley WF Jr. Changes in growth and serum growth hormone and plasma somatomedin-C levels during suppression of gonadal sex steroid secretion in girls with central precocious puberty. J Clin Endocrinol Metab 1988;66:3–9.
8. Rappaport R, Prevot C, Brauner R. Somatomedin-C and growth in children with precocious puberty: a study of the effect of the level of growth hormone secretion. J Clin Endocrinol Metab 1987;65:1112–7.
9. Cara JF, Burstein S, Cuttler L, Moll GW Jr, Rosenfield RL. Growth hormone deficiency impedes the rise in plasma insulin-like growth factor I levels associated with precocious puberty. J Pediatr 1989;115:64–8.
10. Attie KM, Ramirez NR, Conte FA, Kaplan SL, Grumbach MM. The pubertal growth spurt in eight patients with true precocious puberty and growth hormone deficiency: evidence for a direct role of sex steroids. J Clin Endocrinol Metab 1990;71:975–83.

11. DiMartino-Nardi J, Wu R, Fishman K, Saenger P. The effect of long-acting analog of luteinizing hormone-releasing hormone on growth hormone secretory dynamics in children with precocious puberty. J Clin Endocrinol Metab 1991;73:902–6.
12. Sklar CA, Rothenberg S, Blumberg D, Oberfield SE, Levine LS, David R. Suppression of the pituitary-gonadal axis in children with central precocious puberty: effects on growth, growth hormone, insulin-like growth factor-I, and prolactin secretion. J Clin Endocrinol Metab 1991;73:734–8.
13. DiMartino-Nardi J, Wu R, Varner R, Wong WL, Saenger P. The effect of luteinizing hormone-releasing hormone analog for central precocious puberty on growth hormone (GH) and GH-binding protein. J Clin Endocrinol Metab 1994;78:664–8.
14. Stanhope R, Pringle PJ, Brook CG. Growth, growth hormone and sex steroid secretion in girls with central precocious puberty treated with a gonadotrophin releasing hormone (GnRH) analogue. Acta Paediatr Scand 1988;77:525–30.
15. Martha PM Jr, Rogol AD, Carlsson LM, Gesundheit N, Blizzard RM. A longitudinal assessment of hormonal and physical alterations during normal puberty in boys. I. Serum growth hormone-binding protein. J Clin Endocrinol Metab 1993;77:452–7.
16. Postel-Vinay MC, Tar A, Hocquette JF, Clot JP, Fontoura M, Brauner R, Rappaport R. Human plasma growth hormone (GH)-binding proteins are regulated by GH and testosterone. J Clin Endocrinol Metab 1991;73:197–202.
17. Bjarnason R, Boguszewski M, Dahlgren J, Gelander L, Kristrom B, Rosberg S, Carlsson B, Albertsson-Wikland K, Carlsson LM. Leptin levels are strongly correlated with those of GH-binding protein in prepubertal children. Eur J Endocrinol 1997;137:68–73.
18. Kratzsch J, Dehmel B, Pulzer F, Keller E, Englaro P, Blum WF, Wabitsch M. Increased serum GHBP levels in obese pubertal children and adolescents: relationship to body composition, leptin and indicators of metabolic disturbances. Int J Obes Relat Metab Disord 1997;21:1130–6.
19. Oliveira SB, Donnadieu M, Chaussain JL. Changes in growth hormone-binding protein in girls with central precocious puberty treated with a depot preparation of luteinizing hormone-releasing hormone analogue. Horm Res 1993;39:42–6.
20. Kobayashi Y, Murata A, Yasuda T, Minagawa M, Wataki K, Ohnishi H, Niimi H. Suppression of sex steroids by a gonadotrophin-releasing hormone agonist increases serum growth hormone-binding protein activity in girls with central idiopathic precocious puberty. Clin Endocrin 1994;40:351–5.
21. Conn PM, Crowley WF Jr. Gonadotropin-releasing hormone and its analogues. N Engl J Med 1991;324:93–103.
22. Boepple PA, Mansfield MJ, Landy H, Crowley WF Jr. GnRH agonist therapy of central precocious puberty: should the goal be complete suppression? In: Grave GD, Cutler GB Jr, eds. Sexual precocity: etiology, diagnosis, and management. New York: Raven Press, 1993:11–26.
23. Marshall WA, Tanner JM. Variations in pattern of pubertal changes in girls. Arch Dis Child 1969;44:291–303.
24. Ross JL, Cassorla FG, Skerda MC, Valk IM, Loriaux DL, Cutler GB Jr. A preliminary study of the effect of estrogen dose on growth in Turner's syndrome. N Engl J Med 1983;309:1104–6.
25. Cuttler L, Van Vliet G, Conte FA, Kaplan SL, Grumbach MM. Somatomedin-C levels in children and adolescents with gonadal dysgenesis: differences from age-matched normal females and effect of chronic estrogen replacement therapy. J Clin Endocrinol Metab 1985;60:1087–92.

26. Klein KO, Baron J, Colli MJ, McDonnell DP, Cutler GB Jr. Estrogen levels in childhood determined by an ultrasensitive recombinant cell bioassay. J Clin Invest 1994;94:2475–80.
27. Klein KO, Baron J, Barnes KM, Pescovitz OH, Cutler GB Jr. Use of an ultrasensitive recombinant cell bioassay to determine estrogen levels in girls with precocious puberty treated with a luteinizing hormone-releasing hormone agonist. J Clin Endocrinol Metab 1998;83:2387–9.
28. Saggese G, Pasquino AM, Bertelloni S, Baroncelli GI, Battini R, Pucarelli I, Segni M, Franchi G. Effect of combined treatment with gonadotropin releasing hormone analogue and growth hormone in patients with central precocious puberty who had subnormal growth velocity and impaired height prognosis. Acta Paediatr 1995;84:299–304.
29. Pasquino AM, Municchi G, Pucarelli I, Segni M, Mancini MA, Troiani S. Combined treatment with gonadotropin-releasing hormone analog and growth hormone in central precocious puberty. J Clin Endocrinol Metab 1996;81:948–51.
30. Kamp GA, Manasco PK, Barnes KM, Jones J, Rose SR, Hill SC, Cutler GB Jr. Low growth hormone levels are related to increased body mass index and do not reflect impaired growth in luteinizing hormone-releasing hormone agonist-treated children with precocious puberty. J Clin Endocrinol Metab 1991;72:301–7.
31. Rose SR, Municchi G, Barnes KM, Kamp GA, Uriarte MM, Ross JL, Cassorla F, Cutler GB Jr. Spontaneous growth hormone secretion increases during puberty in normal girls and boys. J Clin Endocrinol Metab 1991;73:428–35.
32. Boepple PA, Mansfield MJ, Link K, Crawford JD, Crigler JF Jr, Kushner DC, Blizzard RM, Crowley WF Jr. Impact of sex steroids and their suppression upon skeletal growth and maturation. Am J Physiol 1988;255:E559–66.

Part II

Sex Steroids and GHRH-Somatostatin

12

Hypothalamic Control Mechanisms of Sexually Dimorphic Growth Hormone Secretory Patterns in the Rat

GLORIA S. TANNENBAUM

In most mammalian species, including humans, the secretion of growth hormone (GH) is characterized by multiple episodic bursts. There is, however, a marked sex difference both in the pattern of GH secretion and in the rate of somatic growth [see (1) for review]. The male GH secretory profile, particularly in the rat, is characterized by high-amplitude GH bursts at regular 3.3-h intervals, separated by low or undetectable basal plasma GH levels (2); in contrast, females exhibit irregular, more frequent, lower-amplitude GH pulses superimposed on an elevated GH baseline (3). These gender-related differences in GH secretory patterns are of biological importance in the determination of sex-specific patterns of growth (4) and metabolism (5) in this species. Although the sexual dimorphism in GH secretion has been the subject of intense investigation in recent years (6–21), the underlying mechanisms are still far from clear. This chapter focuses on the roles of the hypothalamic neuropeptides, somatostatin and GH-releasing hormone, and on the modulatory influence of the sex steroids, in genesis of the sexually dimorphic GH secretory patterns in the rat.

Sexually Dimorphic Patterns of Hypothalamic Somatostatin and GHRH Release

Extensive experimental evidence indicates that the episodic pattern of GH release from the pituitary gland is generated by an exquisite interplay between at least two hypothalamic hormones, the stimulatory GH-releasing hormone (GHRH), found in the arcuate nucleus, and its inhibitory counterpart, somatostatin (SRIF), principally derived from periventricular neurons. These neuropeptides are released at the level of the median eminence into the hypophyseal portal circulation to reach somatotrophs in the anterior pituitary, thereby controlling the synthesis or release of GH. Furthermore, a large body of physiological and morphological evidence indicates that GHRH and SRIF not only influence GH secretion at the level of the pituitary gland but also interact within the hypothalamus for GH control (22).

There is compelling evidence for gender-related differences in GH control by hypothalamic SRIF and GHRH. At the level of the hypothalamus, male and female rats show different mRNA (10,14,15) and peptide (9) levels of both SRIF and GHRH as well as different GH autofeedback mechanisms (12). At the level of the pituitary, male somatotrophs exhibit increased GH responsiveness to GHRH (6,13), greater GH secretory capacity (7,8), and elevated mRNA levels of both GH (19) and GHRH receptor (18) compared to their female counterparts. Moreover, many of these parameters can be altered by modulating the sex-steroid environment (11,16,17,20).

We (13,23) and others (6) have postulated that the sexual dimorphism of GH secretion in the rat may be the result of a gender difference in the pattern of hypothalamic SRIF/GHRH signaling to pituitary somatotrophs. Our hypotheses are shown schematically in Figure 12.1. In the male rat (Fig. 12.1, left panel), we propose that SRIF and GHRH are released rhythmically from the hypothalamus into the hypophyseal portal blood, in reciprocal 3- to 4-h cycles, about 180° out of phase, to generate the ultradian rhythm of GH secretion as observed in the peripheral blood. The bulk of evidence now suggests that a decrease in SRIF concentrations associated with an increase in GHRH levels induces GH peaks in males of this species, as well as possibly in the human (24). This phasic release of SRIF likely plays a key role in determining the pulsatility and periodicity of GH secretion.

In females (Fig. 12.1, right panel), it is postulated that the pattern of hypothalamic SRIF secretion into hypophyseal portal blood is continuous, rather than cyclical, as in the male. In the case of GHRH, the female pattern consists of steady-state release, which occurs at a higher level than that of the male, and additional episodic GHRH bursting, which appears to occur at approximately 1-h intervals, is superimposed. The net result of these neuropeptide patterns of release is the more erratic GH secretory profile observed in the peripheral plasma of females. We concluded from these studies that the sexual dimorphism of GH secretion in the rat is primarily caused by the gender difference in mode of hypothalamic SRIF signaling, with females exhibiting tonic, rather than episodic, SRIF secretion compared to males.

Sexually Dimorphic Expression of Somatostatin Receptor Subtypes

In recent years, five different SRIF receptor subtypes, designated sst1 to sst5, have been cloned and characterized (25). The mRNAs for all five SRIF receptors are widely expressed in the brain (26–31) and pituitary (32–35). We have recently found that GHRH neurons in the arcuate nucleus of the male rat hypothalamus express both sst1 and sst2 SRIF receptor genes (36), suggesting that

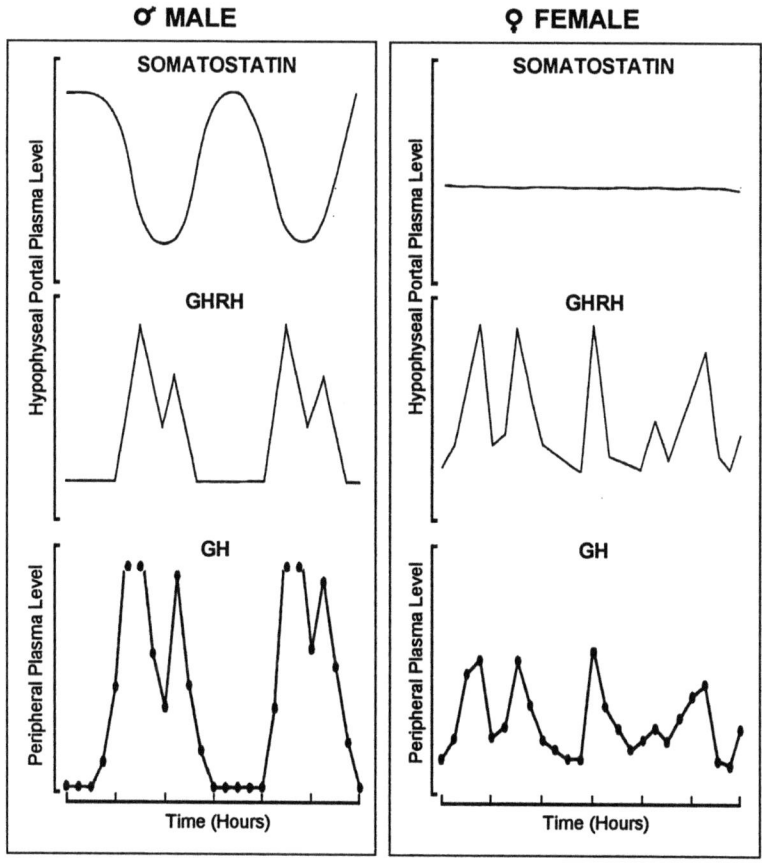

FIGURE 12.1. Hypotheses for the hypothalamic control of pulsatile GH secretion in male (*left panel*) and female (*right panel*) rats. Postulated patterns of hypothalamic somatostatin and GHRH secretion into hypophyseal portal blood, with the net result on pituitary GH release as observed in the peripheral circulation, are shown. Redrawn from Painson and Tannenbaum (13) and Tannenbaum and Ling (23).

SRIF may directly modulate GHRH release into the hypophyseal portal blood and thereby influence GH secretion through interaction with both sst1 and sst2 receptor subtypes. Furthermore, GH has been shown to regulate the expression of sst1 mRNA on arcuate neurons (37). At the level of the pituitary, both the mRNA and protein for the sst2 receptor are present in a large percentage of somatotrophs (32,34,35), and there is pharmacological evidence implicating the sst2 receptor subtype in the regulation of GH release by somatotrophs (38–40). Together, these results suggest that both sst1 and sst2 SRIF receptor subtypes are involved in the neuroendocrine regulation of GH secretion.

We tested the hypothesis that the sexual dimorphism of GH secretion may result, at least in part, from gender-related differences in the transduction of SRIF neuroendocrine actions in the hypothalamus or pituitary. To accomplish this, we compared the distributional pattern and level of expression of sst1 and sst2 receptors in the brain and pituitary gland of adult male and female rats using in situ hybridization histochemistry (41). In the brain, there was a marked sex-related difference in sst1 expression in the arcuate nucleus of the hypothalamus; both the number and labeling density of sst1 mRNA-expressing cells, as revealed by computer-assisted image analysis, were two- to threefold greater in males than in females (see Fig. 12.2). This effect was regionally selective in that there was no

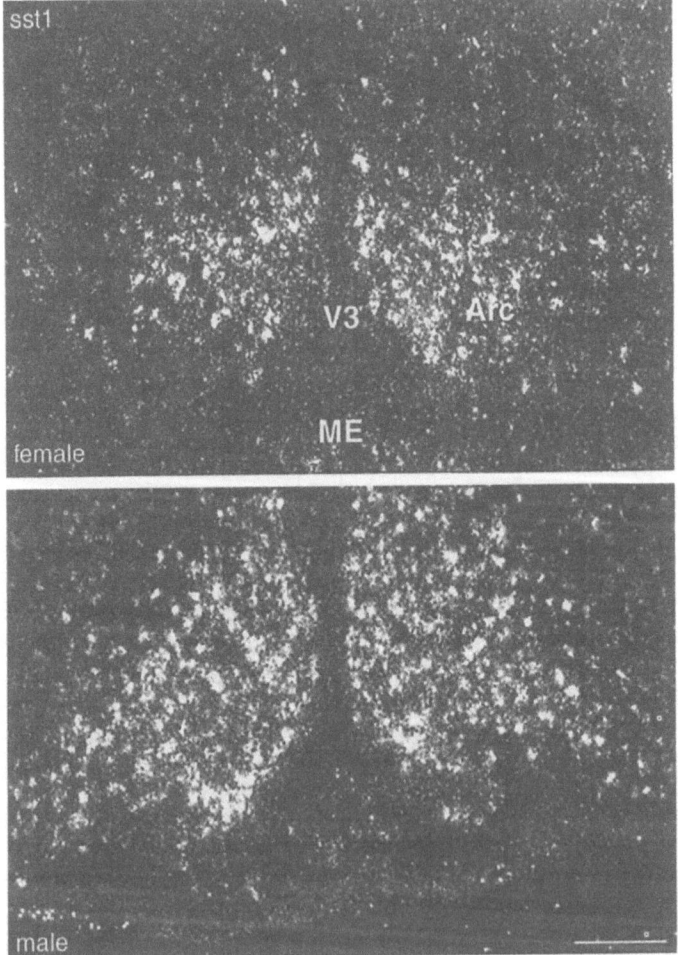

FIGURE 12.2. Comparative distribution of sst1 mRNA in the arcuate nucleus of female and male rats as detected by in situ hybridization. Note that both the number of sst1-hybridizing cells and intensity of individual cell labeling are greater in the arcuate nucleus (*Arc*) of male than female rats. *ME*, median eminence; *V3*, third ventricle. Scale bar, 225 µm. Reprinted with permission from Zhang, Beaudet, and Tannenbaum (41).

significant difference between males and females in the density of sst1 hybridization in other areas of the brain examined, including the cerebral cortex, medial habenula, and ventromedial hypothalamic nucleus. In contrast, no gender-related differences in sst2 mRNA-expressing cells in the arcuate nucleus were found. At the level of the anterior pituitary, the labeling density of sst2 mRNA in males was significantly higher than that of females (Fig. 12.3), but no sex-related difference in pituitary sst1 mRNA was observed. These results demonstrate a sexual dimorphism in the expression of two SRIF receptor subtypes, sst1 and sst2, at the level of the arcuate nucleus and anterior pituitary, respectively. Such dimorphism suggests a differential involvement of sst1 and sst2 in GH regulation with respect to gender, and may imply roles for sst2 and sst1 in transducing SRIF actions on pituitary somatotrophs and GHRH-containing arcuate neurons, respectively, to generate the lower basal and higher GH pulse levels characteristic of the male rat.

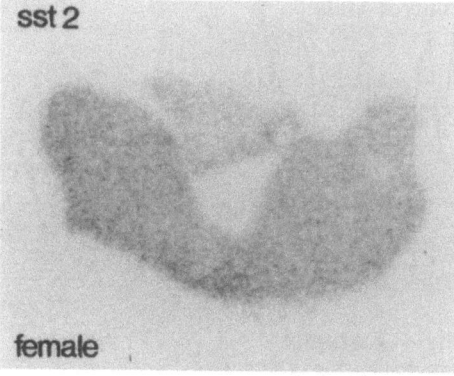

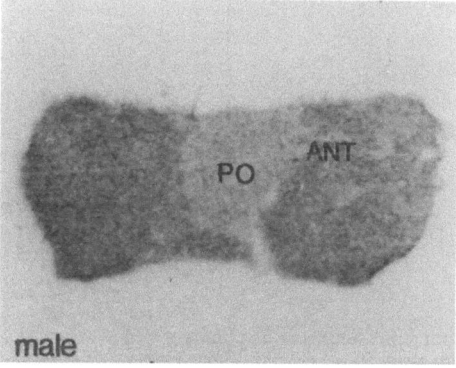

FIGURE 12.3. Film autoradiograms of female and male pituitaries hybridized with a sst2 antisense riboprobe. In both male and female, the anterior pituitary (*ANT*) is more intensely labeled than the posterior pituitary (*PO*). Note the higher expression of sst2 mRNA in the anterior lobe of males than females. Scale bar, 1 mm. Reprinted in part with permission from Zhang, Beaudet, and Tannenbaum (41).

Sex-Steroid Modulation of Sexually Dimorphic GH Secretory Patterns

The sex steroids appear to play a major role in mediating the sexual dimorphism of GH secretion. Considerable evidence indicates that the gender-related differences in GH secretory pattern and growth rate are attributable, at least in part, to the influence of the gonadal steroid milieu during the neonatal and prepubertal periods (1,42). However, recent data indicate that the effects of the gonadal steroids are not limited to the critical neonatal period of imprinting. As shown in Figure 12.4C, short-term exposure of adult male rats to 17-β-estradiol subcutaneous implants causes a marked disruption of the male ultradian rhythm of GH secretion, converting it to a female-like pattern;

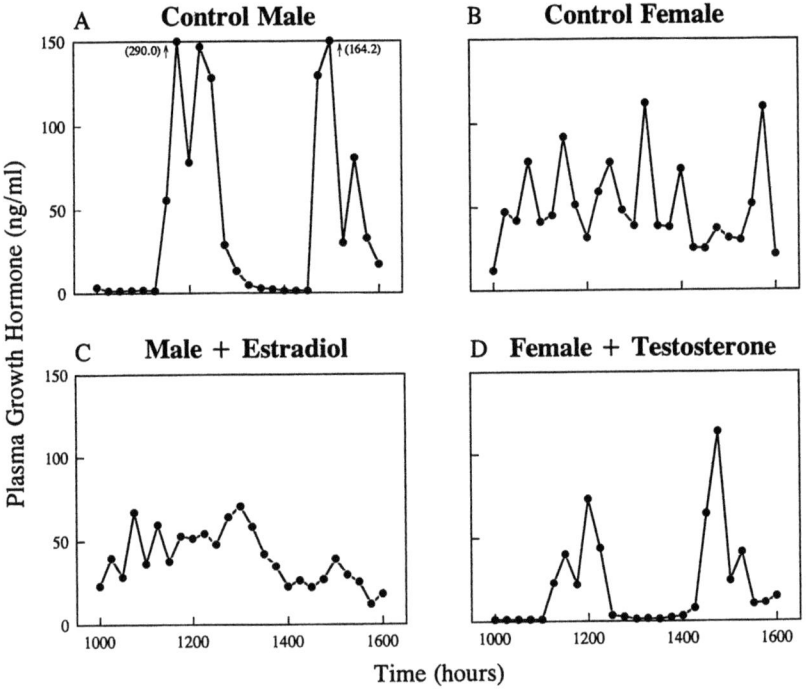

FIGURE 12.4. Individual representative 6-h plasma GH profiles in control male (*A*) and female (*B*) rats treated with either 17-β-estradiol (*C*) or testosterone (*D*), respectively. Exposure of adult male rats to estradiol (s.c. implant for 4 days) completely feminized the male ultradian GH rhythm, whereas testosterone administration to adult females (1 mg s.c. 18 h earlier) converted the erratic female GH secretory pattern to the male-like ultradian rhythm. Based in part on Painson et al. (16) and Tannenbaum (22).

we have observed this dramatic feminization of the GH secretory pattern within 12 h of estradiol treatment (16). Exposure to estradiol during adult life also feminized the male pattern of GH responsiveness to GHRH, as well as rate of somatic growth (16). These findings suggest that estradiol is involved in modulating GH baseline release and pulse frequency.

Conversely, acute exposure to testosterone during adult life profoundly masculinizes the GH pulse-generating circuits of female rats. As illustrated in Figure 12.4D, testosterone treatment (18 h after administration) significantly decreased the GH nadir and the GH pulse frequency, thereby converting the erratic female GH secretory pattern to the very regular male-like ultradian GH rhythm. These results suggest that the enhanced orderliness of the GH release process in males, compared to females, is regulated by testosterone. We postulate that the effects of testosterone are mediated at the level of the hypothalamus by modulating the somatostatinergic branch of the GH neuroendocrine axis, thereby inducing a regularity in SRIF secretion that governs overall periodicity.

A growing body of experimental evidence suggests a hypothalamic site of action for sex-steroid effects on GH secretion. In particular, the gonadal steroids appear to be intimately involved in the regulation of hypothalamic SRIF. Female rats subjected to ovariectomy exhibit an increase in SRIF immunoreactivity in the median eminence (43) and a decrease in SRIF mRNA levels in the periventricular region of the hypothalamus (44); both effects can be reversed by estradiol treatment (43,44). Testosterone, administered either neonatally (45) or in adulthood (10), stimulates SRIF gene expression in the periventricular nucleus. Both estrogen and androgen receptors have been detected in some hypothalamic neurons (46,47).

Regulation of SRIF receptors also appears to be dependent on circulating gonadal steroids. Estradiol can positively modulate ^{125}I-labeled SRIF-binding sites on both rat arcuate perikarya (48) and pituitary membranes (49,50). Furthermore, treatment with estrogen and testosterone increases the expression of three SRIF receptor subtypes (sst1, sst2, and sst3), in pituitary GH_4C_1 cells (51). In normal rats, estradiol treatment increased pituitary mRNA expression of sst1, sst2, and sst3, and decreased expression of sst5 (52), although in another study (53) pituitary sst2 mRNA level was not influenced by gonadal steroids. Some of these effects may be direct, through interaction with estrogen or androgen receptors located in sst-expressing cells, as both estrogen and androgen receptors have been detected in the rat arcuate nucleus and anterior pituitary (46). Furthermore, [^{3}H]estradiol has been localized to GHRH neurons (54), a subset of which express sst1 mRNA (36). Alternatively, the sexual dimorphism in sst receptor expression observed in both brain and pituitary may merely reflect the distinct male/female patterns of endogenous SRIF release (6,13,23) (Fig. 12.1), which, in turn, may be dependent on the activation of hypothalamic SRIF neurons by gonadal steroids (10,15). Although the mechanisms by which gonadal steroids may be regulating sst receptor expression in brain or pituitary have yet to be fully elucidated, undoubtedly the influence of the sex steroids on the somatostatinergic branch of the GH neuroendocrine axis plays an important role in both the generation and maintenance of the sexually dimorphic patterns of GH secretion and, ultimately, body growth.

Acknowledgments. This work was supported by the Medical Research Council of Canada. The author is the recipient of a "Chercheur de carrière" award from the Fonds de la recherche en santé du Québec. The secretarial assistance of Julie Temko in preparing this manuscript is gratefully acknowledged.

References

1. Jansson J-O, Edén S, Isaksson O. Sexual dimorphism in the control of growth hormone secretion. Endocr Rev 1985;6:128–50.
2. Tannenbaum GS, Martin JB. Evidence for an endogenous ultradian rhythm governing growth hormone secretion in the rat. Endocrinology 1976;98:562–70.
3. Edén S. Age- and sex-related differences in episodic growth hormone secretion in the rat. Endocrinology 1979;105:555–60.
4. Robinson ICAF, Clark RG. The secretory pattern of GH and its significance for growth in the rat. In: Isaksson O, Binder C, Hall K, Hökfelt B, eds. Growth hormone—basic and clinical aspects. Amsterdam: Elsevier, 1987:109–27.
5. Waxman DJ, Pampori NA, Ram PA, Agrawal AK, Shapiro BH. Interpulse interval in circulating growth hormone patterns regulates sexually dimorphic expression of hepatic cytochrome P450. Proc Natl Acad Sci USA 1991;88:6868–72.
6. Clark RG, Robinson ICAF. Growth hormone responses to multiple injections of a fragment of human growth hormone-releasing factor in conscious male and female rats. J Endocrinol 1985;106:281–9.
7. Hoeffler JP, Frawley LS. Capacity of individual somatotropes to release growth hormone varies according to sex: Analysis by reverse hemolytic plaque assay. Endocrinology 1986;119:1037–41.
8. Ho KY, Leong DA, Sinha YN, Johnson ML, Evans WS, Thorner MO. Sex-related differences in GH secretion in the rat using the reverse hemolytic plaque assay. Am J Physiol 1986;250:E650–4.
9. Jansson J-O, Ishikawa K, Katakami H, Frohman LA. Pre- and postnatal developmental changes in hypothalamic content of rat growth hormone-releasing factor. Endocrinology 1987;120:525–30.
10. Chowen-Breed JA, Steiner RA, Clifton DK. Sexual dimorphism and testosterone-dependent regulation of somatostatin gene expression in the periventricular nucleus of the rat brain. Endocrinology 1989;125:357–62.
11. Hertz P, Silbermann M, Even L, Hochberg Z. Effects of sex steroids on the response of cultured rat pituitary cells to growth hormone-releasing hormone and somatostatin. Endocrinology 1989;125:581–5.
12. Carlsson LMS, Clark RG, Robinson ICAF. Sex difference in growth hormone feedback in the rat. J Endocrinol 1990;126:27–35.
13. Painson J-C, Tannenbaum GS. Sexual dimorphism of somatostatin and growth hormone-releasing factor signaling in the control of pulsatile growth hormone secretion in the rat. Endocrinology 1991;128:2858–66.

14. Maiter D, Koenig JI, Kaplan LM. Sexually dimorphic expression of the growth hormone-releasing hormone gene is not mediated by circulating gonadal hormones in the adult rat. Endocrinology 1991;128:1709–16.
15. Argente J, Chowen JA, Zeitler P, Clifton DK, Steiner RA. Sexual dimorphism of growth hormone-releasing hormone and somatostatin gene expression in the hypothalamus of the rat during development. Endocrinology 1991;128:2369–75.
16. Painson J-C, Thorner MO, Krieg RJ, Tannenbaum GS. Short-term adult exposure to estradiol feminizes the male pattern of spontaneous and growth hormone-releasing factor-stimulated growth hormone secretion in the rat. Endocrinology 1992;130:511–9.
17. Chowen JA, Argente J, Gonzàlez-Parra S, Garcia-Segura LM. Differential effects of the neonatal and adult sex steroid environments on the organization and activation of hypothalamic growth hormone-releasing hormone and somatostatin neurons. Endocrinology 1993;133:2792–802.
18. Ono M, Miki N, Murata Y, Osaki E, Tamitsu K, Ri T, et al. Sexually dimorphic expression of pituitary growth hormone-releasing factor receptor in the rat. Biochem Biophys Res Commun 1995;216:1060–6.
19. Gonzàlez-Parra S, Chowen JA, Garcia-Segura LM, Argente J. Ontogeny of pituitary transcription factor-1(Pit-1), growth hormone (GH) and prolactin (PRL) mRNA levels in male and female rats and the differential expression of Pit-1 in lactotrophs and somatotrophs. J Neuroendocrinol 1996;8:211–25.
20. Borski RJ, Tsai W, DeMott-Friberg R, Barkan AL. Regulation of somatic growth and the somatotropic axis by gonadal steroids: primary effect on insulin-like growth factor I gene expression and secretion. Endocrinology 1996;137:3253–9.
21. Jaffe CA, Ocampo-Lim B, Guo W, Krueger K, Sugahara I, DeMott-Friberg R, et al. Regulatory mechanisms of growth hormone secretion are sexually dimorphic. J Clin Invest 1998;102:153–64.
22. Tannenbaum GS. Multiple levels of cross-talk between somatostatin (SRIF) and growth hormone (GH)-releasing factor in genesis of pulsatile GH secretion. Clin Pediatr Endocrinol 1994;3:97–110.
23. Tannenbaum GS, Ling N. The interrelationship of growth hormone (GH)-releasing factor and somatostatin in generation of the ultradian rhythm of GH secretion. Endocrinology 1984;115:1952–7.
24. Vance ML, Kaiser DL, Evans WS, Furlanetto R, Vale W, Rivier J, et al. Pulsatile growth hormone secretion in normal man during a continuous 24-hour infusion of human growth hormone releasing factor (1-40). J Clin Invest 1985;75:1584–90.
25. Reisine T, Bell GI. Molecular biology of somatostatin receptors. Endocr Rev 1995;16:427–42.
26. Breder CD, Yamada Y, Yasuda K, Seino S, Saper CB, Bell GI. Differential expression of somatostatin receptor subtypes in brain. J Neurosci 1992;12:3920–34.
27. Bruno JF, Xu Y, Song J, Berelowitz M. Tissue distribution of somatostatin receptor subtype messenger ribonucleic acid in the rat. Endocrinology 1993;133:2561–7.
28. Señaris RM, Humphrey PPA, Emson PC. Distribution of somatostatin receptors 1, 2 and 3 mRNA in rat brain and pituitary. Eur J Neurosci 1994;6:1883–96.
29. Pérez J, Rigo M, Kaupmann K, Bruns C, Yasuda K, Bell GI, et al. Localization of somatostatin (SRIF) SSTR-1, SSTR-2 and SSTR-3 receptor mRNA in rat brain by *in situ* hybridization. Naunyn-Schmiedeberg's Arch Pharmacol 1994;349:145–60.
30. Raulf F, Perez J, Hoyer D, Bruns C. Differential expression of five somatostatin receptor subtypes, SSTR1-5, in the CNS and peripheral tissue. Digestion 1994;55:46–53.

31. Beaudet A, Greenspun D, Raelson J, Tannenbaum GS. Patterns of expression of SSTR1 and SSTR2 somatostatin receptor subtypes in the hypothalamus of the adult rat: relationship to neuroendocrine function. Neuroscience 1995;65:551–61.
32. Day R, Dong W, Panetta R, Kraicer J, Greenwood MT, Patel YC. Expression of mRNA for somatostatin receptor (sstr) types 2 and 5 in individual rat pituitary cells. A double labeling in situ hybridization analysis. Endocrinology 1995;136:5232–5.
33. O'Carroll A-M, Krempels K. Widespread distribution of somatostatin receptor messenger ribonucleic acids in rat pituitary. Endocrinology 1995;136:5224–7.
34. Kumar U, Laird D, Srikant CB, Escher E, Patel YC. Expression of the five somatostatin receptor (SSTR1-5) subtypes in rat pituitary somatotrophes: quantitative analysis by double-label immunofluorescence confocal microscopy. Endocrinology 1997;138:4473–6.
35. Mezey E, Hunyady B, Mitra S, Hayes E, Liu Q, Schaffer J, et al. Cell specific expression of the sst2A and sst5 somatostatin receptors in the rat anterior pituitary. Endocrinology 1998;139:414–9.
36. Tannenbaum GS, Zhang W-H, Lapointe M, Zeitler P, Beaudet A. Growth hormone-releasing hormone neurons in the arcuate nucleus express both sst1 and sst2 somatostatin receptor genes. Endocrinology 1998;139:1450–3.
37. Guo F, Beaudet A, Tannenbaum GS. The effect of hypophysectomy and growth hormone replacement on sst1 and sst2 somatostatin receptor subtype messenger ribonucleic acids in the arcuate nucleus. Endocrinology 1996;137:3928–35.
38. Raynor K, Murphy WA, Coy DH, Taylor JE, Moreau J-P, Yasuda K, et al. Cloned somatostatin receptors: identification of subtype-selective peptides and demonstration of high affinity binding of linear peptides. Mol Pharmacol 1993;43:838–44.
39. Hoyer D, Lubbert H, Bruns C. Molecular pharmacology of somatostatin receptors. Naunyn-Schmiedebergs Arch Pharmacol 1994;350:441–53.
40. Shimon I, Taylor JE, Dong JZ, Bitonte RA, Kim S, Morgan B, et al. Somatostatin receptor subtype specificity in human fetal pituitary cultures: differential role of sstr2 and sstr5 for growth hormone, thyroid-stimulating hormone, and prolactin regulation. J Clin Invest 1997;99:789–98.
41. Zhang W-H, Beaudet A, Tannenbaum GS. Sexually dimorphic expression of sst1 and sst2 somatostatin receptor subtypes in the arcuate nucleus and anterior pituitary of adult rats. J Neuroendocrinol 1999;11:129–36.
42. Jansson J-O, Ekberg S, Isaksson O, Mode A, Gustafsson J-A. Imprinting of growth hormone secretion, body growth, and hepatic steroid metabolism by neonatal testosterone. Endocrinology 1985;117:1881–9.
43. Gross DS. Role of somatostatin in the modulation of hypophysial growth hormone production by gonadal steroids. Am J Anat 1980;158:507–19.
44. Werner H, Koch Y, Baldino F Jr, Gozes I. Steroid regulation of somatostatin mRNA in the rat hypothalamus. J Biol Chem 1988;263:7666–71.
45. Chowen JA, Argente J, Torre-Saleman I, Gonzalez-Parra S, Garcia-Segura LM. Effects of the neonatal sex steroid environment on growth hormone-releasing hormone and somatostatin gene expression. J Pediatr Endocrinol 1993;6:211–8.
46. Simerly RB, Chang C, Muramatsu M, Swanson LW. Distribution of androgen and estrogen receptor mRNA-containing cells in the rat brain: an *in situ* hybridization study. J Comp Neurol 1990;294:76–95.
47. Herbison AE. Sexually dimorphic expression of androgen receptor immunoreactivity by somatostatin neurones in rat hypothalamic periventricular nucleus and bed nucleus of the stria terminalis. J Neuroendocrinol 1995;7:543–53.

48. Slama A, Videau C, Kordon C, Epelbaum J. Estradiol regulation of somatostatin receptors in the arcuate nucleus of the female rat. Neuroendocrinology 1992;56:240–5.
49. Kimura N, Hayafuji C, Konagaya H, Takahashi K. 17β-estradiol induces somatostatin (SRIF) inhibition of prolactin release and regulates SRIF receptors in rat anterior pituitary cells. Endocrinology 1986;119:1028–36.
50. Kimura N, Hayafuji C, Kimura N. Characterization of 17β-estradiol-dependent and -independent somatostatin receptor subtypes in the rat anterior pituitary. J Biol Chem 1998;264:7033–40.
51. Xu Y, Berelowitz M, Bruno JF. Dexamethasone regulates somatostatin receptor subtype messenger ribonucleic acid expression in rat pituitary GH_4C_1 cells. Endocrinology 1995;136:5070–5.
52. Kimura N, Tomizawa S, Arai KN, Kimura N. Chronic treatment with estrogen up-regulates expression of sst_2 messenger ribonucleic acid (mRNA) but down-regulates expression of sst_5 mRNA in rat pituitaries. Endocrinology 1998;139:1573–80.
53. Senaris RM, Lago F, Diéguez C. Gonadal regulation of somatostatin receptor 1, 2 and 3 mRNA levels in the rat anterior pituitary. Mol Brain Res 1996;38:171–5.
54. Shirasu K, Stumpf WE, Sar M. Evidence for direct action of estradiol on growth hormone-releasing factor (GRF) in rat hypothalamus: localization of [^{3}H]estradiol in GRF neurons. Endocrinology 1990;127:344–9.

13

Sex Differences in Growth Hormone-Releasing Hormone (GHRH) and Somatostatin Neurons

DONALD K. CLIFTON AND ROBERT A. STEINER

Male rats grow more rapidly than female rats. Most of the difference in growth rates between the sexes is thought to be the result of differences in gonadal hormone secretion that occur early in development and continue on into adulthood. Gonadal steroids have anabolic and growth-promoting effects on a wide variety of tissues throughout the body, and some of the growth-promoting effects of gonadal steroids are undoubtedly caused by direct actions of these hormones on peripheral tissues such as bone, muscle, and adipose tissue. Gonadal steroids also act on the brain where they influence the regulated secretion of growth hormone (GH) from the pituitary. Therefore, in addition to their direct action on peripheral tissues, gonadal steroids can *indirectly* regulate growth through their effects on circulating GH concentrations.

The release of GH from the pituitary occurs in discrete pulses (1,2), and in the rat this pulsatile mode of secretion has an important impact on growth rates (3,4). GH pulses are generated by the coordinated activity of neurons residing in the hypothalamus. Although there may be many neuronal phenotypes that participate in the process of GH pulse generation, at least two are clearly involved—somatostatin-containing neurons in the periventricular nucleus and growth hormone-releasing hormone (GHRH) containing neurons in the arcuate nucleus. Evidence in the rodent suggests that somatostatin neurons are active between GH pulses, suppressing the release of GH, whereas GHRH neurons are relatively quiescent (5,6). During the generation of a pulse, GHRH neurons become active, and the activity of somatostatin neurons subsides (6). Both the frequency and amplitude of GH pulses are thought to be dependent on the coordinated interaction of somatostatin and GHRH neurons.

Patterns of GH secretion in the rat are sexually dimorphic (7). GH pulses in males are higher in amplitude and occur less frequently than those in females. In addition, males secrete less GH between pulses than females do. Because GH secretory patterns have been shown to influence the rate of growth, sexual differentiation of pulsatile GH secretion is thought to be an important determinant of the sex differences in growth rates. To evaluate the role of somatostatin and GHRH neurons in the sexual differentiation of the GH control system, we have investigated the effects of gender and gonadal steroid treatment on the expression of neuropeptide mRNAs by hypothalamic somatostatin and GHRH neurons in adult rats and have identified potential target sites for gonadal steroids among the neural elements that regulate GH secretion.

The Expression of Neuropeptide mRNAs in Somatostatin and GHRH Neurons Is Sexually Differentiated

To learn more about the molecular basis for the sexual differentiation of GH secretion, we focused initially on somatostatin neurons in the adult rat (8). Using in situ hybridization, we found that somatostatin mRNA levels in the periventricular nucleus are higher in males than females (at proestrus), which is consistent with the putative role of these cells in the neuroendocrine regulation of GH secretion. We found that the expression of somatostatin mRNA in cortical neurons did not appear to be sexually differentiated, reinforcing the argument that cortical somatostatin neurons are not involved in the regulation of GH and suggesting that somatostatin neurons in the periventricular nucleus represent a physiologically distinct population.

In a subsequent study, we examined the sexual differentiation of GHRH neurons (9). We found that the expression of GHRH mRNA in neurons of both the arcuate nucleus and ventromedial nucleus is higher in males than in females (at proestrus), as is somatostatin mRNA in the periventricular nucleus. It is tempting to speculate that the higher expression of both somatostatin and GHRH mRNA in males compared to females is related to differences in the patterns of GH secretion between the sexes. Perhaps males need to secrete more somatostatin between GH pulses to maintain lower interpulse levels of GH secretion and therefore require higher levels of somatostatin mRNA to support the increased rate of secretion. Likewise, it may be that males need to release more GHRH in the process of generating the high-amplitude GH pulses characteristic of that sex and that high levels of GHRH mRNA reflect this increased secretory demand. Whether these simplistic interpretations of differences in mRNA levels reflects reality awaits further investigation.

GHRH neurons are known to express other neuropeptides in addition to GHRH (10). One of these, galanin, has been implicated in the regulation of pulsatile GH secretion (11). Therefore, we sought to determine whether the expression of galanin mRNA in GHRH neurons is also sexually differentiated (12). Because work by others had demonstrated that the infusion of antiserum to galanin converts the male GH pulse pattern to one that appears more like that of the female (11), we hypothesized that galanin mRNA would be higher in males than females. In this study, we used double-label in situ hybridization with a digoxigenin-labeled riboprobe to identify GHRH neurons and an ^{35}S-labeled probe to measure galanin mRNA levels in the same GHRH neurons. As postulated, galanin mRNA levels in GHRH neurons proved to be higher in males compared to females, making it likely that galanin also plays a role in the sexual differentiation of GH secretion.

The Sexual Differentiation of Somatostatin and GHRH mRNA Expression Is Mediated by Testosterone Acting on Androgen Receptors

Sexual differentiation of the GH system could be either the result of organizational processes (programmed at an early age) or caused by the activational effects of hormones circulating in the adult. We hypothesized that circulating gonadal steroids are responsible for sex-related differences in somatostatin mRNA, GHRH mRNA, and galanin mRNA in GHRH neurons. We tested this hypothesis by measuring the levels of these mRNAs in normal (testes-intact), castrated, and castrated plus gonadal steroid-treated males (8,12–14). In these studies we found that castration reduced the expression of all three mRNAs and that replacement with testosterone restored mRNA levels to those found in intact rats. To determine whether the effects of testosterone are mediated either by androgen receptors or, following conversion to estradiol, by estrogen receptors, we compared the effects of 5-α dihydrotestosterone (a nonaromatizable androgen) and 17-β-estradiol on levels of somatostatin and GHRH mRNA. The administration of 17-β-estradiol to castrated males had no effect on either somatostatin or GHRH mRNA, whereas dihydrotestosterone partially restored GHRH mRNA levels and completely restored somatostatin mRNA levels to those found in intact males. These results suggest that testosterone acts directly on androgen receptors to regulate the neuroendocrine GH axis and does not require aromatization to estradiol and subsequent activation of androgen receptors in the rodent.

Although studies such as these make it clear that androgen receptors are involved in the neuroendocrine regulation of the GH system in the rat, they do not identify the neuronal targets at which testosterone acts to accomplish this regulation. Interpretation of the data is complicated by the existence of a short negative feedback loop between GH secretion and the hypothalamic neurons that regulate GH secretion. For example, it is possible that testosterone directly stimulates only somatostatin neurons. This action would cause increased somatostatin release into the portal system and a subsequent inhibition of GH secretion, resulting in an increase in GHRH neuronal activity (caused by decreased negative feedback of GH on GHRH). Likewise, if testosterone were to act only on GHRH neurons to stimulate GHRH secretion, this would cause GH secretion to increase, leading to a stimulation of SS activity. Therefore, testosterone acting on either SS or GHRH neurons could *indirectly* cause an increase in the secretion of the other neuropeptide. However, it is also conceivable that testosterone acts *directly* on both SS and GHRH neurons to increase the release of SS and GHRH into the portal circulation. One way to ascertain immediate sites of testosterone's action is to examine the expression of androgen receptor in the neurons known to be involved in the regulation of GH secretion.

Somatostatin Neurons Express Detectable Androgen Receptors But GHRH Neurons Do Not

The most likely neuronal targets for the effects of testosterone effects on the GH axis are somatostatin neurons in the periventricular nucleus and GHRH neurons in the arcuate nucleus. To test whether these neurons express androgen receptor mRNA, we applied double-label in situ hybridization histochemistry (15) to brain sections from 70-day-old male ($n = 4$) and female ($n = 4$) rats killed between 1000 and 1030 h. A ^{35}S-labeled 572-bp cRNA probe was used to measure androgen-receptor mRNA. This probe was complementary to the region of the androgen-receptor mRNA that encodes the amino terminus and part of the DNA-binding domain. Somatostatin and GHRH mRNA-containing neurons were identified through the use of digoxigenin-labeled cRNA probes. Signal to background ratios (SBR) were calculated for digoxigenin-labeled neurons by counting the number of silver grains over the cell body and dividing by the background silver grain level (16). The SBR was then used to assess whether cells were double labeled. SBRs were determined for every digoxigenin-labeled neuron within defined regions of the hypothalamus, and, in the in situ hybridization assay in which somatostatin neurons were being studied, SBRs for five randomly chosen cortical somatostatin cells per tissue section were also determined.

Because the determination of the percentage of neurons coexpressing androgen-receptor mRNA is dependent on the SBR criterion that is applied, we have displayed the percentage of cells accepted as double labeled as a function of the SBR (Figs. 13.1 and 13.2). As shown in Figure 13.1, the distribution of somatostatin mRNA-containing neurons in the periventricular nucleus that also contain androgen-receptor mRNA is similar for males and females. Even if a relatively conservative SBR of 5 were used as the criterion for determining coexpression, approximately two-thirds of the somatostatin cells in this region would be considered to coexpress mRNA for the androgen receptor in both sexes. The observation that periventricular somatostatin neurons contain androgen-receptor mRNA is consistent with previous reports of the immunohistochemical identification of androgen receptors (protein) in this same population of neurons (17,18). However, our finding that there is no difference in androgen-receptor expression between the sexes is at variance with the latter work of Herbison, who reported that the percentage of somatostatin neurons containing immunoreactive androgen receptors is higher in males than in females (18). There are several possible explanations for this discrepancy, including (1) sex differences in the translation or posttranslational processing of the receptor, and (2) sexually differentiated rates of receptor turnover. The exact mechanism responsible for differential androgen-receptor

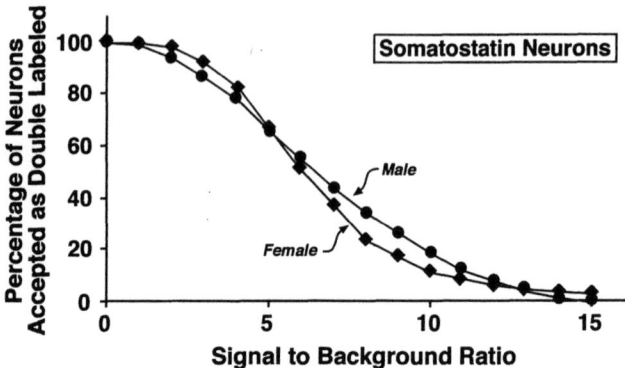

FIGURE 13.1. The frequency distribution of neurons defined as double labeled for androgen-receptor and somatostatin mRNA in the periventricular nucleus of male and female rats plotted as a function of the criterion used to identify double-labeled cells. The signal to background ratio was calculated by dividing the number of silver grains (representing androgen-receptor mRNA) over cell bodies containing somatostatin mRNA by the background silver grain level.

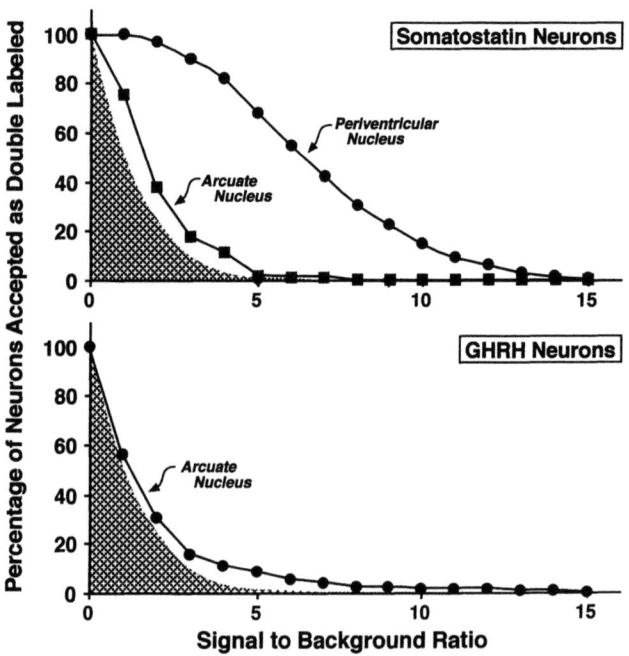

FIGURE 13.2. The frequency distribution of neurons accepted as double labeled for androgen-receptor mRNA and either somatostatin (*top*) or GHRH (*bottom*) mRNA plotted as a function of the criterion used to identify double-labeled cells. Data from both male and female rats were pooled. The *shaded area* represents the frequency distribution obtained by counting silver grains over somatostatin cells in the cortex, an area where somatostatin neurons would not be expected to be androgen sensitive.

antigenic content remains to be clarified; nevertheless, it is clear that levels of androgen-receptor mRNA in these neurons is not influenced by animal gender.

The coexpression of androgen-receptor mRNA in somatostatin (or GHRH) neurons did not appear to be sexually differentiated in any of the areas studied; therefore, we combined the results of males and females as shown in Figure 13.2. Although there was strong evidence for androgen-receptor mRNA expression in periventricular somatostatin neurons, the expression of androgen-receptor mRNA in somatostatin and GHRH neurons located in the arcuate nucleus and somatostatin neurons in the cortex was much less apparent. Again, when an SBR criterion of 5 is applied, 66% of the somatostatin neurons in the periventricular nucleus appear to contain androgen-receptor mRNA, but less than 2% of the somatostatin neurons in the arcuate nucleus and cortex and less than 10% of the GHRH neurons in the arcuate nucleus appear to do so. We included measurements of androgen-receptor mRNA in cortical somatostatin neurons because we did not expect these neurons to be responsive to androgens; the low SBR values in the cortex are consistent with our expectation. Although it is conceivable that neurons with SBRs between 1 and 5 are actually expressing low levels of the androgen-receptor mRNA, it is also probable that many of the neurons with low SBRs are "false positives." Based on these observations, we conclude that although most of the somatostatin neurons in the periventricular nucleus are androgen sensitive, few (if any) of the somatostatin or GHRH neurons in the arcuate nucleus are.

A Model for Sex Differences in Somatostatin and GHRH Neurons

With this information in hand, we can begin to assemble a rough model for sexually differentiated neuropeptide gene expression in somatostatin and GHRH neurons of the rat. It is clear that, in the male, testosterone acts directly on somatostatin neurons in the periventricular nucleus to increase the expression of somatostatin mRNA. Testosterone also increases GHRH mRNA levels, but this effect is most likely indirect. At least some of the effects of testosterone on GHRH neurons may be mediated by somatostatin neurons acting through the GH negative feedback (or other) system. The logic for this proposition depends on the assumption that testosterone-induced SS mRNA expression is associated with an increase in somatostatin release into the portal system. If this were the case, then increased somatostatin should inhibit GH secretion from the pituitary. Lower levels of circulating GH would reduce the negative feedback influence of GH on GHRH neurons, resulting in increased GHRH neuronal activity. GH feedback provides an attractive explanation for the regulation of GHRH mRNA, but GH feedback can not explain the effects of testosterone on galanin mRNA expression in GHRH neurons.

Unlike its inhibitory effect on GHRH mRNA, GH stimulates galanin gene expression in GHRH neurons (19). Therefore, if the effects of testosterone were mediated by GH feedback, one would expect testosterone to inhibit, rather than stimulate, galanin mRNA levels in GHRH neurons. Some other mechanisms must be responsible for mediating testosterone's effects on galanin in GHRH neurons. Because the receptors that mediate the effects of testosterone on galanin in GHRH neurons have not been characterized, it is possible that estrogen rather than androgen receptors are involved in this process. In fact, the galanin gene contains an estrogen-sensitive AP-1 site, and galanin expression is regulated by estrogen in other neurons both inside and outside the hypothalamus (20,21). Thus, one would predict that GHRH neurons express estrogen receptors, although we have not tested this.

In addition to the possibility that somatostatin regulates GHRH mRNA levels through GH feedback mechanisms, it is also probable that the testosterone signal is transmitted to GHRH neurons through afferent connections from androgen receptor-containing neurons located elsewhere in the hypothalamus. Again, somatostatin neurons in the periventricular nucleus are likely candidates for performing this function because GHRH neurons are heavily innervated by somatostatin fibers (22,23), although in one study the origin of these fibers has been questioned (24). Neurons containing substance P, metenkephalin, thyrotropin-releasing hormone, and corticotropin-releasing hormone also make direct contacts with GHRH neurons (25), but we do not know whether these neurons contain androgen receptors.

Summary and Conclusion

We have made progress toward understanding the mechanisms that lead to sex differences in the expression of neuropeptides in somatostatin and GHRH neurons, yet there are still pieces to this puzzle that remain to be put into place. Most of the differences appear to result from the activational effects of circulating testosterone. Testosterone increases somatostatin mRNA levels by binding to androgen receptors on somatostatin neurons in the periventricular nucleus. Testosterone also increases GHRH mRNA by acting via androgen receptors, but the target neurons mediating this effect remain unidentified. In addition, although testosterone stimulates galanin gene expression, this may be mediated by estrogen receptors or other pathways.

Our work has focused on the sexual differentiation of mRNA expression in the output neurons of the hypothalamic system that regulates GH secretion. This is only one aspect of a complex system that regulates growth, and most certainly mRNA regulation does not tell the entire story. For example, even though estrogen does not appear to influence either somatostatin or GHRH mRNA levels in adult male rats, the administration of estradiol causes males to exhibit female-like GH pulse patterns (26). If and how sex differences in neuropeptide gene expression are involved in the generation of such sexually dimorphic patterns of GH secretion and the possible interspecies differences in such neuroregulation remain to be elucidated.

Acknowledgments. The study of androgen-receptor mRNA expression in GHRH and somatostatin neurons was conducted by Dr. Yvonne Chan while she was a postdoctoral fellow in our laboratory.

References

1. Martin JB, Renaud LP, Brazeau PJ Jr. Pulsatile growth hormone secretion: suppression by hypothalamic ventromedial lesions and by long-acting somatostatin. Science 1974;186:538–40.
2. Tannenbaum G, Martin JB. Evidence for an endogenous ultradian rhythm governing growth hormone secretion in the rat. Endocrinology 1976;98:562–70.
3. Jansson J-O, Albertsson-Wikland K, Eden S, Throngren K-H, Isaksson O. Circumstantial evidence for a role of the secretory pattern of growth hormone in control of body growth. Acta Endocrinol Copenh 1982;99:24–30.
4. Clark RG, Jansson J-O, Isaksson O, Robinson ICAF. Intravenous growth hormone: growth responses to patterned infusions in hypophysectomized rats. J Endocrinol 1985;104:53–61.
5. Tannenbaum GS. Physiological role of somatostatin in regulation of pulsatile growth hormone secretion. Adv Exp Med Biol 1985;188:229–59.
6. Plotsky PM, Vale W. Patterns of growth hormone-releasing factor and somatostatin secretion into the hypophysial-portal circulation of the rat. Science 1985;230:461–3.
7. Eden S. Age- and sex-related differences in episodic growth hormone secretion in the rat. Endocrinology 1979;105:555–60.
8. Chowen-Breed JA, Steiner RA, Clifton DK. Sexual dimorphism and testosterone-dependent regulation of somatostatin gene expression in the periventricular nucleus of the rat brain. Endocrinology 1989;125:357–62.
9. Argente J, Chowen JA, Zeitler P, Clifton DK, Steiner RA. Sexual dimorphism of growth hormone-releasing hormone and somatostatin gene expression in the hypothalamus of the rat during development. Endocrinology 1991;128:2369–75.
10. Meister B, Hökfelt T. Peptide- and transmitter-containing neurons in the mediobasal hypothalamus and their relation to GABAergic systems: possible roles in control of prolactin and growth hormone secretion. Synapse 1988;2:585–605.
11. Maiter DM, Hooi SC, Koenig JI, Martin JB. Galanin is a physiological regulator of spontaneous pulsatile secretion of growth hormone in the male rat. Endocrinology 1990;126:1216–22.
12. Delemarre-van-de-Waal HA, Burton KA, Kabigting EB, Steiner RA, Clifton DK. Expression and sexual dimorphism of galanin messenger ribonucleic acid in growth hormone-releasing hormone neurons of the rat during development. Endocrinology 1994;134:665–71.
13. Zeitler P, Argente J, Chowen-Breed JA, Clifton DK, Steiner RA. Growth hormone-releasing hormone messenger ribonucleic acid in the hypothalamus of the adult male rat is increased by testosterone. Endocrinology 1990;127:1362–8.
14. Argente J, Chowen-Breed JA, Steiner RA, Clifton DK. Somatostatin messenger RNA in hypothalamic neurons is increased by testosterone through activation of androgen receptors and not by aromatization to estradiol. Neuroendocrinology 1990;52:342–9.
15. Marks DL, Wiemann JN, Burton KA, Lent KL, Clifton DK, Steiner RA. Simultaneous visualization of two cellular mRNA species in individual neurons by use of a new double in situ hybridization method. Mol Cell Neurosci 1992;3:395–405.

16. Chan YY, Steiner RA, Clifton DK. Regulation of hypothalamic neuropeptide-Y neurons by growth hormone in the rat. Endocrinology 1996;137:1319–25.
17. Huang X, Harlan RE. Androgen receptor immunoreactivity in somatostatin neurons of the periventricular nucleus but not in the bed nucleus of the stria terminalis in male rats. Brain Res 1994;652:291–6.
18. Herbison AE. Sexually dimorphic expression of androgen receptor immunoreactivity in rat hypothalamic periventricular nucleus and bed nucleus of the stria terminalis. J Neuroendocrinol 1995;7:543–53.
19. Chan YY, Grafstein-Dunn E, Delemarre-Van de Waal HA, Burton KA, Clifton DK, Steiner RA. The role of galanin and its receptor in the feedback regulation of growth hormone secretion. Endocrinology 1996;137:5303–10.
20. Marks DL, Lent KL, Rossmanith WG, Clifton DK, Steiner RA. Activation-dependent regulation of galanin gene expression in gonadotropin-releasing hormone neurons in the female rat. Endocrinology 1994;134:1991–8.
21. Tseng JY, Kolb PE, Raskind MA, Miller MA. Estrogen regulates galanin but not tyrosine hydroxylase gene expression in the rat locus ceruleus. Brain Res Mol Brain Res 1997;50:100–6.
22. Liposits Z, Merchenthaler I, Paull WK, Flerkó B. Synaptic communication between somatostatinergic axons and growth hormone-releasing factor (GRF) synthesizing neurons in the arcuate nucleus of the rat. Histochemistry 1988;89:247–52.
23. Horváth S, Palkovits M, Görcs T, Arimura A. Electron microscopic immunocytochemical evidence for the existence of bidirectional synaptic connections between growth hormone-releasing hormone- and somatostatin-containing neurons in the hypothalamus of the rat. Brain Res 1989;481:8–15.
24. Willoughby JO, Brogan M, Kapoor R. Hypothalamic interconnections of somatostatin and growth hormone releasing factor neurons. Neuroendocrinology 1989;50:584–91.
25. Halász B. Neuroendocrinology in 1992. Neuroendocrinology 1993;57:1196–207.
26. Painson JC, Thorner MO, Krieg RJ, Tannenbaum GS. Short-term adult exposure to estradiol feminizes the male pattern of spontaneous and growth hormone-releasing factor-stimulated growth hormone secretion in the rat. Endocrinology 1992;130:511–9.

14

Sex-Steroid Interaction with Somatostatin and GHRH in the Rat

Franco Sanchez-Franco, Fernando Pazos, Gumersindo Fernandez, Judith Lopez, Nuria Palacios, and Lucinda Cacicedo

Gonadal steroids modulate endocrine systems at both hypothalamic and pituitary levels (1). In particular, in the rat, testosterone (T) and estradiol (E_2) have been shown to influence the synthesis and secretion of growth hormone (GH), as well as the pituitary cell response to somatostatin (SS) and GH-releasing hormone (GHRH) (2,3). These effects are exerted directly on the somatotroph (4) or through the alteration of the hypothalamic SS or GHRH (5,6), suggesting that in the rat hypothalamic GRF and SS neurons are targets for gonadal steroids. Previous studies done in vivo by in situ hybridization have indicated that T stimulates SS and GHRH gene expression (5,6), but not all studies have been in agreement with these findings (7). Also, differences in SS and GHRH concentrations in hypothalamic tissue have been reported in different developmental stages by several authors (1,8–12).

To further understand the interaction between gonadal steroids and SS or GHRH, we have studied the direct action of T and E_2 on the synthesis and secretion of the two neuropeptides in fetal rat hypothalamic cells in primary culture. Also, we have focused on the interaction between the gonadal and GH axes during puberty and aging in the male rat, two developmental stages during which a rapid coactivation or a parallel decline of the two axes occurs. In these settings, interference with the normal physiological development of either axis might influence behavior of the other.

Action of Gonadal Steroids on SS and GHRH in Fetal Rat Hypothalamic Cells

Fetal rat hypothalamic cells from 17-day-old embryos were maintained in primary culture. On day 20 of culture, each culture dish was incubated for 48 h with 2 ml of minimal essential medium (MEM) containing different concentrations of T or E_2 or with MEM alone in the control group of plates. As

TABLE 14.1. Action of testosterone (T) on IR-SS in fetal rat hypothalamic cells in primary culture.

Testosterone (ng/dl)	IR-SS (pg/plate)		
	Cell extract	Media	Total
Control	480 ± 40	1520 ± 99	2000 ± 70
10	440 ± 30	1405 ± 98	1845 ± 60
20	425 ± 35	1311 ± 125**	1726 ± 80*
40	350 ± 40**	1250 ± 99**	1600 ± 70**

IR-SS, immunoreactive somatostatin. The values represent the mean ± SEM ($n = 5$).
*, $p < 0.05$ vs. control.
** $p < 0.01$ vs. Control.

shown in Table 14.1, 48-h treatment with T had a significant inhibitory effect on immunoreactive SS (IR-SS) released into the media, on SS content in the cells, and on the total IR-SS in the plate. The total IR-SS in the plate (media and cell content) reflects the SS synthesis. This action of T occurred in a dose-related manner between T concentrations of 10 and 40 ng/dl, the highest concentration of T used in the study.

The T action on GRF is represented in Table 14.2. Total (media and cell content) immunoreactive GHRH (IR-GHRH) per plate significantly increased in the presence of increasing doses of T. A similar effect was observed when cell extracts and media were analyzed separately, indicating that GHRH synthesis is increased in fetal hypothalamic cells under the direct action of T. The E_2 action on IR-SS is represented in Table 14.3. Increasing concentrations of E_2 produced a continuous and significant decrease of IR-SS in the media and a progressive, not significant, increase in intracellular accumula-

TABLE 14.2. Action of testosterone on IR-GHRH in fetal rat hypothalamic cells in primary culture.

Testoterone (ng/dl)	IR-GRF (pg/plate)		
	Cell extract	Media	Total
Control	280 ± 20	88 ± 8	368 ± 16
10	326 ± 40*	96 ± 6	422 ± 14**
20	364 ± 16***	100 ± 9	464 ± 26**
40	425 ± 22***	110 ± 16	535 ± 27***

IR-GHRH, immunoreactive growth hormone-releasing hormone. The values represent the mean ± SEM ($n = 5$).
*, $p < 0.05$ vs. control.
**, $p < 0.01$ vs. control.
***, $p < 0.001$ vs. control.

TABLE 14.3. Action of estradiol (E_2) on IR-SS in fetal rat hypothalamic cells in primary culture.

Estradiol (ng/dl)	IR-SS (pg/plate)		
	Cell extract	Media	Total
Control	680 ± 40	1360 ± 45	2040 ± 42
0.1	701 ± 80	1179 ± 60**	1880 ± 51
1.0	750 ± 60	1100 ± 80***	1850 ± 62
10.0	854 ± 45**	995 ± 49***	1849 ± 31

The values represent the mean ± SEM ($n = 5$).
**, $p < 0.01$ vs. control.
***, $p < 0.001$ vs. control.

tion of IR-SS, with no significant difference in the total IR-SS content per plate. These data indicate that under these experimental conditions E_2 inhibits the release of SS with no influence on its synthesis. The effect of E_2 on IR-GHRH was inconsistent, but showed a tendency to decrease the IR-GHRH content per plate.

These results demonstrate, in agreement with those reported by others doing in situ hybridization studies in adult male rats (4), that T has a direct stimulatory action on GHRH synthesis. However, these data do not support the finding, suggested by others (3), that T increases SS gene expression and translation and therefore SS synthesis. Developmental differences in the regulation of the neuropeptides influencing GH secretion in the rat have been demonstrated (10). Possibly, differences in the milieu of the cells or in the methodology can explain these disagreements.

Role of Androgens in Regulating the Somatotropic Axis During Puberty

To assess the role of androgens in the regulation of the GH axis, the physiological state of puberty was chosen because the pubertal growth spurt is characterized by a rapid combined activation of both the gonadal and the GH axes. In these studies we have used the androgen receptor-blocking agent flutamide, which is devoid of androgen effect. Flutamide (100 mg/kg) was administered s.c. to 30-day-old male rats for 4 weeks. Long-term flutamide administration to pubertal male rats significantly decreased the abundance of SS mRNA in the hypothalamus ($p < 0.01$). This inhibition on the SS mRNA message coincided with a significant increase ($p < 0.05$) of the IR-SS content in the whole hypothalamus as expressed per milligram of protein. These data suggest that posttranscriptional mechanisms could be implicated in this action and indicate that androgens potentiate SS secretion as a mechanism of regulating GH secretion. This stimulatory action on SS secretion can be deduced from the increase in hypothalamic IR-SS and the significant decrease of SS mRNA that occur in the absence of androgenic activity.

A slight, but not significant, alteration of hypothalamic IR-GHRH was produced by the treatment with flutamide. This response resembles that observed in fetal hypothalamic cells in primary culture, but in the opposite direction, because it is the consequence of the blockade of the androgen receptor. These data are in accordance with previous in vivo SS mRNA studies (5,10). The complexity of interpretation of these results also derives from the fact that serum E_2 concentration is increased as a consequence of high T levels, and might be seen as an E_2 effect rather than a blockade of T action.

To further understand the mechanism of the action of androgens on SS and GRF, a gonadotropin-releasing hormone (GnRH) agonist (Leuproreline, 65.5 µg rat^{-1} day^{-1}) was given s.c. to 30-day-old prepubertal male rats for 4 weeks. This low-T, low-E_2 model produced a slight not significant decrease in the accumulation of hypothalamic SS mRNA, and there was no modification of the IR-SS content in the whole hypothalamus. There was no significant alteration of IR-GHRH in the hypothalamus in these experimental circumstances.

The data obtained from this experiment support the importance of E_2, because the alterations seen in the low T activity with high E_2 levels are not evident in this low-T, low-E_2 model. This interpretation is supported by earlier data obtained in testicular-feminized male rats, deficient in intracellular androgen receptors, which had elevated basal GH levels (13). These observations presented so far underscore the influence of androgens on SS, and potentially on GHRH, although some actions of androgens are inconsistent.

The results obtained in these studies prompted us to design a third model to further understand the action of gonadal steroids on the GH secretagogues. Surgically castrated 30-day-old prepubertal male rats were treated with the GnRH antagonist Antide. As shown in Figure 14.1, surgical castration alone did not significantly alter SS mRNA content or the IR-SS in the hypothalamic extracts, whereas IR-GRF was significantly decreased ($p < 0.05$). There were no significant differences in hypothalamic SS and GRF between the surgically castrated animals treated with Antide and those not exposed to the antagonist.

Surgical castration in the male rat is a complex model for physiological studies. Nevertheless, together with our previous studies, we can infer that the decrease in gonadal steroids is not relevant to physiological regulation of SS even during maximal activation of the GH axis, such as in the peripubertal stage. No significant alterations are observed in hypothalamic SS mRNA or IR-SS levels in the absence of T and E_2. As in the case of the GnRH agonist Leuproreline, there is a decrease of hypothalamic IR-GHRH. To explain the elevation of baseline plasma GH levels that occurs in castrated adult male rats (1), we suggest that the decrease in hypothalamic IR-GHRH content is caused by an increase in its release with no alteration of SS secretion.

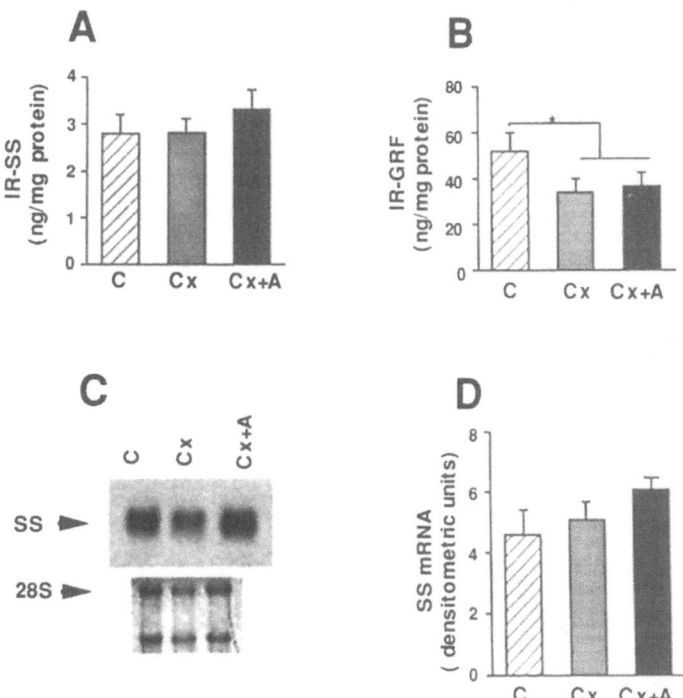

FIGURE 14.1. Effect of surgical castration (Cx) or surgical castration and treatment with the GnRH-receptor antagonist Antide (100 μg rat^{-1} day^{-1} s.c.) (Cx + A) on hypothalamic SS and GRF content in prepubertal 30-day-old male rats. Control animals (C) received only vehicle. A, IR-SS content in hypothalamic extracts; B, IR-GRF content in hypothalamic extracts; C, northern blot anlysis of hypothalamic SS mRNA of RNA pools of seven rats per group; D, densitometric quantitation of the autoradiogram adjusted for the internal control of charge (ribosomal 28S). Data are expressed as arbitrary units after correction for 28S ribosomal band; each point represents the mean ± SEM of three determinations of SS mRNA levels from independent experiments. *, $p < 0.05$.

Role of Androgens on the Regulation of Somatotropic Axis During Aging

Aging is another physiological condition in which the GH and the gonadotropic axes decline in a parallel manner. We therefore considered aging to be another physiological situation for the study of their interaction. In fact, in the male rat GH gene expression and GH secretion decline with senescence, independently of the weight and body composition alterations that occur with aging, as has been demonstrated in our laboratory (14–16). In a similar manner, the gonadal function of the male rat progressively and slowly declines with aging (17,18).

Different experimental models were designed to study the influence of the gonadal steroids on the GH axis during aging. Surgical castration was performed on 90-day-old and 2-year-old male rats, and the consequences on GH and hypothalamic SS and GHRH were studied. In the old rats, surgical castration provoked no alteration of GH mRNA accumulation in the pituitary or IR-GH in the pituitary and the serum, whereas in the young adult 90-day-old rats a slight significant decrease ($p < 0.05$) was observed in GH mRNA and IR-GH levels in the pituitary, with no significant modification of IR-GH in the serum. Exogenous testosterone injected s.c. to these animals returned these alterations to the normal levels. In the old rats, T treatment significantly increased ($p < 0.05$) GH mRNA and IR-GH levels in the pituitary and the serum IR-GH above those found in the aged control rats.

With regard to hypothalamic SS, the group of aging rats showed no significant modifications of SS mRNA or IR-SS content after castration or castration plus T administration. In the 90-day-old group, castration provoked a mild significant decrease ($p < 0.05$) of hypothalamic SS mRNA and IR-SS content, which returned to normal with the administration of exogenous T. There was no significant alteration in the IR-GHRH.

When chemical castration was performed in young adult 90-day-old male rats by s.c. administration of the GnRH agonist Leuproreline (65.5 µg rat^{-1} day^{-1} for 30 days), there was a mild, albeit not significant, decrease of GH mRNA, and at the same time a significant increase of IR-GH content ($p < 0.05$) in the pituitary and a significant decrease ($p < 0.001$) of IR-GH in the serum. However, in 90-day-old rats a significant decrease of hypothalamic SS mRNA ($p < 0.05$) and IR-SS content ($p < 0.001$) was seen with this treatment. No modifications of the IR-GHRH occurred in these circumstances.

Summary

These studies indicate that fetal neuronal hypothalamic cells are responsive to sex-steroids, which therefore can regulate SS and GHRH in vivo. During pubertal development there is some interplay between the gonadal and the GH–IGF-I axes, but both axes function with evident independence. In aging male rats, exogenous T can influence GH without significantly altering hypothalamic SS and GRF.

Acknowledgments. This work was supported by grants from FIS (94/0308 and 96/1574) and from Ministerio de Educación y Ciencia (PB 94-1271 and 95-0258).

References

1. Jansson JO, Eden S, Isaksson O. Sexual dimorphism in the control of growth hormone secretion. Endocr Rev 1985;6:128–50.
2. Kerrigan KR, Rogol AD. The impact of gonadal steroid hormone action on growth hormone secretion during childhood and adolescence. Endocr Rev 1992;13:281–98.
3. Wehrenberg WB, Giustina A. Basic counterpoint: mechanisms and pathways of gonadal steroid modulation of growth hormone secretion. Endocr Rev 1992;13:299–308.
4. Hertz P, Silberman M, Even L, Hochberg Z. Effects of sex steroids on the response of cultured rat pituitary cells to growth hormone-releasing hormone and somatostatin. Endocrinology 1989;125:581–5.
5. Chowen-Breed JA, Steiner RA, Clifton DK. Sexual dimorphism and testosterone-dependent regulation of somatostatin gene expression in the periventricular nucleus of the rat brain. Endocrinology 1989;125:357–62.
6. Zeitler P, Argente J, Choween-Breed JA, Clifton DK, Steiner RA. Growth hormone-releasing hormone messenger ribonucleic acid in the hypothalamus of the adult male rat is increased by testosterone. Endocrinology 1990;127:1362–8.
7. Fernandez G, Sanchez-Franco F, Frailes MT, Tolon RM, Lorenzo MJ, Lopez J, Cacicedo L. Regulation of somatostatin and growth hormone-releasing factor by gonadal steroids in fetal rat hypothalamic cells in culture. Regul Pept 1992;42:135–44.
8. Frohman LA, Downs TR, Chomzynski P. Regulation of growth hormone secretion. Front Neuroendocrinol 1992;13:344–406.
9. Gevers EF, Wit JM, Robinson ICAF. Effect of gonadectomy on growth and GH responsiveness in dwarf rats. J Endocrinol 1995;145:69–79.
10. Werner H, Koch Y, Baldino F, Gozes I. Steroid regulation of somatostatin mRNA in the rat hypothalamus. J Biol Chem 1988;263:7666–72.
11. Jansson JO, Frohman LA. Differential effects of neonatal and adult androgen exposure on the growth hormone secretory pattern in male rats. Endocrinology 1987;120:1551–7.
12. Gabriel SM, Millard WJ, Koening JI. Sexual and developmental differences in peptides regulating growth hormone secretion in the rat. Neuroendocrinology 1989;50:299–307.
13. Bardin CW, Bullock LP, Sherins RJ, Blackburn WR. Androgen metabolism and mechanism of action in male pseudo-hermaphroditism: a study of testicular feminization. Recent Prog Horm Res 1973;29:65–87.
14. De Gennaro V, Zoli M, Cocchi D, Maggi A, Marrama P, Aguati LF, Muller EE. Reduced growth hormone releasing factor (GHRF)-like immunoreactivity and GHRF gene expression in the hypothalamus of aged rats. Peptides 1989;10:705–8.
15. Corpas E, Harman SM, Blackman MR. Human growth hormone and human aging. Endocr Rev 1993;14:20–39.
16. Velasco B, Cacicedo L, Escalada J, Lopez J, Sanchez-Franco F. Growth hormone gene expression and secretion in aging rats is age dependent and not age-associated weight increase related. Endocrinology 1998;139:1314–20.
17. Meites J, Steger RW, Huang HH. Relation of neuroendocrine system to reproductive decline in aging rats and human subjects. Fed Proc 1980;39:3168–72.
18. Wise PM, Kashon ML, Krajnak KM, Rosewell KL, Cai A, Scarbrough K, Harney JP, McShane T, Lloyd JM, Weiland NG. Aging of the female reproductive system: a window into brain aging. Rec Prog Horm Res 1997;52:279–305.

15

Sex-Steroid Effects on Primate Somatotrophs

KENNETH C. COPELAND

No perfect animal model for studies of pubertal growth has been identified, for a variety of reasons that include the dissimilarities between animals and the human in terms of biochemical or physical markers, seasonal breeding, or the absence of a well-defined growth spurt coincident with sexual maturation. In the 1930s, Spence and Yerkes (1) described physical changes specific for puberty in nonhuman primates, with similarities to those of man. Based on the seminal auxological and anthropometric works of Yerkes, Gavin, Tanner, and others (1–4) and the development of hormone radioimmunoassay methodology in the 1950s, studies using primates as models for pubertal growth began to emerge. Although numerous descriptive studies of hormonal and physical events of human puberty were published in the 1960s and 1970s, intervention studies remained scarce because of the inherent ethical restraints on the use of human subjects, thus intensifying the search for appropriate nonhuman models.

The following chapter is largely a historical account of studies using nonhuman primate models to explore the process of pubertal growth and the relationships between gonadal steroids and the growth hormone (GH)–insulin-like growth factor (IGF)-I axis. Most of the data reported derive from in vivo studies and encompass a time span of approximately 20 years. The report focuses first on the appropriateness and limitations of various primate models for studies of pubertal growth and second reviews intervention studies in primates designed to examine the relationships between gonadal steroids and the GH–IGF-I axis.

Nonhuman Primate Models for Pubertal Growth

Puberty in all primates is characterized by obvious physical and biochemical changes heralding the development of sexual maturity. Ages of puberty for most primate species are defined, and correlations between physical changes or events (e.g. testicular enlargement or menarche) and pituitary and gonadal hormonal events (e.g. increases in gonadotropins and testosterone or estradiol) have been documented in several primates (5–10). These hypothalamopituitary-gonadal events comprising "gonadarche" appear to be universal among both human and nonhuman primates. Adrenarche, the period of maturation of the androgen-secreting zone of the adrenal cortex (5), occurs in humans before gonadarche and is responsible in part for the appearance of pubic and axillary hair, particularly in women. Of nonhuman primates studied, only the chimpanzee appears to undergo a biochemical adrenarche similar to that of humans (7,11–13). By contrast, the rhesus monkey (11,13) and baboon (14) do not.

Several studies of various nonhuman primates, particularly the chimpanzee (9) and baboon (10), have documented correlations between pubertal increases in gonadal steroid hormone production and a period of accelerated weight gain (Figs. 15.1 and 15.2). Accelerated linear growth coincident with gonadal maturation has been more difficult to document, in large part because most relevant studies have been cross-sectional in nature. Of all nonhuman primates examined, the baboon (Fig. 15.2) has perhaps the most pronounced rate of increase in crown-rump length (CRL) at the time of puberty (10). Accelerated limb growth at puberty also has been reported in nonhuman primates (15), although relevance to the human is questionable because human pubertal growth is predominantly truncal rather than limb (16).

As in the human, increases in serum IGF-I concentrations at puberty have been reported in virtually every nonhuman primate examined. Pubertal increases in IGF-I concentrations are greater in the female chimpanzee than in the male (Fig. 15.3) (9), a pattern similar to that of humans (17) and dissimilar to that of baboons (Fig. 15.4) (18) and monkeys (19), in both of which primate species serum IGF-I values are greater in males than in females.

IGF-I circulates in blood bound to at least six binding proteins, responsible not only for circulatory transport but also for modulation of biological action (20). Although differences in binding protein characteristics and profiles in various nonhuman species have been described, the binding protein profiles in both chimpanzee (9) and baboon (21) are remarkably similar to those of the human. Like the human, nonhuman primate binding proteins interfere with direct IGF quantification, especially during treatment with agents that influence binding protein expression, synthesis, or secretion (e.g., estrogen). Indeed, estrogen treatment of female baboons is associated with a profound alteration in the IGF-I-binding protein profile (21). Binding protein interference appears to be corrected by assay methodologies that involve preincubation of plasma or serum with acid or acid–ethanol (9,21) to precipitate protein-bound IGF-I.

Interventional Studies

Although descriptive studies have documented puberty-associated increases in GH and IGF-I in both humans (22–24) and nonhuman primates (9,18,19),

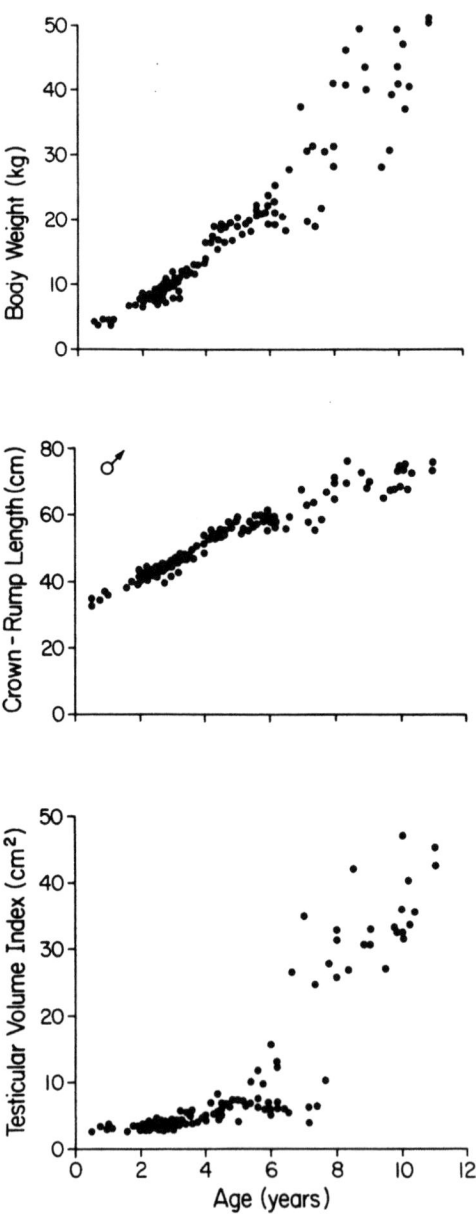

FIGURE 15.1. Physical growth characteristics, including body weight (*top panel*), crown-rump length (CRL; *middle panel*), and testes size (testicular volume index, TVI; *bottom panel*) vs. age in male chimpanzees. Adapted from Copeland et al. (9).

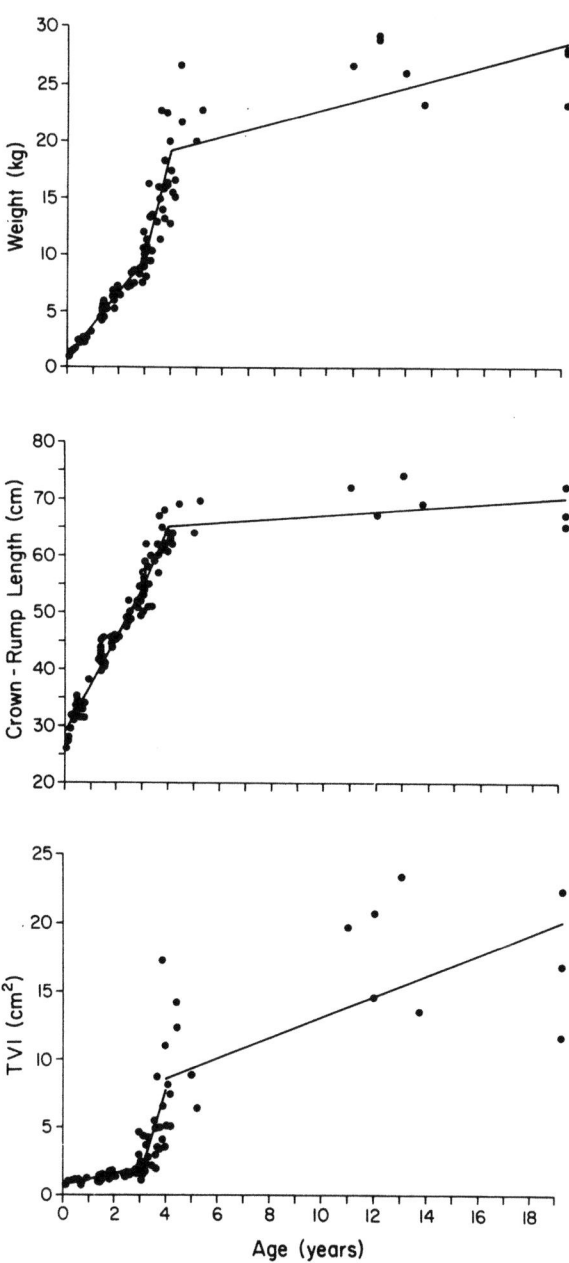

FIGURE 15.2. Physical growth characteristics, including body weight (*top panel*), CRL (*middle panel*), and testes size (testicular volume index, TVI; *bottom panel*) vs. age in male baboons. Slopes were calculated by linear regression analysis and are presented for 0–3 years, 3–4 years, and 4 years and older. Slopes between 3–4 years are significantly different for body weight ($p < 0.05$), CRL ($p < 0.06$), and testes size ($p < 0.05$). Adapted from Castracane et al. (10).

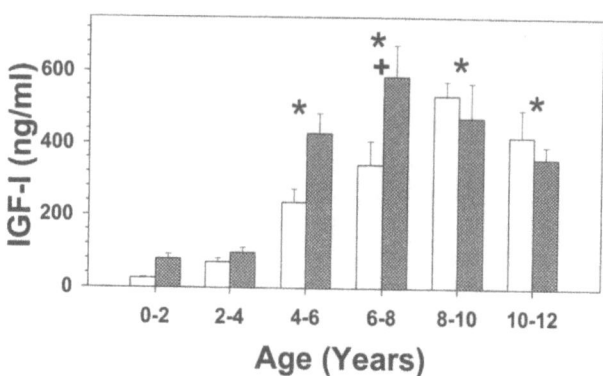

FIGURE 15.3. Plasma IGF-I concentrations vs. age in male (*open bars*) and female (*hatched bars*) chimpanzees. *, different from animals 4 years and younger, $p < 0.05$; +, higher in females compared to males, all ages considered together, $p = 0.01$, and animals 0–2, 4–6, and 6–8 years considered separately, $p < 0.05$). Adapted from Copeland et al. (9).

attempts to replicate such increases in GH and IGF-I during sex-steroid treatment often have been futile, yielding contradictory results. Estrogen treatment of primates (25) and human subjects under a variety of conditions including acromegaly (26–28), tall stature (29) and postmenopause (30) and in normal childhood (31) has been associated with increases in growth hormone secretion, yet typically with reductions in serum IGF-I concentrations.

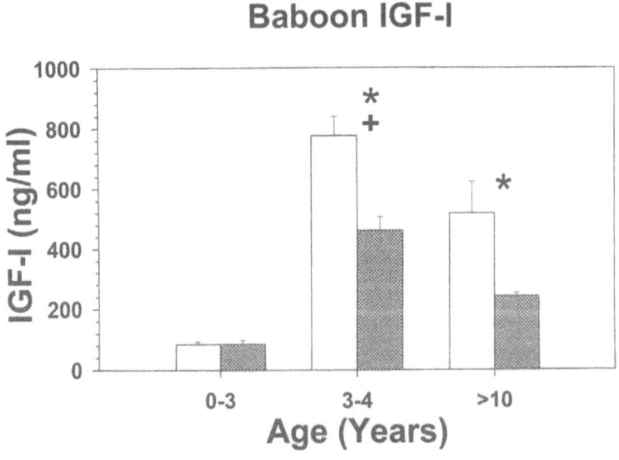

FIGURE 15.4. Plasma IGF-I concentrations vs. age in male (*open bars*) and female (*hatched bars*) baboons. *, different from animals 0–3 years, $p < 0.05$; +, higher in males compared to females, ages 3–4 years, $p < 0.001$). Adapted from Copeland et al. (18).

These estrogen-induced inhibitory effects on IGF-I appear not to reflect assay technique because similar results have been reported using bioassay (26,27) and radioimmunoassay (28,30) methodology. Fewer reports have described an increase in IGF-I concentrations following sex-steroid treatment, primarily in younger and less mature human subjects (32,33). GH and IGF-I concentrations are elevated in children with precocious puberty (34), although effects of treatment with GnRH agonists on IGF-I levels have not always been consistent (34–37), as discussed in Chapter 11. Differences postulated to explain these paradoxical IGF-I responses to therapeutic interventions have included (1) sex-steroid dose, (2) route of sex-steroid administration, and (3) age or developmental stage effect.

Dose Effect of Estrogen

In 1984, using a female baboon model, we first sought to examine the hypothesis that "physiologic estrogen stimulates, while pharmacologic estrogen inhibits IGF-I production." Sixteen chronically castrated adult female baboons were implanted with blank ($n = 5$) or crystalline estradiol-containing ($n = 11$) silastic capsules that remained in place for 7 days. Blood samples were drawn sequentially on days 0, 4, 7, and 13. Serum estradiol concentrations increased into the physiological adult female range (96 ± 9.2 pg/ml), and significantly greater than those of the control blank capsule-treated animals (28 ± 3.4 pg/ml; $p < 0.0005$). IGF-I concentrations increased significantly by 4 and 7 days and returned to baseline levels by day 13, indicating a stimulatory effect of "physiologic" estrogen delivered subcutaneously (Fig. 15.5).

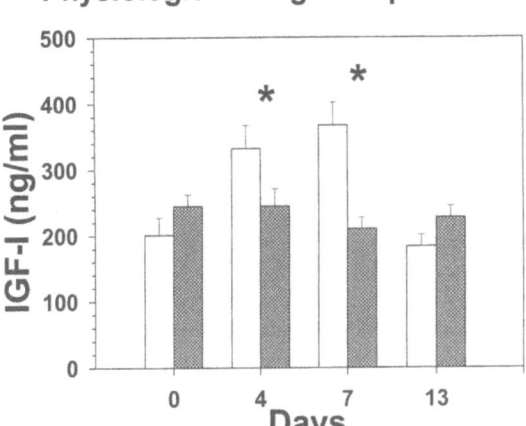

FIGURE 15.5. Mean plasma IGF-I concentrations in 16 chronic castrate adult female baboons implanted with estradiol ($n = 11$; *open bars*) or blank ($n = 5$; *hatched bars*) capsules between days 0 and 7 (*, $p < 0.02$). Adapted from Copeland et al. (21).

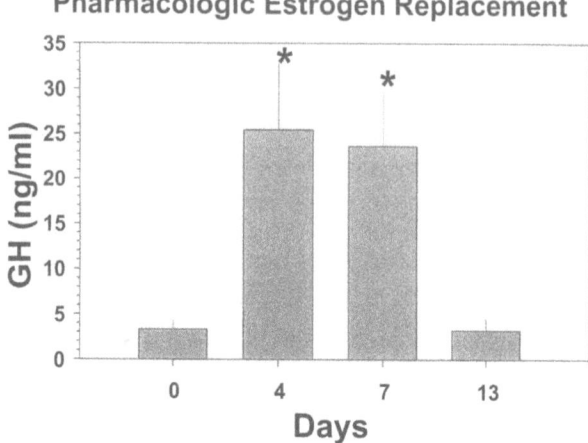

FIGURE 15.6. Mean serum GH concentrations in six intact adult female baboons treated with six daily i.m. injections of 1 mg estradiol benzoate, days 1–6 (*, $p < 0.05$). Adapted from Copeland et al. (21).

To investigate the effects of pharmacological estrogen treatment, six normal adult female baboons were given six consecutive daily intramuscular (i.m.) injections of 1 mg estradiol benzoate in oil, with blood samples obtained on days 0, 4, 7, and 13. This treatment resulted in peak serum estradiol concentrations well into the pharmacological range (1950 ± 320 pg/ml) by day 7. This treatment was associated with significant increases in GH (Fig. 15.6) and IGF-I (Fig. 15.7) concentrations by days 4 and 7, indicating a stimulatory effect of pharmacological estrogen treatment on both GH and IGF-I production (21).

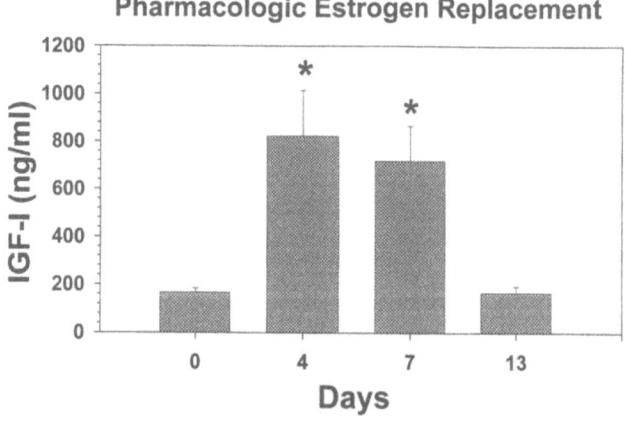

FIGURE 15.7. Mean plasma IGF-I concentrations in six intact adult female baboons treated with six daily i.m. injections of 1 mg estradiol benzoate, days 1–6 (*, $p < 0.01$). Adapted from Copeland et al. (21).

These studies excluded a simple paradoxical dose effect because under these experimental conditions both high- and low-dose estrogen stimulated IGF-I production. Chromatographic analysis of baboon plasma following treatment with pharmacological estrogen indicated a marked alteration in the relationship between IGF-I and its binding proteins (21). Estrogen treatment resulted in a profound increase in IGF binding over a broad range of molecular weights (analyzed by neutral gel chromatography), including not only samples eluting in the 150-kDa region [encompassing the IGF-binding protein (IGF-BP-3) tertiary complex], but also lower molecular weight proteins (likely including both IGF-BP-1 and IGF-BP-2).

Route of Administration

In humans, high-dose i.v. estrogen results in a prompt reduction in IGF-I concentrations, in contrast with no change in levels following prolonged (6-month) low-dose (5 µg ethinyl estradiol/day) oral estrogen treatment in the same subjects (38). Studies comparing effects of oral and transcutaneous estrogen exposure are inconsistent; oral estrogens typically reduced circulating IGF-I concentrations, but transcutaneous estrogen treatment resulting variously in either increased (39), decreased (40,41), or unchanged (42) circulating IGF-I concentrations. Similar studies comparing effects of parenteral with enteral with transcutaneous delivery of sex steroids have not been reported in nonhuman primates.

Age or Developmental Stage

Several recent studies in primates have examined the effects of age, reproductive maturity, and developmental stage on IGF-I responses to sex-steroid exposure, both in vitro (43) and in vivo (44–46). Crawford and Handelsman (47) examined the effect of castration on IGF-I and IGF-BP-3 concentrations in both prepubertal and sexually mature male baboons. Castration of prepubertal animals prevented the puberty-associated augmentations in IGF-I and IGF-BP-3, increases that were restored by subcutaneous delivery of testosterone by pellet. By contrast, castration of sexually mature baboons was not associated with a decline in circulating IGF-I concentrations. These data suggested to the authors a predominant influence of androgens on IGF-I in the young male baboon, a phenomenon they termed "uniquely pubertal." More recently, Wilson (48) investigated the effects of estradiol treatment in female rhesus monkeys, both with and without cotreatment with IGF-I. He observed that estradiol treatment resulted in an increase in IGF-BP-3 regardless of age. By contrast, estradiol treatment resulted in differential age effects on IGF-I, with suppressive effects in adults and stimulatory effects in adolescent animals (48). These data suggested to the author an "age-dependent uncoupling of estradiol regulation of the GH–IGF-I axis."

Conclusions

Substantial data support the use of nonhuman primates as models for the study of pubertal growth and the relationships between gonadal steroid hormones and the GH–IGF-I axis. The chimpanzee is most similar to the human in terms of biochemical changes at puberty, including the presence of an adrenarche and higher IGF-I concentrations in females than in males. In terms of anthropometric data, a linear growth spurt, as well as a profound increase in the rate of weight gain at puberty, has been documented most convincingly in the baboon. Studies involving gonadal steroid treatment of primates indicate that dose, route of administration, and age or stage of sexual maturity all have modulating effects on GH, IGF-I, and IGF-binding protein responses.

Acknowledgments. The author expresses his appreciation to Drs. Dan Castracane, Thomas Koehl, and Jorg Eichberg for collaboration and advice, to Drs. Louis Underwood and Judson Van Wyk for IGF-I materials in the 1970s and 1980s and priceless mentorship, to Ms. Carrol Cribbs for secretarial help, and to Dr. Philip Beckett and Ms. Lucia Copeland for editorial assistance.

References

1. Spence KW, Yerkes RM. Weight, growth and age in chimpanzee. Am J Phys Anthropol 1937;22:229–46.
2. Gavan JA. Growth and development of the chimpanzee: a longitudinal and comparative study. Hum Biol 1953;25:93–143.
3. Tanner JM. Growth at adolescence, 2nd ed. Oxford: Blackwell, 1962:223–39.
4. Gavan JA, Swindler DR. Growth rates and phylogeny in primates. Am J Phys Anthropol 1966;24:181–90.
5. Winter JSD, Faiman C, Reyes FI, Hobson WC. Gonadotrophins and steroid hormones in the blood and urine of prepubertal girls and other primates. Clin Endocrinol Metab 1978;7:513–30.
6. Winter JSD, Faiman C, Hobson WC, Reyes FI. The endocrine basis of sexual development in the chimpanzee. J Reprod Fertil 1980;28(suppl):131–8.
7. Hobson WC, Fuller GB, Winter JSD, Faiman C, Reyes FI. Reproductive and endocrine development in the great apes. In: Graham CE, ed. Reproductive biology of the great apes. New York: Academic Press; 1981:83–103.
8. Glassman DM, Coelho AM Jr, Carey KD, Bramblett CA. Weight growth in Savannah baboons: a longitudinal study from birth to adulthood. Growth 1984;48:425–33.
9. Copeland KC, Eichberg JW, Parker CR Jr, Bartke A. Puberty in the chimpanzee: somatomedin-C and its relationship to somatic growth and steroid hormone concentrations. J Clin Endocrinol Metab 1985;60:1154–60.
10. Castracane VD, Copeland KC, Reyes P, Kuehl TJ. Pubertal endocrinology of yellow baboon (*Papio cynocephalus*): plasma testosterone, testis size, body weight, and crown-rump length in males. Am J Primatol 1986;11:263–70.
11. Cutler GB Jr, Glenn M, Bush M, Hodgen GD, Graham CE, Loriaux DL. Adrenarche: a survey of rodents, domestic animals and primates. Endocrinology 1978;103:2112–8.

12. Cutler GB Jr, Loriaux DL. Adrenarche and its relationship to the onset of puberty. Fed Proc 1980;39:2384–90.
13. Smail PJ, Faiman C, Hobson WC, Fuller GB, Winter JSD. Further studies on adrenarche in nonhuman primates. Endocrinology 1982;111:844–8.
14. Castracane VD, Cutler GB, Loriaux DL. Pubertal endocrinology of the baboon: adrenarche. Am J Physiol 1981;241:E305–9.
15. Watts ES, Gavan JA. Postnatal growth of nonhuman primates: the problem of the adolescent spurt. Hum Biol 1982;54:53–70.
16. Tanner JM, Whitehouse RH, Hughes PCR, Carter BS. Relative importance of growth hormone and sex steroids for the growth at puberty of trunk length, limb length, and muscle width in growth-hormone deficient children. J Pediatr 1976;89:1000–5.
17. Underwood LE, Copeland KC, Clemmons, Chatelain PG, Blethen SL, Van Wyk JJ. Radioimmunoassay of somatomedin-C: clinical applications. In: Waldhausch WK, ed. Diabetes. Amsterdam: Excerpta Medica, 1979:278–82.
18. Copeland KC, Kuehl TJ, Castracane VD. Pubertal endocrinology of the baboon: elevated somatomedin-C/insulin-like growth factor I at puberty. J Clin Endocrinol Metab 1982;55:1198–201.
19. Styne DM. Serum insulin-like growth factor 1 concentrations in the developing rhesus monkey. J Med Primatol 1991;20:334–7.
20. Baxter RC. Insulin-like growth factor binding proteins in the human circulation: a review. Horm Res 1994;42:140–4.
21. Copeland KC, Johnson DM, Kuehl TJ, Castracane VD. Estrogen stimulates growth hormone and somatomedin-C in castrate and intact female baboons. J Clin Endocrinol Metab 1984;58:698–703.
22. Luna AM, Wilson DM, Wibbelsman CJ, Brown RC, Nagashima RJ, Hintz RL, Rosenfeld RG. Somatomedins in adolescence: a cross-sectional study of the effect of puberty on plasma insulin-like growth factor I and II levels. J Clin Endocrinol Metab 1983;57:268–71.
23. Kerrigan JR, Rogol AD. The impact of gonadal steroid hormone action on growth hormone secretion during childhood and adolescence. Endocr Rev 1992;13:281–98.
24. Clark PA, Rogol AD. Growth hormones and sex steroid interactions at puberty. Endocrinol Metab Clin North Am 1996;25:665–81.
25. Koritnik DR, Cronin MJ, Orth DN, Evans, Thorner MO. Pituitary response to intravenous hypothalamic releasing peptides in cynomolgus monkeys treated with contraceptive steroids. J Clin Endocrinol Metab 1987;65:37–45.
26. Wiedemann E, Schwartz E. Suppression of growth hormone-dependent human serum sulfation factor by estrogen. J Clin Endocrinol Metab 1972;34:51–4.
27. Wiedemann E, Schwartz E, Frantz AG. Acute and chronic estrogen effects upon serum somatomedin activity, growth hormone, and prolactin in man. J Clin Endocrinol Metab 1976;42:942–52.
28. Clemmons DR, Underwood LE, Ridgway EC, Kliman B, Kjellberg RN, Van Wyk JJ. Estradiol treatment of acromegaly. Reduction of immunoreactive somatomedin-C and improvement in metabolic status. Am J Med 1980;69:571–4.
29. Gourmelen M, LeBouc Y, Girard F, Binoux M. Serum levels of insulin-like growth factor (IGF) and IGF binding protein in constitutionally tall children and adolescents. J Clin Endocrinol Metab 1984;59:1197–200.
30. Lieberman SA, Mitchell AM, Marcus R, Hintz RL, Hoffman AR. The insulin-like growth factor I generation test: resistance to growth hormone with aging and estrogen replacement therapy. Horm Metab Res 1994;26:229–33.

31. Veldhuis JD, Metzger DL, Martha PM Jr, Mauras N, Kerrigan JR, Keenan B, et al. Estrogen and testosterone, but not a nonaromatizable androgen, direct network integration of the hypothalamo-somatotrope (growth hormone)-insulin-like growth factor I axis in the human: evidence from pubertal pathophysiology and sex-steroid hormone replacement. J Clin Endocrinol Metab 1997;82:3414–20.
32. Ross JL, Cassorla FG, Skerda MC, Valk IM, Loriaux DL, Cutler GB Jr. A preliminary study of the effect of estrogen dose on growth in Turner's syndrome. N Engl J Med 1983;309:1104–6.
33. Cuttler L, Van Vliet G, Conte FA, Kaplan, SL, Grumbach MM. Somatomedin-C levels in children and adolescents with gonadal dysgenesis: differences from age-matched normal females and effect of chronic estrogen replacement therapy. J Clin Endocrinol Metab 1985;60:1087–92.
34. Harris DA, Van Vliet G, Egli CA, Grumbach MM, Kaplan SL, Styne DM, Vainsel M. Somatomedin-C in normal puberty and in true precocious puberty before and after treatment with a potent luteinizing hormone-releasing hormone agonist. J Clin Endocrinol Metab 1985;61:152–9.
35. Mansfield MJ, Rudlin CR, Crigler JF Jr, Karol KA, Crawford JD, Boepple PA, et al. Changes in growth and serum growth hormone and plasma somatomedin-C levels during suppression of gonadal sex steroid secretion in girls with central precocious puberty. J Clin Endocrinol Metab 1988;66:3–9.
36. Juul A, Scheike T, Nielsen, Krabbe S, Müller J, Skakkebæk NE. Serum insulin-like growth factor I (IGF-I) and IGF-binding protein 3 levels are increased in central precocious puberty: effects of two different treatment regimens with gonadotropin-releasing hormone agonists, without or in combination with an antiandrogen (cyproterone acetate). J Clin Endocrinol Metab 1995;80:3059–67.
37. Kanety H, Karasik A, Pariente C, Kauschansky A. Insulin-like growth factor-I and IGF binding protein-3 remain high after GnRH analogue therapy in girls with central precocious puberty. Clin Endocrinol 1996;45:7–12.
38. Copeland KC. Effects of acute high dose and chronic low dose estrogen on plasma somatomedin-C and growth in patients with Turner's syndrome. J Clin Endocrinol Metab 1988;66:1278–82.
39. Weissberger AJ, Ho KKY, Lazarus L. Contrasting effects of oral and transdermal routes of estrogen replacement therapy on 24-hour growth hormone (GH) secretion, insulin-like growth factor I, and GH-binding protein in postmenopausal women. J Clin Endocrinol Metab 1991;72:374–81.
40. Friend KE, Hartman ML, Pezzoli SS, Clasey JL, Thorner MO. Both oral and transdermal estrogen increase growth hormone release in postmenopausal women—a clinical research center study. J Clin Endocrinol Metab 1996;81:2250–6.
41. Helle SI, Omsjø IH, Hughes SCC, Botta L, Hüls G, Holly JMP, Lønning PE. Effects of oral and transdermal oestrogen replacement therapy on plasma levels of insulin-like growth factors and IGF bonding proteins 1 and 3: a cross-over study. Clin Endocrinol 1996;45;727–32.
42. Bellantoni MF, Vittone J, Campfield AT, Bass KM, Harman SM, Blackman MR. Effects of oral versus transdermal estrogen on the growth hormone/insulin-like growth factor I axis in younger and older postmenopausal women: a clinical research center study. J Clin Endocrinol Metab 1996;81:2848–53.
43. Bethea CL, Freesh F. Estrogen action on growth hormone in pituitary cell cultures from adult and juvenile macaques. Endocrinology 1991;129:2110–8.

44. Wilson ME. Gonadal steroids influence serum somatomedin-C concentrations in prepubertal female rhesus monkeys. Endocrinology 1986;119:666–71.
45. Wilson ME, Gordon TP, Tanner JM. Constant low-dose oestradiol replacement accelerates skeletal maturation and growth in ovariectomized adolescent rhesus monkeys. J Clin Endocrinol Metab 1993;137:519–27.
46. Wilson ME. Administration of IGF-I affects the GH axis and adolescent growth in normal monkeys. J Clin Endocrinol Metab 1997;153:327–35.
47. Crawford BA, Handelsman DJ. Androgens regulate circulating levels of insulin-like growth factor (IGF)-I and IGF binding protein-3 during puberty in male baboons. J Clin Endocrinol Metab 1996;82:65–72.
48. Wilson ME. Regulation of the growth hormone-insulin-like growth factor I axis in developing and adult monkeys is affected by estradiol replacement and supplementation with insulin-like growth factor I. J Clin Endocrinol Metab 1998;83:2018–28.

16

Sex-Steroid Effects on Perifused Pituitary

Richard J. Krieg, Jr., Paul M. Martha, Jr., James R. Kerrigan,
Judy M. Batson, Timothy E. Sayles, Steven J. Kraus,
Dennis W. Matt, and William S. Evans

Sex steroids are known to affect growth and growth hormone (GH) secretion throughout mammalian life. Early studies on the imprinting effects of testosterone (T) on growth and GH secretion in rats (1) showed that neonatal gonadectomy (GX) caused significant disruption of GH secretion in adult male rats. Treatment of neonatally GX male rats with T during neonatal and adult life restored a significant component of the GH pulsatile pattern. The presence of the ovary prevented the imprinting effect of T that would occur if female rats had otherwise undergone neonatal GX (2). These imprinting effects most likely involve the brain as the primary target.

Considerable evidence exists for the effects of steroids on GH secretion during puberty. Some of the most recent reports (3) showed that important parameters of pulsatile GH in boys were increased by T treatment. Interestingly, T administration also augmented GH release that was stimulated by L-arginine and GH-releasing hormone (GHRH) but not by GHRH alone. These data also tend to support the probability that T primarily affects central nervous control systems and possibly somatostatin in particular; see Chapter 3.

Recent data in adult animals show that the orderliness of the GH secretory pattern (which is of significant importance to the stimulation of body growth) is related to gender and the presence of the gonad (4). Previous studies had shown that the continued presence of T in adult life was necessary for the high-amplitude pulsatile GH secretion seen in male rats (5).

Although the effects of T have been found to be mostly stimulatory, the spectrum of effects that estrogen (E_2) exerts on GH secretion has been somewhat more equivocal. Consistent with the decreased level of growth in female mammals, early results showed that E_2 treatment caused a decrease in the pituitary GH content of male rats (6). Contrasting results in humans showed that serum concentrations of free E_2 correlated best with the (significantly greater) 24-h integrated GH concentrations observed in women (7).

With the apparent weight of these results favoring a significant effect, certainly of T, on central nervous system control of GH secretion, the studies described here were aimed at clarifying what effects in vivo gonadal steroid treatment and testicular feminization have on the specific ability of perifused pituitary tissue to secrete GH in vitro.

Materials and Methods

Experiment 1: Hormone Secretion by Perifused Pituitary Tissue Preparations

We first investigated the secretion of pituitary hormone by perifused hemipituitaries, pituitary fragments, and dispersed pituitary cells using male and female Long-Evans rats. Daily vaginal smears were performed to establish the stage of the female estrous cycle.

For each repetition, 12 male and 9 female rats were selected. Anterior pituitaries were collected after anesthesia and decapitation. Pituitary tissue from 4 male animals (8 halves) were used for each perifusion column. The female pituitaries were separated such that 2 randomly selected halves from diestrous II, proestrous, and estrous animals (i.e., a total of 6 halves) were pooled for each perifusion column.

One column of tissue from male or female rats was loaded with Bio-Gel and hemipituitaries. Another column of tissue from male or female rats was loaded with Bio-Gel and tissue that had been cut into fragments (each approximately 0.5 mm square). The final pair of columns was loaded with Bio-Gel and dispersed cells from male or female pituitaries that had been prepared as described previously (8). The weights of the pituitary tissue contained in each column were determined before processing the tissue and loading of the columns.

All six columns were simultaneously perifused with Krebs-Ringer-HEPES buffer (KRGH) with 0.2% glucose, 0.5% BSA (bovine serum albumin), and 0.03 g/100 ml bacitracin at a mean flow rate of about 0.43 ml/min. Eluate was collected at 5-min fractions on a fraction collector. All columns were run for 2.5 h of equilibration, followed by a 1-h collection period to establish basal hormone secretory rates. Tissues were then exposed to 2.5-min pulses of four concentrations of gonadotropin-releasing hormone (GnRH) at 1-hour intervals. The four concentrations (0.3, 1.0, 3.0, and 10.0 nanomolar [nM]) were applied in a different sequence for each of the three repetitions of the experiment. The eluate samples were stored frozen until measurement of luteinizing hormone (LH) by radioimmunoassay (RIA). The amounts of hormone secreted were expressed as a *secretory rate* (*SR*) in "pg/min/mg pituitary tissue." This was calculated as $SR = [LH](f)/(mg\ pituitary)$, where [LH] was the concentration of LH in ng/ml as determined by RIA, f was the flow rate in ml/min, and "mg pituitary" was the predetermined weight of the tissue placed in each column. Basal secretory rates and the responses of the various pituitary tissue preparations to releasing hormone stimulation were compared statistically.

Experiment 2: Effect of In Vivo Sex-Steroid Treatment of Gonadectomized Male Rats on GH Secretion by Perifused Pituitary Cells

Forty-eight male and 12 female Sprague-Dawley rats were used in these experiments, 36 male rats were gonadectomized (GX). At the time of GX, 3 weeks before pituitary cell perifusion, 12 GX rats received subcutaneous implants of crystalline T (Sigma, St. Louis, MO) in silastic capsules (40 mm). Another 12 GX rats received subcutaneous implants of crystalline 17-β-estradiol (Sigma) in silastic capsules (10 mm). The remaining 12 GX rats were untreated. Female rats were used specifically on the diestrus II stage of the estrous cycle. Animals were killed by decapitation and plasma was collected and frozen for the measurement of steroid levels by RIA. Anterior pituitary tissue was quickly collected, and dispersed pituitary cells were prepared as described, using tissue from 4 animals for each column. The numbers of cells and their viability were determined using a hemocytometer and the trypan blue exclusion test. Three repetitions of each group were run.

Cells were perifused with Medium 199 (Gibco, Grand Island, NY) containing 0.25% BSA and antibiotics at a flow rate of approximately 0.43 ml/min. After a 4-h equilibration period, cells were exposed to separate 2.5-min pulses of eight concentrations of hGHRH-40 (0.03, 0.1, 0.3, 1.0, 3.0, 10.0, 30.0, and 100.0 nM). The pulses were given in random order, so that each repetition exposed the cells to a different sequence. Eluate fractions were stored frozen until GH was quantitated by RIA. The GH response was expressed as a SR in "ng/min/10 million (10^7) cells." The complete response to each hGHRH concentration was calculated as the area under the curve, above baseline.

Experiment 3: Effect of In Vivo T Treatment of Gonadectomized Female Rats on GH Secretion by Perifused Pituitary Cells

Thirty-six female and 12 male Sprague-Dawley rats were used in these experiments; 24 female rats were ovariectomized (OVX). At the time of OVX, 3 weeks before pituitary cell perifusion, 12 of these rats (OVX + T) received subcutaneous implants of crystalline T (Sigma) in silastic capsules (25 mm). The other 12 OVX female rats remained untreated. Changes in body weight were recorded for the three groups of female rats over the period of OVX or OVX+T.

Perifusion of the dispersed pituitary cells was performed using tissue from three animals per column and four repetitions. Five concentrations (0.01, 0.1,

1.0, 10.0, and 100 nM) of hGHRH-40 (Bachem, Torrance, CA) were applied as 2.5-min pulses separated by 30-min intervals. GH secretion was expressed as SR as described, and the complete response to each concentration of hGHRH-40 was calculated as the area under the curve above baseline.

Experiment 4: GH Secretion by Perifused Dispersed Pituitary Cells from Male, Female, and Testicular Feminized Male Rats

Normal male, normal female, and testicular-feminized (Tfm) King-Holtzman rats (Stanley-Gumbreck line; Intorgene, Oklahoma City, OK) were raised in our own colony and used at ages between 113 and 124 days. Dispersed pituitary cells were prepared as previously described (8), and columns of cells from each of the three groups were run simultaneously in each of seven repetitions of the experiment. After a 4-h equilibration period of perifusion with M199 (Gibco), cells were exposed to eight concentrations (0.03, 0.1, 0.3, 1.0, 3.0, 10.0, 30.0, and 100.0 nM) of hGHRH-40 (Bachem) in a random order for each repetition. Eluate fractions were assayed for rat GH by RIA, and the data were expressed as SR and areas under the curve as described.

Results

Experiment 1: Hormone Secretion by Perifused Pituitary Tissue Preparations

Figure 16.1 shows representative LH responses to GnRH of the different pituitary preparations from male and female rats. As expected for LH, significantly greater responses were observed in preparations from female rats. It is clearly seen in subsequent experiments that the reverse is true for GH, i.e., GH responses were significantly larger in pituitary tissue from male rats.

Table 16.1 shows that the basal secretory rate was greatest in the pituitary fragment group from either male or female rats. The duration of the LH response to GnRH tended to be higher in the hemipituitary and fragment groups. The hemipituitary group actually had the longest duration, as evidenced by the fact that two particular concentrations of GnRH produced significantly longer durations in the (male) hemipituitary group than the fragment group (3nM, 40 ± 4.9 versus 28 ± 4.5 min; and 10nM, 54 ± 4.5 min versus 42 ± 3.5 min, respectively). The quickness of the GnRH response in the dispersed cell preparation from female rats was substantiated by the significantly lower "time to peak" LH levels. The rapid recovery of the dispersed cells was demonstrated by the significantly shorter overall duration of their response.

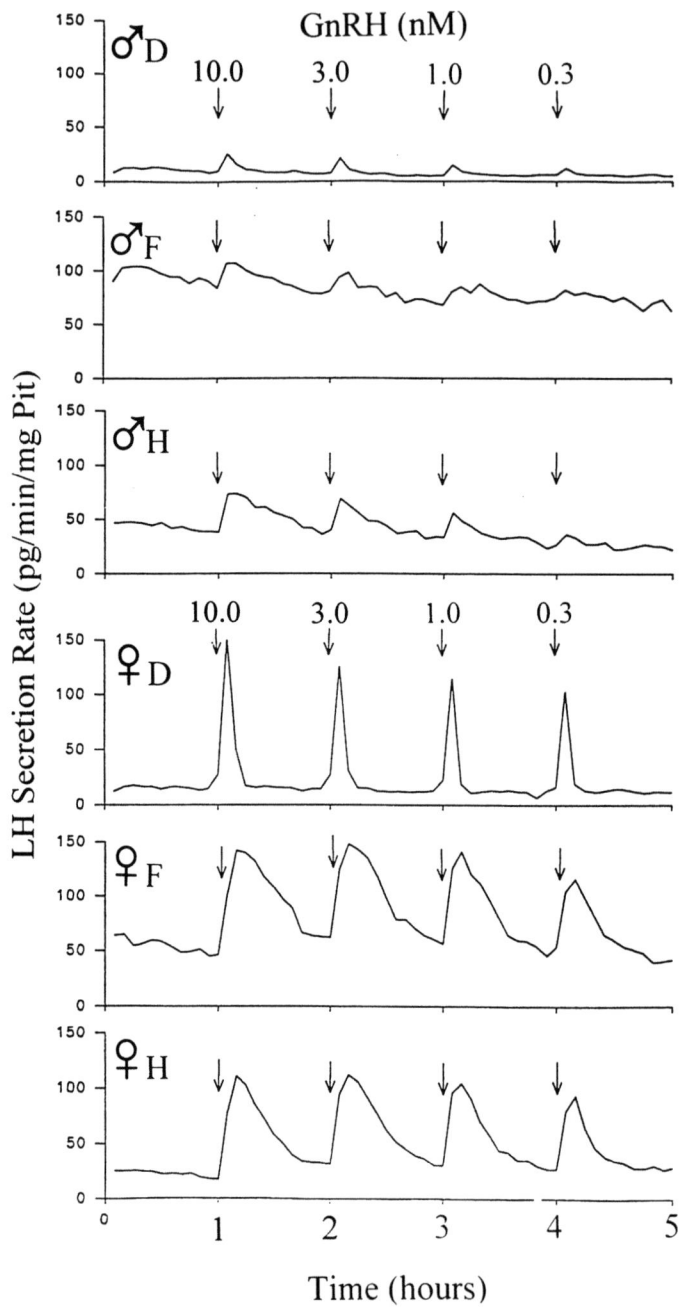

FIGURE 16.1. Representative LH release by dispersed anterior pituitary cells (D), pituitary fragments (F), and hemipituitaries (H) from male rats (labeled ♂D, ♂F, and ♂H, respectively) and female rats (labeled ♀D, ♀F, and ♀H, respectively) in response to four concentrations of gonadotropin-releasing hormone (GnRH).

TABLE 16.1. Secretory characteristics of dispersed pituitary cells (DIS), pituitary fragments (FRG), and hemipituitaries (HEM) from male or female rats during in vitro perifusion.

Group	Basal LH secretion (pg/min^{-1}/mg^{-1} Pit)	GnRH-stimulated LH secretory response	
		Overall duration (min)	Overall time to peak (min)
Male.DIS	10.85 ± 0.27[a]	22.50 ± 1.44[a,b]	5.83 ± 0.83[a]
Male.FRG	90.26 ± 1.28[b]	31.67 ± 3.86[b,c]	6.25 ± 0.65[a]
Male.HEM	37.68 ± 1.02[c]	35.42 ± 4.75[c]	5.83 ± 0.56[a]
Fem.DIS	14.48 ± 0.51[d]	16.67 ± 1.28[a]	5.00 ± 0.00[a]
Fem.FRG	58.53 ± 1.80[e]	46.67 ± 4.23[d]	10.42 ± 0.42[b]
Fem.HEM	23.07 ± 0.98[f]	49.50 ± 4.10[d]	9.17 ± 0.56[b]

Values are expressed as mean ± SEM. Different superscripts (a–f) denote significant differences ($p < 0.05$).

The quick rise to peak and short duration of the response in the dispersed cell preparation were a closer approximation to an expected physiological response, with no sequestration of releasing hormone. This effect was the rationale behind the use of dispersed pituitary cells in the following experiments.

Experiment 2: Effect of In Vivo Sex-Steroid Treatment of Gonadectomized Male Rats on GH Secretion by Perifused Pituitary Cells

Figure 16.2 shows the responses of perifused dispersed pituitary cells from the five groups of animals to eight concentrations of hGHRH-40. The largest responses were clearly produced by cells from intact male animals. The next greatest response to hGHRH was by cells from T-treated GX male rats (GXT), with cells from untreated GX rats responding slightly less well. Clearly the smallest responses were produced by cells from intact female and estradiol-treated GX male animals (GXE$_2$). These data are delineated by the overall responses to hGHRH shown in Table 16.2. The magnitude of responses followed this order: intact male > GXT > GX > intact female = GXE$_2$.

Experiment 3: Effect of In Vivo T Treatment of Gonadectomized Female Rats on GH Secretion by Perifused Pituitary Cells

Intact male rats weighed 353.3 ± 7.6 g and all the intact female rats weighed 254 ± 5.3 g at the beginning of the experiment. During the 3-week period after OVX, intact female rats gained 22.0 ± 3.3 g, while the OVX females gained

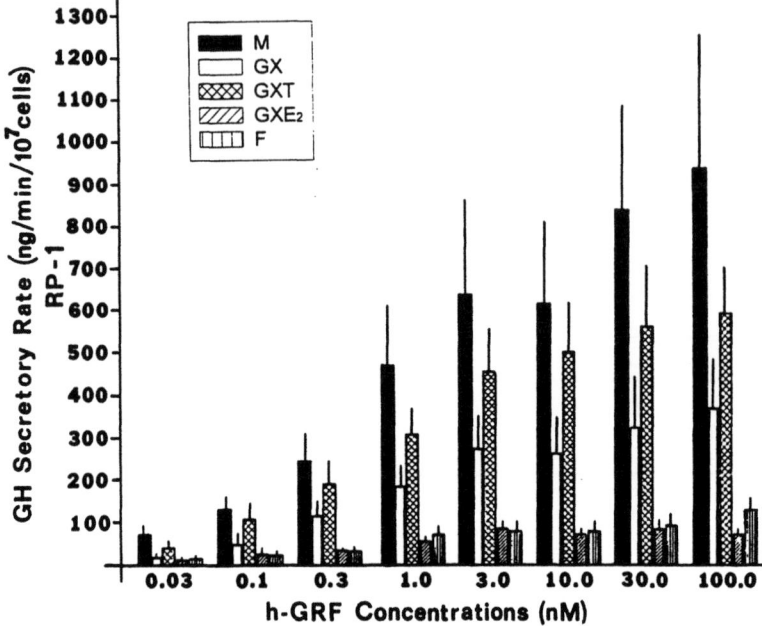

FIGURE 16.2. Responses of the five groups in experiment 2 to eight concentrations of hGHRH-40 (h-GRF). Testosterone augmentation of the GH response in cells from GX (gonadectomized) male rats and the profound depression of the GH response in cells from GX male rats is conspicuous. Reproduced with permission from Evans et al. (8).

89.2 ± 3.8 g, and the OVX females treated with T (OVX+T) gained 86.0 ± 3.7 g. Thus, T did not effect weight gain during the period of treatment.

Figure 16.3 shows the response of dispersed pituitary cells from intact male, intact female, OVX, and OVX+T rats to five concentrations of hGHRH. Cells from all three groups of female rats showed profoundly lower responses than cells from male rats. Three weeks of in vivo T treatment clearly had no effect on the ability of dispersed pituitary cells from female rats to secrete GH in response to hGHRH.

TABLE 16.2. Basal and hGHRH-40-stimulated GH secretory rates of perifused dispersed pituitary cells from five specific groups of animals.

Group	Basal	Overall hGHRH-40-stimulated
Intact male	5.4 ± 1.3[a,b]	496.0 ± 92.0[a]
Castrate male	5.7 ± 1.8[a,b]	203.0 ± 37.3[b]
Castrate + T	1.4 ± 2.7[a]	348.0 ± 52.8[c]
Castrate + E_2	12.2 ± 8.3[a,b]	58.1 ± 6.8[d]
Diestrus day 2	4.3 ± 0.6[b]	68.6 ± 9.5[d]

In vivo testosterone significantly augmented the response of cells from castrate male rats, while in vivo 17-β-estradiol (E_2) profoundly suppressed the response. Values are in ng min^{-1}/10^{-7} cells and, like the superscripts, are expressed as in Table 16.1. Reproduced with permission from Evans et al. (8).

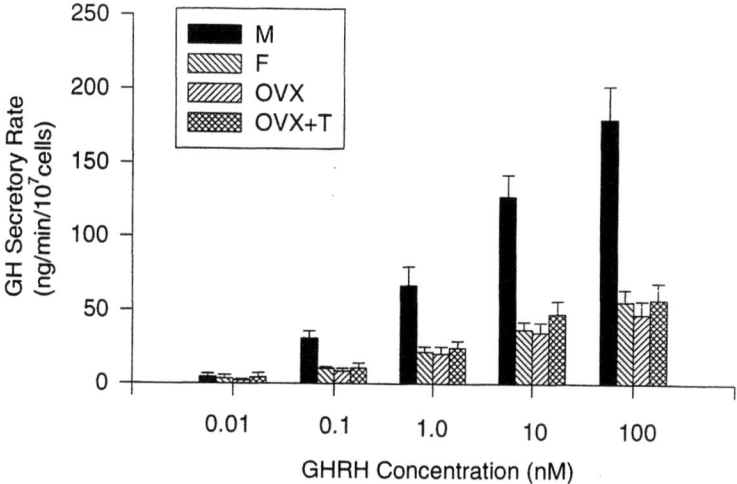

FIGURE 16.3. Responses of the four groups in experiment 3 to five concentrations of hGHRH-40. Lack of testosterone augmentation of the GH response by cells from OVX (ovariectomized) rats is evident.

Experiment 4: GH Secretion by Perifused Dispersed Pituitary Cells from Male, Female, and Testicular-Feminized Male Rats

Circulating levels of T and 17-β-estradiol in male, female, and Tfm rats are shown in Table 16.3. E_2 levels in Tfm rats were comparable to those in female rats and threefold higher than those in males. T was significantly higher in Tfm rats than in male rats by more than threefold. Figure 16.4 shows the response of perifused dispersed pituitary cells from the three groups to eight concentrations of hGHRH-40. The difference between the cells from male and female rats is conspicuous, as is the intermediate position of the response of cells from Tfm animals. Table 16.4 clearly shows the intermediate position of the response from Tfm cells. An interesting and significant correlation of the

TABLE 16.3. Circulating levels of testosterone and 17-β-estradiol in male, female, and Tfm rats. Values and superscripts are expressed the same as in Table 16.1.

Group	Testosterone (ng/ml)	17-β-Estradiol (pg/ml)
Male	1.12 ± 0.13[a]	9.25 ± 4.18
Female	ND	26.90 ± 10.97
Tfm	3.99 ± 0.78[b]	32.25 ± 7.22

ND, nondetectable (<0.20 ng/ml).
Reproduced with permission from Kerrigan et al. (10).

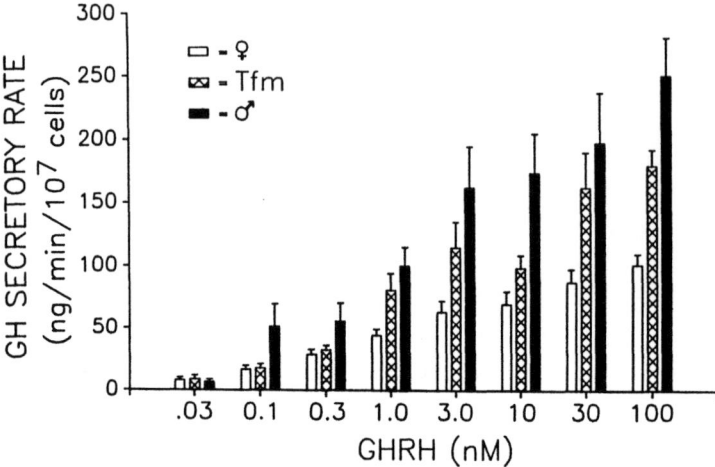

FIGURE 16.4. Responses of dispersed pituitary cells from male, female, and testicular-feminized male (Tfm) rats to eight concentrations of hGHRH-40. These results show the intermediate response of cells from Tfm rats. Reproduced with permission from Batson et al. (11).

body weights of all the animals with the in vitro response of the perifused pituitary cells is shown in Figure 16.5.

Discussion

The correlation of GH secretory rate with body weight seen in the Tfm studies is indicative of the relationship between the potential for growth dependence and the ability of pituitary cells to secrete GH, as well as the ability of the pituitary gland to mediate sex steroid effects. The actions of sex steroids on the pituitary gland directly, as opposed to on hypothalamic control mechanisms, is an impor-

TABLE 16.4. GH secretory rates of pituitary cells from male, female, and Tfm rats.

Group	Basal	GHRH-Stimulated
Male	22.0 ± 5.0[a]	124.6 ± 9.8[a]
Female	6.2 ± 2.4[b]	52.4 ± 8.0[b]
Tfm	9.5 ± 2.2[b]	87.6 ± 3.9[c]

Values are in ng min^{-1}/10^{-7} cells and, like the superscripts, are expressed as in Table 16.1. Reproduced with permission from Batson et al. (11).

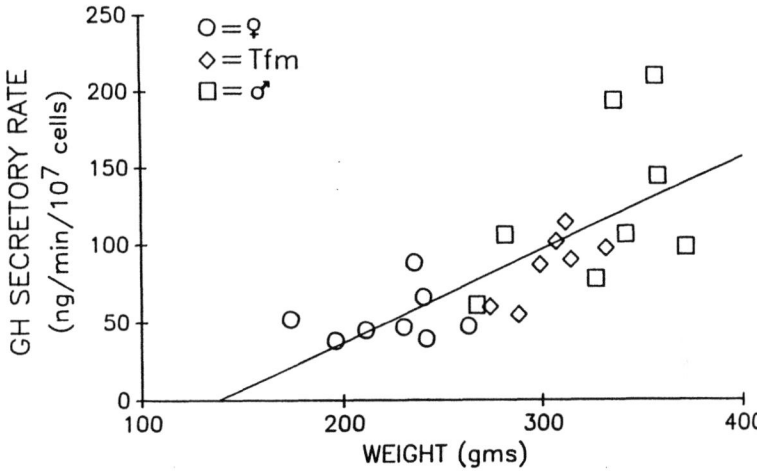

FIGURE 16.5. Relationship of body weight to secretory capacity of dispersed pituitary cells from male, female, and Tfm rats. Reproduced with permission from Batson et al. (11).

tant distinction, and the results of these studies can contribute to its resolution. Although T in vivo can clearly induce the pituitary gland of GX male rats to secrete more GH, it cannot accomplish the same end in OVX female animals. These sex differences may be related to the initial imprinting effects of T in the sense that a hypothalamic effect might be required through changes in GHRH or somatostatin. This hypothalamic effect of T would be possible in a hypothalamus that is "programmed" to augment GH secretion (i.e., a male hypothalamus), but not in one that is not imprinted in the same way (i.e., a female hypothalamus). The effects of E_2 might be more directly related to the pituitary gland. Clearly, E_2 decreased the ability of pituitary cells to release GH. E_2 is also known to favor the production and secretion of prolactin from pituitary cells at the expense of GH (9). However, the potential involvement of hypothalamic somatostatin (12) or changes in pituitary somatostatin receptors in estrogen actions (13) should not be overlooked.

The response in Tfm animals is probably caused by the incomplete androgen resistance that exists in Tfm rats, as opposed to nearly complete resistance in Tfm mice and some Tfm humans (14). The high levels of T probably stimulated some GH production while the female levels of E_2 suppressed it. The question of imprinting in these animals has not been studied.

In conclusion, these in vitro results support the ability of T, in the rat, to augment GH secretion, probably through an action on the hypothalamus. The potent action of E_2 to decrease GH secretory capacity in our model is somewhat at odds with other findings in humans, and the site of E_2 action remains to be clearly established.

References

1. Jansson J-O, Ekberg S, Isaksson O, Mode A, Gustafsson J-A. Imprinting of growth hormone secretion, body growth, and hepatic steroid metabolism by neonatal testosterone. Endocrinology 1985;117:1881–9.
2. Jansson J-O, Frohman LA. Inhibitory effect of the ovaries on neonatal androgen imprinting of growth hormone secretion in female rats. Endocrinology 1987;121:1417–23.
3. Giustina A, Scalvini T, Tassi C, Desenzani P, Poiesi C, Wehrenberg WB, et al. Maturation of the regulation of growth hormone secretion in young males with hypogonadotropic hypogonadism pharmacologically exposed to progressive increments in serum testosterone. J Clin Endocrinol Metab 1997;82:1210–9.
4. Gevers E, Pincus SM, Robinson ICAF, Veldhuis JD. Differential orderliness of the GH release process in castrate male and female rats. Am J Physiol 1998;274:R437–44.
5. Jansson J-O, Frohman LA. Differential effects of neonatal and adult androgen exposure on the growth hormone secretory pattern in male rats. Endocrinology 1987;120:1551–7.
6. Birge CA, Peake GT, Mariz IK, Daughaday WH. Radioimmunoassayable growth hormone in the rat pituitary gland: effects of age, sex and hormonal state. Endocrinology 1967;81:195–204.
7. Ho KY, Evans WS, Blizzard RM, Veldhuis JD, Merriam GR, Samojlik E, et al. Effects of sex and age on the 24-hour profile of growth hormone secretion in man: importance of endogenous estradiol concentrations. J Clin Endocrinol Metab 1987;64:51–8.
8. Evans WS, Krieg RJ, Limber ER, Kaiser DL, Thorner MO. Effects of in vivo gonadal hormone environment on in vitro hGRF-40-stimulated GH release. Am J Physiol 1985;249:E276–80.
9. Goth MI, Lyons CE Jr, Ellwood MR, Barrett JR, Thorner MO. Chronic estrogen treatment in male rats reveals mammosomatotropes and allows inhibition of prolactin secretion by somatostatin. Endocrinology 1996;137:274–80.
10. Kerrigan JR, Martha PM Jr, Krieg RJ Jr, Rogol AD, Evans WS. Somatostatin inhibition of growth hormone secretion by somatotropes from male, female, and androgen receptor-deficient rats: evidence for differing sensitivities. Endocrinology 1989;125:3078–83.
11. Batson JM, Krieg RJ Jr, Martha PM Jr, Evans WS. Growth hormone (GH) response to GH-releasing hormone by perifused pituitary cells from male, female, and testicular feminized rats. Endocrinology 1989;124:444–8.
12. Painson J-C, Thorner MO, Krieg RJ, Tannenbaum GS. Short term adult exposure to estradiol feminizes the male pattern of spontaneous and growth hormone-releasing factor-stimulated growth hormone secretion in the rat. Endocrinology 1992;130:511–9.
13. Djordjijevic D, Zhang J, Priam M, Viollet C, Gourdji D, Kordon C, et al. Effect of 17β-estradiol on somatostatin receptor expression and inhibitory effects on growth hormone and prolactin release in rat pituitary cell cultures. Endocrinology 1998;139:2272–7.
14. Quigley CA, DeBellis A, Marschke KB, El-Awady MK, Wilson EM, French FS. Androgen receptor defects: historical, clinical, and molecular perspectives, Endocr Rev 1995;16:271–321.

17

Somatotrope Heterogeneity and Its Involvement in Growth Hormone (GH) Regulation

JUSTO PASTOR CASTAÑO, JOSÉ LUIS RAMÍREZ,
JOSÉ CARLOS GARRIDO-GRACIA, AND FRANCISCO GRACIA-NAVARRO

Cells that constitute the population of somatotropes within the anterior pituitary gland are not all alike, but differ from one another in a number of characteristics that include not only ultrastructure but also functional and molecular attributes. This concept, known as somatotrope heterogeneity, emerged from pioneering studies by Hopkins and Farquhar (1) on the unequal incorporation of labeled amino acids by dispersed rat pituitary cells, which revealed that individual somatotropes could differ in their protein biosynthetic rates. Subsequent studies by Hymer's group (2) confirmed and extended these observations by showing that rat somatotropes could be separated by use of density gradients into subpopulations that differed in ultrastructure and secretory capacity. Through the application of the reverse hemolytic plaque assay by Frawley and Neill (3) to the analysis of growth hormone (GH) secretion at the single-cell level, our understanding of somatotrope heterogeneity received a definitive impulse. Nevertheless, despite the valuable information gained via these technical and theoretical advances, there are still a number of aspects of somatotrope heterogeneity that await elucidation.

This chapter summarizes the results from some of our recent studies on two new aspects of somatotrope heterogeneity that may help us to better understand its functional basis as well as its significance for somatotrope function (4–11). The first aspect relates to the role of the dynamic nature of somatotrope heterogeneity in changes in the GH axis. More specifically, we have investigated whether somatotrope subpopulations undergo morphofunctional changes associated with growth and aging, and what mechanisms may underlie the dynamics of such heterogeneity. In a second set of studies, we examined the contribution of functional heterogeneity of somatotrope subpopulations in response to primary regulatory factors to the complex activities of the entire somatotrope population. In particular, we addressed the question of whether different, and even opposite, responses of subpopulations to a given secretagogue involve different signaling mechanisms.

Isolation and Characterization of Somatotrope Subpopulations

Our approach to analyzing these aspects of heterogeneity has been to study somatotrope subpopulations separated by centrifugation of pituitary cells in Percoll density gradients (4). Application of this technique to pig or rat pituitary cells yields two major subpopulations of low-density (LD) and high-density (HD) somatotropes (4–7), which are likely to correspond to those originally separated from rat pituitary (2). In prepubertal female pigs, LD and HD somatotropes represented 49% and 40%, respectively, of the total population of somatotropes (5), whereas in adult male rats they accounted for 41% and 35% of total somatotropes, respectively (7). Visual examination of these subpopulations at the electron microscopic level demonstrated that, in both species, LD somatotropes contain fewer secretory granules and more rough endoplasmic reticulum than HD somatotropes. Morphometrical evaluation confirmed the existence of such quantitative differences in ultrastructure between LD and HD somatotropes (5,7). Furthermore, culture of these subpopulations revealed that differences between LD and HD somatotropes also include GH secretory ability because, regardless of the animal model, LD somatotropes release significantly less GH than their HD counterparts under basal conditions (5,8). Therefore, Percoll gradient separation shows that the population of somatotropes from both male rat and female pig pituitary comprises two major subpopulations that display distinctive features in ultrastructure and secretory capacity.

Dynamics of Somatotrope Subpopulations During Postnatal Growth and Aging

GH secretion is a truly dynamic process in that the levels of GH released by somatotropes vary over time. Thus, in addition to the typical pulsatile pattern of GH release over the course of minutes or hours, mean plasma levels of GH also show strong longer-term fluctuations that are closely associated with processes such as postnatal growth or aging in which GH plays important roles (12–14). Consequently, we reasoned that if heterogeneity is to be important for somatotrope function, it should be related to the dynamics of the GH axis. To address this question, changes in the relative proportions, ultrastructure, and secretory characteristics of somatotrope subpopulations were evaluated during postnatal growth in female pigs (neonate, prepubertal, and mature animals; 1, 5–6, and 18–24 months old, respectively) (4–6) and during aging in male rats (adult, old, and senescent rats; 5, 19, and 26 months old, respectively) (7,8). In pigs, LD and HD somatotropes represented the major subpopulations of the somatotrope population throughout the period of active growth. However, the relative proportions of LD and HD somatotropes changed during this period. Specifically, the subpopulation with higher levels of GH secretion, i.e., HD somatotropes, was more abundant in neonatal and young than mature animals, and its numerical importance decreased while that of the less-secretory subpopulation, i.e., LD somatotropes, increased with age (4–6). In addition, evaluation of GH release from cultured LD and HD somatotropes indicated that the amount of hormone secretion decreased with age in both subpopulations, but that this decline was more pronounced in HD than in LD cells (6).

A strikingly similar course of changes was observed in rat somatotrope subpopulations during aging. Thus, the proportion of LD somatotropes increased with aging as that of HD reciprocally decreased (7). Moreover, there was a marked age-associated decrease in the secretory ability of both subpopulations, which was accompanied by a reduction in the content of GH and its mRNA within individual somatotropes (7,8). Therefore, the dynamics of LD and HD somatotropes during porcine growth and rat aging revealed an age-related decrease in the relative proportion of HD cells, coupled with an increase in LD somatotropes, as well as a common reduction in the secretory capacity of both subsets. These findings parallel the reported decline of GH secretion in these and other species during postnatal development and aging (12–14). Hence, dynamic changes in the somatotrope subpopulations associated with age in both species may provide a cellular basis to further understand the age-related decline in GH release. In this model, HD cells would constitute a somatotrope subtype with high levels of GH content and release closely associated with the period of active growth. As the animal grows and ages, LD somatotropes, which release and contain less GH, would progressively prevail over HD somatotropes in the pituitary.

However, in spite of the age-related changes in relative proportions and secretory capacities displayed by LD and HD somatotropes, it is noteworthy that somatotropes within each subpopulation essentially retained their distinct morphofunctional characteristics throughout the animal's lifespan (5–8). To be more specific, irrespective of the animal's age or the species considered, HD cells always contained more secretory granules and less rough endoplasmic reticulum than LD somatotropes. Likewise, HD somatotropes released more GH than LD cells under basal culture conditions in both species at all ages. Bearing in mind the dynamic changes shown by somatotrope subpopulations during growth and aging, these results suggest that somatotropes from a subpopulation (e.g., HD) may modify their functional and morphological characteristics and become members of another subpopulation (e.g., LD) with different properties, and vice versa. This hypothesis can be better understood in the context of a secretory cycle through which somatotropes would alternate between different phases or functional states, which at any point in time would be perceived as distinct subpopulations. In this scenario, the physiological status of the animal would dictate the specific balance among the different subpopulations, and HD and LD somatotropes could show interconversion to adjust the response of the entire somatotrope population so as to ensure that the appropriate hormonal requirements are fulfilled.

Dynamics of Somatotrope Subpopulations Induced by GRF and SRIF

Evidence in support of the notion of dynamic interconversions between somatotrope subpopulations has been provided by experiments carried out in vivo in rats (9). The two hypothalamic factors that primarily control somatotrope function, GH-releasing factor (GRF) and somatostatin (SRIF), were tested as possible modulators of postulated interconversions between LD and HD somatotropes. To this end, adult male rats were injected with GRF or SRIF daily for a week. Then, pituitary cells were dispersed and separated on Percoll gradients, and the relative proportions and morphofunctional features of LD and HD somatotropes were evaluated. GRF, but more notably SRIF, induced an increase in the proportion of HD somatotropes and a concomitant decrease in LD somatotropes, yet the overall number of somatotropes per gland was not modified. Moreover, despite the changes in proportions of LD and HD cells induced by treatment with GRF or SRIF, the distinctive ultrastructural and secretory features of the somatotropes comprising each subpopulation remained unchanged (9). Therefore, it seems reasonable to suggest that such changes in proportions indeed reflect interconversions between LD and HD somatotropes. These and our previous findings, together with results reported for other pituitary cell types, particularly rat lactotropes (15–17) and amphibian melanotropes (18–20), reinforce the idea that pituitary cells can undergo secretory cycles in which subpopulations represent different phases. This cycle would not however consist of a simple process of degranulation and granule replenishment of the cell cytoplasm. Instead, cells in different functional states (i.e., subpopulations or phases of the putative secretory cycle) would exhibit additional distinctive features, as has been shown in lactotropes and melanotropes and as is described next for porcine somatotropes.

Functional Heterogeneity of Porcine Somatotrope Subpopulations Includes Nonuniform Responsiveness to Regulatory Factors and Molecular Signaling

A detailed analysis of the response of porcine LD and HD somatotropes to GRF and SRIF in vitro has provided new insights into the possible importance of functional heterogeneity in this cell type (5,10,11). Initial studies on the secretory behavior of cultured LD and HD subpopulations obtained from prepubertal female pigs revealed a striking dissimilarity between the effects induced by SRIF in each subpopulation (5). LD somatotropes displayed a typical response in that GRF stimulated GH release and SRIF inhibited this stimulation but did not alter basal GH release. Conversely, response of HD somatotropes was 'atypical' because although GRF also stimulated GH release, SRIF did not inhibit this GRF effect. Instead, SRIF itself stimulated basal GH release from HD somatotropes (5). These results corroborated the principle that porcine somatotropes are functionally heterogeneous, and pointed to these subpopulations as a good model to further examine how GRF and SRIF regulate GH secretion from pig somatotropes. Accordingly, our next step was to investigate the subcellular mechanisms that underlie the response of porcine LD and HD somatotropes to these factors. Two types of experiments were applied to achieve this goal. First, we examined the direct effects of these peptides on the production of the second messengers cAMP and inositol phosphates (IPs) in cultures of LD and HD cells, as well as their direct effects on the concentration of free cytosolic calcium ions ($[Ca^{2+}]_i$) in single LD and HD somatotropes, as measured with Indo-1 microfluorometry. Second, to evaluate the relative contribution of these intracellular messengers to the secretory response of somatotropes, the effects of the peptides on GH release from LD and HD somatotropes were measured after selectively inhibiting key signaling enzymes such as adenylate cyclase (AC) or phospholipase C (PLC), and after blocking extracellular Ca^{2+} influx or depleting intracellular Ca^{2+} stores.

Application of the foregoing approaches revealed that the seemingly similar secretory responses of LD and HD somatotropes to GRF (5) were mediated through partially divergent signaling pathways in each subpopulation (10; and unpublished results). As expected, GRF activated cAMP production in both subpopulations, and this activation was critical for GRF-induced GH release, because AC inhibition abolished this stimulation. In contrast, GRF only activated IP turnover in LD cultures but not in HD cells. As a result, blockade of PLC partially reduced GH release from LD but not from HD somatotropes (unpublished results). A more detailed analysis of this problem was facilitated by the study of GRF-induced changes in $[Ca^{2+}]_i$ in single somatotropes (10). This peptide induced distinct patterns of $[Ca^{2+}]_i$ rise in each subpopulation. In all responsive LD somatotropes, GRF elicited a plateau type of response, whereas in more than 90% of responsive HD somatotropes, GRF caused a transient or spike type of response. Application of Ca^{2+} channel blockers and depletors of intracellular Ca^{2+} pools revealed that the response of LD somatotropes to GRF consists of an initial mobilization of intracellular Ca^{2+} followed by an influx of Ca^{2+} through L-type voltage-sensitive Ca^{2+} channels (VSCC), whereas that of HD somatotropes depends exclusively on extracellular Ca^{2+} entry through L-type VSCC (10). This apparently differential contribution of Ca^{2+} sources to GRF-induced Ca^{2+} responses mirrored their importance for GRF-stimulated GH release. In fact, blockade of Ca^{2+} influx reduced GRF-induced GH secretion in both subpopulations, but depletion of intracellular Ca^{2+} with thapsigargin only reduced GH release in LD cells (unpublished results). Thus, we infer that GRF-induced GH release and $[Ca^{2+}]_i$ increase are mediated through differential activation of two intracellular signaling pathways in LD and HD porcine somatotropes. More specifically, although the AC/cAMP/extracellular Ca^{2+} system is the predominant and requisite cascade employed by GRF in both subpopulations, the peptide also requires activation of the PLC/IP/intracellular Ca^{2+} pathway to exert its full effect in LD somatotropes.

In the case of SRIF, a similar investigative approach was used to examine its mode of signaling on somatotropes (11; and unpublished results). High (10^{-7} M) SRIF doses were found to markedly increase cAMP production in HD cultures but not in LD cells. Additionally, AC inhibition abrogated SRIF-stimulated GH release from HD somatotropes. On the other hand, although IP turnover was stimulated by SRIF in cultures of HD cells and not in LD cells, PLC blockade did not modify SRIF-induced GH secretion from HD somatotropes. Interestingly, SRIF only evoked modest decreases in $[Ca^{2+}]_i$ in single LD somatotropes and small increases in $[Ca^{2+}]_i$ in HD somatotropes, thus suggesting that net $[Ca^{2+}]_i$ increases were not required by SRIF to exert its stimulatory action. Indeed, SRIF-stimulated GH release from HD somatotropes was not affected by blocking extracellular Ca^{2+} influx or by depleting intracellular Ca^{2+} stores. In sum, these results indicate that SRIF exerts a differential and specific stimulatory effect on GH secretion from porcine HD somatotropes that requires AC activation and the subsequent cAMP production. Conversely, and surprisingly, PLC activation, or Ca^{2+} influx or mobilization, does not seem to be involved in this process (11; and unpublished results). Thus, when viewed as a whole, available data indicate that porcine somatotropes are heterogeneous not only in ultrastructure and secretory capacity but also in molecular signaling in response to relevant regulatory factors.

Conclusions and Future Directions

Two main conclusions emerge from the studies summarized here. First, somatotrope heterogeneity should be viewed as a dynamic process, rather than as a static phenomenon, as classically considered. Such a dynamic view of heterogeneity should enable a more comprehensive and integrative analysis of this process and of its significance to changes in the somatotropic axis. Our analyses, and those offered by other scientists, of somatotropes and other pituitary cell types (2,5–9,15–21) strongly support the idea that somatotrope heterogeneity is in fact a reflection of the secretory cycle in this cell type. However, this cycle does not represent a simple process of successive states of granule synthesis and release, but includes a complex variety of subcellular and molecular changes that affect relevant aspects of cell function. In line with this idea, our second conclusion is that molecular heterogeneity of signaling mechanisms underlies the differential responses of somatotrope subpopulations to secretagogues. This thesis partly explains the complex responses of mixed somatotrope populations to regulatory factors. Taking these concepts one step further, we propose that differential and dynamic regulation of functionally distinct somatotrope subpopulations cyclicity may constitute a key mechanism in the control of somatotrope function and, thus, of the GH axis.

Future studies on this subject should aim at elucidating the extra- and intracellular signals and molecular mechanisms that control the dynamics of somatotrope heterogeneity. Among the wide array of factors that could modulate this process, sex steroids constitute primary candidates. Thus, for example, estradiol has been shown to control transitions between rat lactotrope subpopulations, and the steroid milieu is known to influence the subpopulation composition of the lactotrope population (21,22). Moreover, gonadal steroids can modulate the effects of secretagogues, e.g., dopamine, on lactotropes by modifying the signaling proteins and dopamine receptor subtypes involved in the response (23,24). Interestingly, recent studies have shown that estradiol modifies the type of SRIF receptors expressed in the rat pituitary, thereby opening the possibility that sex steroids regulate one of the relevant aspects of heterogeneity presented herein, the somatotrope response to SRIF (25).

Acknowledgments. This work has been supported by grants CRG-971039 (NATO), CVI-0139 (PAI, Junta de Andalucia, Spain), and PB94-0451-CO2-01 (MEC, Spain).

References

1. Hopkins CR, Farquhar MG. Hormone secretion by cells dissociated from rat anterior pituitaries. J Cell Biol 1973;59:276–303.
2. Snyder G, Hymer WC, Snyder J. Functional heterogeneity in somatotrophs isolated from the rat anterior pituitary. Endocrinology 1977;101:788–99.
3. Frawley LS, Neill JD. A reverse hemolytic plaque assay for microscopic visualization of growth hormone release from individual cells: evidence for somatotrope heterogeneity. Neuroendocrinology 1984;39:484–7.
4. Torronteras R, Castaño JP, Ruiz-Navarro A, Gracia-Navarro F. Application of a Percoll density gradient to separate and enrich porcine pituitary cell types. J Neuroendocrinol 1993;5:257–66.
5. Castaño JP, Torronteras R, Ramírez JL, Gribouval A, Sánchez-Hormigo A, Ruiz-Navarro A, Gracia-Navarro F. Somatostatin increases growth hormone (GH) secretion in a subpopulation of porcine somatotropes: evidence for functional and morphological heterogeneity among porcine GH-producing cells. Endocrinology 1996;137:129–36.

6. Castaño JP, Ruiz-Navarro A, Malagón MM, Hidalgo-Díaz C, Gracia-Navarro F. Secretory and morphological heterogeneity of porcine somatotropes during postnatal development. J Neuroendocrinol 1997;9:769–75.
7. Dobado-Berrios PM, Ruiz-Navarro A, Almadén Y, Malagón MM, Garrido JC, Ramírez-Gutiérrez JL, Gracia-Navarro F. Heterogeneity of growth hormone (GH) -producing cells in aging male rats: ultrastructure and GH gene expression in somatotrope subpopulations. Mol Cell Endocrinol 1996;118:181–91.
8. Dobado-Berrios P, Ruiz-Navarro A, López-Pedrera R, González de Aguilar JL, Torronteras R, Hidalgo-Díaz C, Gracia-Navarro F. Heterogeneity of growth hormone (GH) -producing cells in aging male rats. II. In vitro GH releasing activity of somatotrope subpopulations. Mol Cell Endocrinol 1996;123:127–37.
9. Malagón MM, Garrido-Gracia JC, Torronteras R, Dobado-Berrios PM, Ruiz-Navarro A, Gracia-Navarro F. Cell heterogeneity as a reflection of the secretory cell cycle. Ann NY Acad Sci 1998;839:244–8.
10. Ramírez JL, Torronteras R, Malagón MM, Castaño JP, García-Navarro S, González de Aguilar JL, Martínez-Fuentes AJ, Gracia-Navarro F. Growth hormone-releasing factor mobilizes cytosolic free calcium through different mechanisms in two somatotrope subpopulations from porcine pituitary. Cell Calcium 1998;23:207–17.
11. Ramírez JL, Torronteras R, García-Navarro S, Castaño JP, Gracia-Navarro F. Differences in second messengers (Ca^{2+} and cAMP) suggest a dual role for SRIF in regulating GH release from porcine somatotropes. Ann NY Acad Sci 1998;839:375–7.
12. Eden S. Age- and sex-related differences in episodic growth hormone secretion in the rat. Endocrinology 1979;105:555–60.
13. Klindt J, Stone RT. Porcine growth hormone and prolactin: concentrations in the fetus and secretory patterns in the growing pig. Growth 1984;48:1–15.
14. Takahashi S, Gottschall PE, Quigley KL, Goya RG, Meites J. Growth hormone secretory patterns in young, middle-aged and old female rats. Neuroendocrinology 1987;46:137–42.
15. Castaño JP, Kineman RD, Frawley LS. Dynamic fluctuations in the secretory activity of individual lactotropes as demonstrated by a modified sequential plaque assay. Endocrinology 1994;135:1747–52.
16. Castaño JP, Kineman RD, Frawley LS. Dynamic monitoring and quantification of gene expression in single, living cells: a molecular basis for secretory cell heterogeneity. Mol Endocrinol 1996;10:599–605.
17. Lledo PM, Guérineau N, Mollard P, Vincent JD, Israel JM. Physiological characterization of two functional states in subpopulations of prolactin cells from lactating rats. J Physiol 1991;437:477–94.
18. González de Aguilar JL, Malagón MM, Vázquez-Martínez RM, Lihrmann I, Tonon MC, Vaudry H, Gracia-Navarro F. Two frog melanotrope cell subpopulations exhibiting distinct biochemical and physiological patterns in basal conditions and under thyrotropin-releasing hormone stimulation. Endocrinology 1997;138:970–7.
19. Gracia-Navarro F, González de Aguilar JL, Vázquez-Martínez RM, Tonon MC, Vaudry H, Malagón MM. Melanotrope cell heterogeneity in the pars intermedia of amphibians. Ann NY Acad Sci 1998;839:223–8
20. Vázquez-Martínez RM, Malagón MM, González de Aguilar JL, Desrues L, Tonon MC, Vaudry H, Gracia-Navarro F. Relationship between melanotrope cell heterogeneity and background adaptation in the frog intermediate lobe. Ann NY Acad Sci 1998;839:431–3.

21. Kukstas LA, Verrier D, Zhang J, Chen C, Israel JM, Vincent JD. Evidence for a relationship between lactotroph heterogeneity and physiological context. Neurosci Lett 1990;120:84–6.
22. Velkeniers B, Kazemzadeh M, Vanhaelst L, Hooghe-Peters EL. Functional heterogeneity with respect to oestrogen treatment in prolactin cell subpopulations separated by Percoll gradient centrifugation. J Endocrinol 1994;141:251–8.
23. Kukstas LA, Domec C, Bascles L, Bonnet J, Verrier D, Israel JM, Vincent JD. Different expression of the two dopaminergic D2 receptors, D2415 and D2444, in two types of lactotroph each characterised by their response to dopamine, and modification of expression by sex steroids. Endocrinology 1991;129:1101–3.
24. Livingstone JD, Lerant A, Freeman ME. Ovarian steroids modulate responsiveness to dopamine and expression of G-proteins in lactotropes. Neuroendocrinology 1998;68:172–9.
25. Djordjijevic D, Zhang J, Priam M, Viollet C, Gourdji D, Kordon C, Epelbaum J. Effect of 17β-estradiol on somatostatin receptor expression and inhibitory effects on growth hormone and prolactin release in rat pituitary cell cultures. Endocrinology 1998;139:2272–7.

Part III

How Sex Steroids Modulate GH Action on Target Tissues

18

Sex-Steroid and GH Interactions in the Regulation of Lipid Metabolism

STAFFAN EDÉN, JAN OSCARSSON, AND MALIN OTTOSSON

Growth hormone (GH) and sex steroids interact at several levels in the control of growth and body composition and reproduction. However, it is not completely clear which interactions are of primary significance for the effects of the respective hormones. For example, sex steroids influence the overall secretion and secretory pattern of GH (see Chapter 12, this volume), which may be of importance for the effects of GH on target tissues (see Chapters 32 and 33). On the other hand, GH may influence the responsiveness of target tissues to sex steroids or gonadotrophins (see Chapter 8).

Both GH and sex steroids have been shown to influence lipid metabolism. In males, body fat is lower and the anatomic distribution of fat among different fat depots seem to be dependent upon sex steroids (1). Lipoprotein metabolism is also sex dependent (2). GH has a marked influence on body fat. GH deficiency is associated with accumulation of body fat (3) whereas acromegalic patients have decreased body fat (4). GH seems to be more effective in its lipolytic actions on visceral fat depots, although all fat depots seem to be responsive to GH (5). Interestingly, changes in body composition in at least male acromegalic subjects seem to be dependent upon intact gonadal function (4), indicating that there are direct interactions between sex steroids and GH at the level of the target tissues.

We have studied the interactions between GH and sex steroids in the regulation of lipid metabolism using different experimental systems. Hypophysectomized rats have been used to study the importance of the changes in GH secretory pattern for the effects of GH and sex steroids in this regulation. In this model, hypophysectomized rats have been given replacement therapy with thyroxine and cortisol. In addition, GH has been given as a subcutaneous continuous infusion to mimic the female secretory pattern of GH, or as intermittent injections, either subcutaneously twice daily (6) or as intermittent pulses (7) to emulate the male pattern. To clarify direct interactions between GH and sex steroids at the level of the target tissue, human adipose tissue has been incubated for several days in a defined medium (8), and hepatic lipoprotein metabolism has been studied in primary cultures of rat hepatocytes (9).

Roles of GH and Sex-Steroids in the Control of Adipose Tissue Metabolism

Adipose tissue mass increases during the first year of life. Thereafter, there is a steady but slow increase in adipose tissue mass, which accelerates during the prepubertal and pubertal period (10). During this time, the sex differences in distribution of body fat become apparent (11). With increasing age, the relative amount of fat increases (12), although total body weight might decrease in the elderly (13).

It is apparent that many of the age- and sex-related differences in body fat accumulation and distribution can be attributed to changes in the tissue effects of and responsiveness to GH and sex steroids. For example, there is an inverse correlation between changes in GH secretion with increasing age and the accumulation of body fat (14). Testosterone seems to be important for the distribution of fat in normal adult men (1). Also, estrogens affect body fat distribution. The typical increase in adipocyte cell size in the femoral subcutaneous fat depot in women disappears after menopause, but this decrease can be prevented by hormonal replacement therapy (15).

The sites of action of GH and sex steroids in the regulation of adipose tissue mass and distribution are not completely clear. Adipose tissues from experimental animals and humans express GH receptors (16,17). We have studied the direct effects of GH in incubated human adipose tissue. In this experimental system, lipoprotein lipase (LPL) activity is induced in the presence of cortisol. Addition of GH markedly inhibits LPL activity within 8 h. This effect of GH does not seem to be exerted at the transcriptional level because steady-state LPL mRNA levels are unaffected by GH (8). Inhibition is accompanied by increased basal rates of lipolysis (Ottosson M, unpublished). Similar effects of GH in vitro on lipolysis have been demonstrated in isolated adipocytes obtained immediately after excision, isolation, and exposure to GH (18) and in cultured human preadipocytes allowed to differentiate into adipocytes (19). Taken together, these results strongly suggest that GH indeed has a direct effect on human adipose tissue metabolism, resulting in an inhibition of lipid accumulation and a stimulation of lipolysis. Testosterone also inhibited LPL activity when added to human adipose tissue in culture, whereas estrogens were without effect (Ottosson M, unpublished). Thus, both GH and testosterone can affect lipid metabolism directly at the level of the target tissue.

In the rat, there is a decrease in catecholamine-induced lipolysis in adipocytes obtained from hypophysectomized rats. GH treatment of these rats resulted in an increase in catecholamine-induced lipolysis, without any differences among the effects of GH on adipocytes obtained from different fat depots. In hypophysectomized female rats, the lipolytic response to noradrenaline or isoproteronol was completely restored by GH treatment, whereas this was not the case in male rats (20). Testosterone was without effect when given alone. However, GH and testosterone treatment in combination had an additive effect on catecholamine-induced lipolysis. Both hormones increased β-adrenergic receptors, but GH was able to increase lipolysis also through post-adrenergic-receptor mechanisms (21).

The studies cited here indicate that GH has a marked direct effect on adipose tissue, in line with the findings of GH-receptor expression in this tissue. Testosterone has an additive effect to that of GH. Indeed, androgen receptors are present in both adipose precursor cells and adipocytes (22). The presence of estrogen receptors in human adipose tissue has been controversial (23–25). As stated earlier, we have been unable to show any direct effects of estrogens on LPL activity in cultured human adipose tissue. However, it seems likely that estrogens may also exert direct effects on adipose tissue growth and differentiation, and further studies are needed to clarify any direct influence of estrogens on various adipose tissue depots.

Roles of GH and Sex Steroids in the Regulation of Hepatic Lipoprotein Metabolism

There are sex differences in lipoprotein concentrations and hepatic lipoprotein metabolism in both humans and experimental animals, that can be ascribed to effects of gonadal steroids. However, comparison of sex-steroid hormone-receptor expression in the liver with the physiological concentrations of sex steroids may suggest that the sex steroids under physiological circumstances act indirectly on the liver. For example, physiological plasma concentrations of estradiol in the rat are probably too low to induce a translocation of the estrogen receptor to the nucleus in rat liver cells (26). Moreover, it has been difficult to obtain reproducible effects of estrogens in vitro on cultured cells of hepatic origin (27–29). Conversely, pharmacological doses of sex steroids, which may be a result of oral administration of sex steroids, may affect the liver directly (see Chapter 6). For example, low doses of estradiol in intact female rats increase serum high-density lipoprotein (HDL) concentrations, whereas high doses have the opposite effect (30), the latter probably a "pharmacological" effect.

In the hypophysectomized rat model, we have been unable to demonstrate any effects of physiological replacement doses of sex steroids on serum lipoprotein concentrations, in line with the contention that the sex steroids act indirectly in this context. The concentrations and fatty acid composition of plasma phospholipids are sexually differentiated. In gonadectomized rats, phospholipid levels and fatty acid composition are normalized with sex hormone replacement therapy, whereas sex steroids are without effect in hypophysectomized rats (31). When rats were treated with GH as a continuous infusion, phospholipid concentrations and fatty acid composition were "feminized," whereas GH given twice daily had no such effect (32,33). Similarly, the sex difference in HDL and apolipoprotein E concentrations was shown to be dependent upon sex differences in GH secretory patterns (9,34–36). There are exceptions to the tenet that sex steroids act on the liver via an influence on the secretory pattern of GH. Liver fatty acid-binding protein is two to three times more abundant in the female rat liver than in the male rat liver, independent of the secretory pattern of GH (37).

In summary, in the rat, sex-steroid hormones affect hepatic lipoprotein metabolism by affecting GH secretory pattern. Whether similar regulatory principles are applicable in humans is unclear. As thoroughly discussed elsewhere in this volume, sex steroids affect GH secretion in humans, but it is evident that their influence is much less marked compared with that in the rat. There is circumstantial evidence that the mode of GH administration may affect lipoprotein metabolism also in GH-deficient humans. For example, continuous adminsitration of GH for 2 weeks increased apolipoprotein E concentrations more markedly than treatment with the same dose given as one daily subcutaneous injection (38). Thus, it is possible that the pattern of GH secretion is also important in the regulation of hepatic lipoprotein metabolism in humans.

In conclusion, sex steroids and GH interact at several levels in the regulation of body growth and metabolism. Sex steroids directly interact with GH at the level of target tissues, for example, adipose tissue. However, many if not most physiological effects of sex steroids on the liver may be mediated indirectly by their influence on GH secretion (Fig. 18.1).

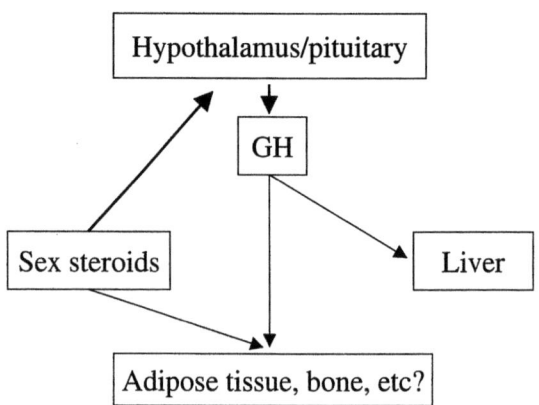

FIGURE 18.1. Proposed schematic model for the interactions between sex steroids and GH in the regulation of lipid metabolism. Sex steroids act on adipose tissue and other target tissues directly and by influencing GH secretion, whereas their effects on liver lipoprotein metabolism are mainly indirect via altering GH secretion.

References

1. Björntorp P. Growth hormone, insulin-like growth factor-I and lipid metabolism: interactions with sex steroids. Horm Res 1996;46:188–91.
2. Svanborg A. Possible relationship between plasma lipids and hormonal steroids. In: Salahanic HA, Kipnis DM, Van Der Wiele RL, eds. Metabolic effects of gonadal hormones and contraceptive steroids. New York: Plenum Press, 1969: 242–6.
3. Rosén T, Bosaeus I, Tölli J, Lindstedt G, Bengtsson B-Å. Increased body fat mass and decreased extracellular fluid volume in adults with growth hormone deficiency. Clin Endocrinol 1993;38:63–71.
4. Rosén T, Bengtsson B-Å, Brummer R J, Bosaeus I, Edén S. Body composition in acromegaly. J Clin Endocrinol 1989;30:121–30.
5. Bengtsson B-Å, Edén S, Lönn L, Kvist H, Stokland A, Lindstedt G, Bosaeus I, Tölli J, Sjöström L, Isaksson OGP. Treatment of adults with growth hormone (GH) deficiency with recombinant human growth hormone. J Clin Endocrinol Metab 1993;76:309–17.
6. Jansson J-O, Albertsson-Wikland K, Edén S, Thorngren K-G, Isaksson O. Circumstantial evidence for a role of the secretory pattern of growth hormone in control of body growth. Acta Endocrinol (Copenh) 1982;99:24–30.
7. Oscarsson J, Carlsson L, Bick T, Lidell A, Olofsson S-O, Edén S. Evidence for the role of the secretory pattern of growth hormone in the regulation of serum concentrations of cholesterol and apolipoprotein E in rats. J Endocrinol 1990;128:433–8.
8. Ottosson M, Vikman-Adolfsson K, Enerbäck S, Elander A, Björntorp P, Edén S. Growth hormone inhibits lipoprotein lipase activy in human adipose tissue. J Clin Endocrinol Metab 1995;80:936–41.
9. Sjöberg A, Oscarsson J, Borén J, Edén S, Olofsson S-O. Mode of growth hormone administration influences triacylglycerol synthesis and assembly of apolipoprotein B-containing lipoproteins in cultured rat hepatocytes. J Lipid Res 1996;37:275–9.
10. Knittle J, Timmers K, Ginsburg-Fellner F. The growth of adipose tissue in children and adolescents. Cross-sectional and longitudinal studies of adipose cell number and size. J Clin Invest 1979;63:929–41.
11. Malina RM. Growth of muscle tissue and muscle mass. In: Falkner F, Tanner JM, eds. Human growth. New York: Plenum Press, 1986:77–99.
12. Bruce A, Andersson M, Arvidsson B, Isaksson B. Body composition. Prediction of normal body potassium, body water and body fat in adults on the basis of body height, body weight and age. Scand J Clin Lab Invest 1980;40:461–73.
13. Steen B. Body composition and aging. Nutr Rev 1988;46:45–51.
14. Corpas E, Harman S, Blackman M. Human growth hormone and human aging. Endocr Rev 1993;14:20–39.
15. Rebuffé-Scrive M, Eldh J, Hafström L-O, Björntorp P. Metabolism of mammary, abdominal and femoral adipocytes in women before and after menopause. Metabolism 1986;35:792–7.
16. Fagin K, Lackey S, Reagan C, DiGirolamo M. Specific binding of growth hormone by rat adipocytes. Endocrinology 1980;107:608–15.
17. DiGirolamo M, Edén S, Enberg G, Isaksson O, Lönnroth P, Hall K, Smith U. Specific binding of human growth hormone but not insulin-like growth factors by human adipocytes. FEBS Lett 1986;205:15–9.

18. Harant I, Beauville M, Crampes F, Riviere D, Tauber M-T, Tauber J-P, Garrigues M. Response of fat cells to growth hormone (GH): effect of long term treatment with recombinant human GH in GH-deficient adults. J Clin Endocrinol Metab 1994;78:1392–5.
19. Wabitsch M, Braun S, Hauner H, Heinze E, Ilondo MM, Shymko R, De Myets P, Teller WM. Mitogenic and antiadipogenic properties of human growth hormone in differentiating human adipocyte precursor cells in primary culture. Pediatr Res 1996;40:450–6.
20. Yang S, Björntorp P, Liu X, Edén S. Growth hormone regulates lipolysis and β-adrenergic receptors in rat adipocytes: no difference in the effects of growth hormone between visceral and subcutaneous fat. J Obes Res 1996;4:471–8.
21. Yang S, Xu X, Edén S. Additive effects of growth hormone and testosterone on lipolysis in adipocytes of hypophysectomized rats. J Endocrinol 1995;147:147–52.
22. De Pergola G, Xu X, Yang S, Giorgino R, Björntorp P. Up-regulation of androgen receptor binding in male rat fat pad adipose precursor cells exposed to testosterone: study in a whole cell assay system. J Steroid Biochem Mol Biol 1990;37:553–8.
23. Brönnegård M, Ottosson M, Boos J, Marcus C, Björntorp P. Lack of evidence for estrogen and progesterone receptors in human adipose tissue. J Steroid Biochem Mol Biol 1994;51:275–81.
24. Mizutani T, Nishikawa Y, Adachi H, Enomoto T, Ikegami H, Kurachi H, Nomura T, Miyake A. Identification of estrogen receptor in human adipose tissue and adipocytes. J Clin Endocrinol Metab 1994;78:950–4.
25. Price TM, O'Brien SN. Determination of estrogen receptor messenger ribonucleic acid (mRNA) and cytochrome P450 aromatase mRNA levels in adipocytes and adipose stromal cells by competetive polymerase chain reaction amplification. J Clin Endocrinol Metab 1993;77:1041–5.
26. Eisenfeld AJ, Aten RF. Estrogen receptor in the mammalian liver. In: Langer M, Chianduss L, Chopra IJ, Martini L, eds. The endocrines and the liver.Serono Symposia No. 51. London: Academic Press, 1982:49–62.
27. Brindley D, Salter A. Hormonal regulation of the hepatic low density lipoprotein receptor and the catabolism of low density lipoproteins. Relationship with the secretion of very low density lipoproteins. Prog Lipid Res 1991;22:287–96.
28. Semenkovich C, Ostlund R. Estrogens induce low density lipoprotein receptor activity and decrease intracellular cholesterol in human hepatoma cell line Hep G2. Biochemistry 1987;26:4987–92.
29. Wade DP, Knight BL, Soutar AK. Hormonal regulation of low-density lipoprotein (LDL) receptor activity in human hepatoma HepG2 cells: insulin increases LDL receptor activity and diminishes its suppression by exogenous cholesterol. Eur J Biochem 1988;174:213–8.
30. Ferreri LF, Naito HK. Effect of estrogen on rat serum cholesterol concentrations: consideration of dose, type of estrogen, and treatment duration. Endocrinology 1978;102:1621–7.
31. Edén S, Oscarsson J, Jansson J-O, Svanborg A. The influence of gonadal steroids and the pituitary on the levels and composition of plasma phospholipids in the rat. Metabolism 1987;36:527–32.
32. Oscarsson J, Jansson J-O, Edén S. Plasma pattern of growth hormone regulates sexual differentiation of phosphatidylcholine in rat plasma. Acta Physiol Scand 1988;133:257–65.

33. Oscarsson J, Edén S. Sex differences in fatty acid composition of rat liver phosphatidyl choline are regulated by the plasma pattern of growth hormone. Biochim Biophys Acta 1988;959:280–7.
34. Oscarsson J, Olofsson S-O, Bondjers G, Edén S. Differential effects of continuous versus intermittent administration of growth hormone to hypophysectomized female rats on serum lipoproteins and their apoproteins. Endocrinology 1989;125:1638–49.
35. Oscarsson J, Olofsson S-O, Vikman K, Edén S. Growth hormone regulation of serum lipoproteins in the rat: different growth hormone regulatory principles for apolipoprotein (apo) B and the sexually dimorphic apo E concentrations. Metabolism 1991;11:1191–8.
36. Sjöberg A, Oscarsson J, Edén S, Olofsson S-O. Continuous but not intermittent administration of growth hormone to hypophysectomized rats increases apolipoprotein-E secretion from cultured hepatocytes. Endocrinology 1994;134:790–8.
37. Carlsson L, Nilsson I, Oscarsson J. Hormonal regulation of liver fatty acid-binding protein in vivo and in vitro: effects of growth hormone and insulin. Endocrinology 1998;139:2699–709.
38. Oscarsson J, Ottosson M, Johansson J-O, Wiklund O, Mårin P, Björntorp P, Bengtsson B-Å. Two weeks of daily injections and continuous infusion of recombinant human growth hormone (GH) in GH-deficient adults. II. Effects on serum lipoproteins and lipoprotein and hepatic lipase activity. Metabolism 1986;45:370–7.

19

Estrogen Replacement Therapy and the Response to Human Growth Hormone

GIAN PAOLO CEDA, GIORGIO VALENTI, AND ANDREW R. HOFFMAN

In young adults, growth hormone (GH) levels are higher in women than in men, and there is a positive correlation between serum estrogen and GH levels. Plasma IGF-I levels fall immediately after the onset of the menopause (1), and mean serum 24-h GH levels are lower than in premenopausal controls. Both estrogen and progesterone enhance the GH response to the somatostatin inhibitor pyridostigmine (2). Oral ethinyl estradiol increases serum GH levels but lowers IGF-I levels (3), suggesting a block in GH action at the liver.

In an earlier study, we obtained similar results in patients administered conjugated estrogens. We also employed an IGF-I generation test to examine the effects of age, gender, and oral estrogen replacement on the responsiveness of serum IGF-I to acute GH administration (4). A single dose of GH (0.1 mg/kg) was given subcutaneously to 31 healthy adults, who were divided into five study groups: young (20–29 years) men, young women in the follicular phase of the menstrual cycle, older (60–69 years) men, older women not on estrogen replacement, and older women on oral estrogen replacement. Blood was sampled over 72 h following GH administration. The older men had lower peak IGF-I levels and a lower IGF-I response to GH than did the younger men. Older women who were taking oral estrogen had lower basal IGF-I than the older women who were not taking hormone replacement therapy. IGF-binding protein (IGF-BP-3) levels were unchanged 24 hours after GH administration. The high-affinity circulating GH-binding protein (GHBP) is thought to be derived from proteolysis of the GH receptor (GHR), and thus GHBP levels may provide an index of GHR concentration and thereby reflect the body's sensitivity to GH. After GH administration, GHBP levels fell by approximately 20%, with no differences among the groups, and there was no correlation between the IGF-I response to GH administration and basal GHBP levels. Thus, oral estrogen therapy in postmenopausal women is accompanied by apparent resistance to GH action.

As IGF-I levels fall, feedback inhibition at the level of the hypothalamopituitary unit declines and serum GH increases. While oral estrogens lower IGF-I levels, transdermal estradiol had no (or a lesser) effect on serum

GH levels (5), and actually increased serum IGF-I in one study (6,7). In another study, transdermal estradiol decreased the GH response to growth hormone-releasing hormone (GHRH), but had no effect on serum IGF-I levels (8). The effects on GH pulsatility have been variable (9,10). We examined the effect of transdermal estradiol plus progestin on the GH–IGF axis (11). Seven postmenopausal women (51 ± 1.3 years old) with normal body mass index (BMI) wore a transdermal estrogen patch designed to deliver 50 μg of estradiol/day and took oral medroxyprogesterone acetate 10 mg/day for the last 10 days of each month. After 6 months of this therapy, there was no change in the GH response to a single injection of GHRH compared with baseline. Fasting plasma IGF-I, IGF-BP-1, and IGF-BP-3 levels also did not change, nor did this treatment alter any serum lipid parameters, IGF-I, or IGF-BP-3. Thus, while oral estrogens exacerbate the relative IGF-I deficiency of aging, probably because of a "first-pass" hepatic effect, our results suggest that transdermal estrogen therapy does not exacerbate the endocrine deficiencies associated with the somatopause.

To clarify the role of the dose of estradiol delivered by transdermal patch in the regulation of GH and IGF-I, we studied the effects of increasing doses of transdermal E_2 for 1 week on IGF-I and IGF-BP secretion, on the GH responses to GHRH, and on plasma lipid levels in a group of elderly women.

Methods

Seventeen healthy women, aged 79 ± 1.5 years, were recruited from the community. All the subjects had a normal BMI (25 ± 1.2 kg/m^2) and had not been treated with hormone replacement therapy for at least 6 months. The subjects were divided into three groups, receiving the following doses of transdermal estradiol for 7 days:

Group 1 (n = 5), 50 μg/day

Group 2 (n = 6), 100 μg/day

Group 3 (n = 6), 200 μg/day

Blood samples were drawn each morning for determination of IGF-I, IGF-BP-3, IGF-BP-1, estradiol (E_2), luteinizing hormone (LH), follicle-stimulating hormone (FSH), and plasma lipid levels. The pituitary response to GHRH was determined by injecting subjects with GHRH at a dose of 1μg/kg b.w. and then measuring serum GH levels.

Results

Transdermal estrogen administration resulted in a dose-responsive increase in circulating estradiol levels, with plasma levels in the 200 μg/day group similar to those normally achieved in the late follicular phase and at midcycle in young menstruating women (Fig. 19.1). Associated with this increase in estradiol was a concomitant and significant decline in serum FSH and LH levels (data not shown).

GH secretory capacity was investigated using the standard GHRH stimulation test. Before the initiation of transdermal estradiol, GHRH administration resulted in a minimal increment in GH levels, consistent with the previously reported decreased responsiveness characteristic of the somatopause. However, when the GHRH test was repeated after 1 week of estradiol treatment, the GH responses tended to be higher in the 100 µg/day group and were significantly higher than those recorded in basal condition in the group of women treated with 200 µg/day of transdermal estradiol.

The differences in the GH secretory pattern are more evident when the pituitary responses are evaluated as area under the curve. In the group receiving 50 µg/day, no difference was observed between the responses obtained before and after treatment, but in the 100 µg/day and 200 µg/day groups, the area under the curve almost doubled after estrogen treatment (Fig. 19.2, top). Serum IGF-I levels did not change in the 50 or 100 µg/day groups (Fig. 19.2, bottom). However, there was a significant decline in IGF-I levels in the group receiving 200 µg/day. No significant changes in IGF-BP-1 or IGF-BP-3 levels were observed in any of the three treatment groups (data not shown). There

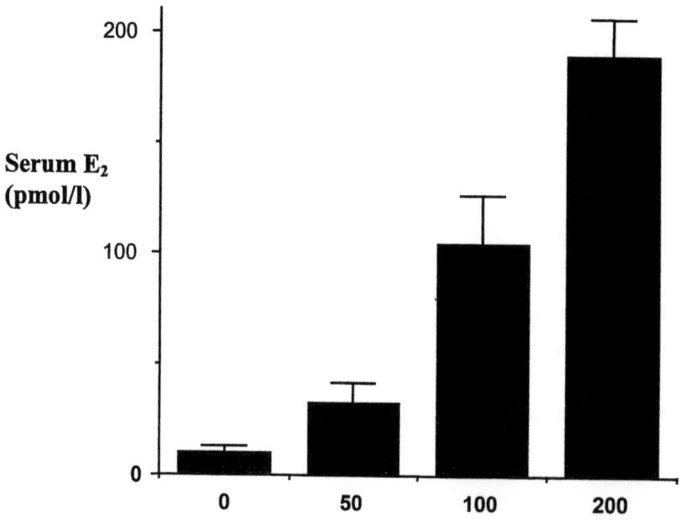

FIGURE 19.1. Serum estradiol concentrations in subjects wearing transdermal estradiol patches for 1 week. Estradiol-17-beta doses are noted in micrograms/day.

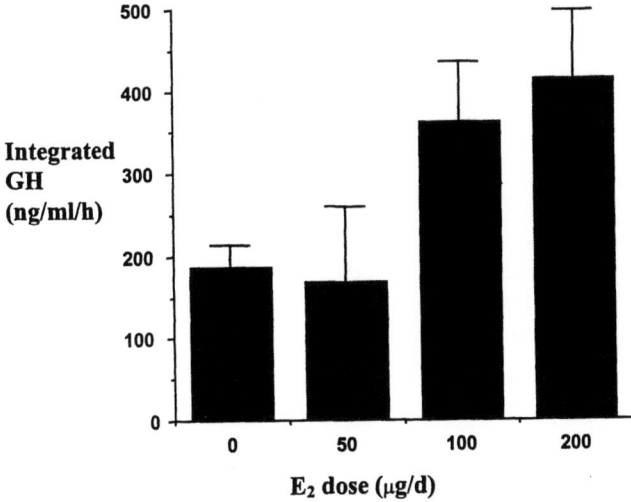

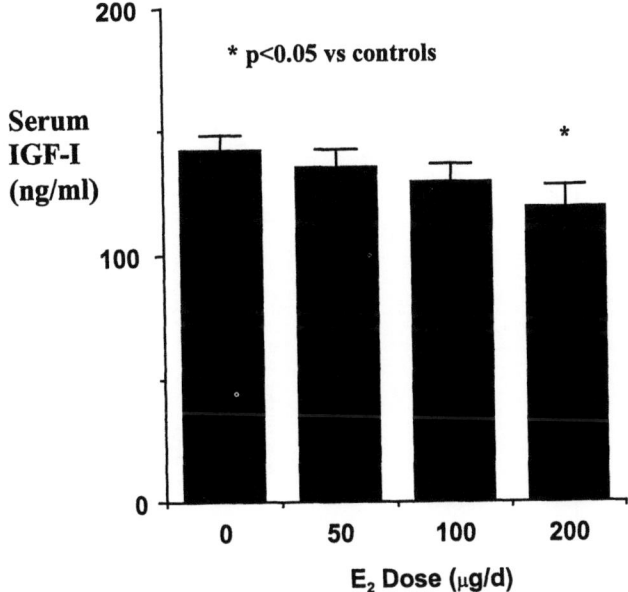

FIGURE 19.2. Response of the GH–IGF axis to treatment with transdermal estradiol *Top:* Integrated GH concentrations measured as area under the serum GH curve after a GHRH stimulation test. The area under the curve was significantly greater than baseline for the subjects who wore the 100- or the 200 μg/day patch. *Bottom:* Serum IGF-I levels after 1 week of transdermal estradiol therapy.

TABLE 19.1. The effects of transdermal E_2 treatment on GH secretion and IGF-I levels.

Source	Length of treatment	E_2 Dose	IGF-I levels	GH levels
Dell'Aglio et al. (1994) (11)	6–12 months	50 µg E_2 + medroxyprogesterone; 10 mg days 15–25	No change No change in BP-3 and BP-1	GHRH test: no change
Weissberger et al. (1991) (6)	2 months	100 µg E_2 + norethistrone; 5 mg days 15–21	Increase	Mean 24-h GH: no change
Bellantoni et al. (1991) (8)	2 months	50, 100, 150 µg E_2 + medroxyprogesterone; 10 mg days 15–28	No change	GHRH test: decreased GH responses
Slowinska-Srzednicka et al. (1993) (7)	3–6 months	100 µg E_2 + chlormadinone acetate; 2 mg days 15–21	Increase	Not studied
Friend et at. (1996) (10)	15 days	200 µg E_2 only	Decrease	Increased integrated GH concentration, pulse amplitude, peak levels
Bellantoni et al. (1996) (9)	6 weeks	100 µg E_2 only	Slight decrease	No change in GH pulsatility and GH response

was no significant effect of estradiol treatment at any dose on serum total cholesterol, HDL, LDL, Apo A1, apo-B, and Lp (a) concentrations. Moreover, one week of estrogen therapy did not lead to any changes in body composition as measured by bioimpedance analysis

Conclusion

The contrasting effects of oral and transdermal estrogens on the GH–IGF axis have been reported in detail (see summary on Table 19.1). In our study, short-term administration of low-dose transdermal E_2, which is commonly used in postmenopausal subjects, did not significantly affect IGF-I or GH secretion. However higher, pharmacological doses of transdermal E_2 (200 μg) did induce effects on IGF-I and GH levels similar to those observed after oral estrogen administration. This increase in GH secretion is most likely secondary to the reduction of serum IGF-I levels, because several studies, both in vitro and in vivo, have clearly demonstrated the ability of IGF-I to inhibit the synthesis and secretion of GH at the pituitary and hypothalamic levels. Moreover, the fact that the IGF-I levels are reduced while the IGF-BP levels are not affected suggests further reduced (free) IGF-I bioactivity. In contrast with the effects on GH and IGF-I levels, high doses of transdermal estradiol did not result in any significant modification of plasma lipid levels.

Important interactions exist between the GH–IGF axis and estrogen. For example, in at least one model system, IGF-I has been shown directly to activate the estrogen receptor (ER), stimulating the expression of estrogen-responsive genes even in the absence of estrogen (12). We hypothesize that a similar anabolic interaction between IGF-I and estrogen may also occur in other tissues, such as muscle, skin, and CNS. According to this proposed model, if the menopausal/somatopausal woman is treated with oral estrogens, serum IGF-I levels fall, and optimal estrogenic activity therefore cannot be achieved in bone, for example. If IGF-I or GH is given with low-dose, "physiologic" transdermal estrogen, it is possible that the increased circulating IGF-I would synergize with estrogen to enhance the overall anabolic effect. We speculate further that GH or IGF-I therapy would synergize with the ability of estrogen hormone replacement therapy to decrease cardiovascular risk factors and mortality in elderly women, given the inferred interactions of sex steroids and GH on lipid metabolism (see Chapter 18).

Reference

1. Romagnoli E, Minisola S, Carnevale V, Scarda A, Rosso R, Scarnecchia L, et al. Effect of estrogen deficiency on IGF-I plasma levels: relationship with bone mineral density in perimenopausal women. Calif Tissue Int 1993;53:1–6.
2. O'Keane V, Dinan TG. Sex steroid priming effects on growth hormone response to pyridostigmine throughout the menstrual cycle. J Clin Endocrinol Metab 1992;5:11–4.

3. Dawson-Hughes B, Stern D, Goldman J, Reichlin S. Regulation of growth hormone and somatomedin-C secretion in postmenopausal women: effect of physiological estrogen replacement. J Clin Endocrinol Metab 1986;63:424–32.
4. Lieberman SA, Mitchell AM, Hintz R, Marcus R, Hoffman AR. The insulin-like growth factor-I generation test: resistance to growth hormone during aging and estrogen replacement therapy. Horm Metab Res 1994;26:229–33.
5. Mercuri N, Petraglia F, Genazzani AD, Amato F, Sgherzi MR, Maietta-Latessa A, et al. Hormonal treatments modulate pulsatile plasma growth hormone, gonadotrophin and osteocalcin levels in postmenopausal women. Maturitas 1993;17:51–62.
6. Weissberger AJ, Ho KKY, Lazarus L. Contrasting effects of oral and transdermal routes of estrogen replacement therapy on 24-hour growth hormone (GH) secretion, insulin-like growth factor I, and GH-binding protein in postmenopausal women. J Clin Endocrinol Metab 1991;72:374–81.
7. Slowinska-Srzednicka J, Zgliczynski S, Chotkowska E, Srzednicki M, Stopinska-Gluszak U, Jeske W, et al. Effects of transdermal 17-beta-oestradiol combined with oral progestogen on lipids and lipoproteins in hypercholesterolaemic postmenopausal women. J Intern Med 1993;234:447–51.
8. Bellantoni MF, Harman SM, Cho DE, Blackman MR. Effects of progestin-opposed transdermal estrogen administration on growth hormone and insulin-like growth factor-I in postmenopausal women of different ages. J Clin Endocrinol Metab 1991;72:172–8.
9. Bellantoni MF, Vittone J, Campfield AT, Bass KM, Harman SM, Blackman MR. Effects of oral versus transdermal estrogen on the growth hormone/insulin-like growth factor I axis in younger and older postmenopausal women: a clinical research center study. J Clin Endocrinol Metab 1996;81:2848–53.
10. Friend KE, Hartman ML, Pezzoli SS, Clasey JL, Thorner MO. Both oral and transdermal estrogen increase growth hormone release in postmenopausal women—a clinical research center study. J Clin Endocrinol Metab 1996;81:2250–6.
11. Dell'Aglio E, Valenti G, Hoffman AR, Zuccarelli A, Passari G, Ceda GP. Lack of effect of transdermal estrogen on the growth hormone-insulin-like growth factor axis. Horm Metab Res 1994;26:211–2.
12. Ma ZQ, Santagati S, Patrone C, Pollio G, Vegeto E, Maggi A. Insulin-like growth factors activate estrogen receptor to control growth and differentiation of the human neuroblastoma cell line SK-ER3. Mol Endocrinol 1994;8:910–8.

20

Influence of Gender on Response to Growth Hormone Substitution Therapy in Adults with Growth Hormone Deficiency

FERDINAND ROELFSEMA AND YVONNE JOHANNA HENRICA JANSSEN

Growth hormone (GH) plays a crucial role in postnatal growth. Disorders in the molecular processing of GH, its secretion, or its peripheral tissue effects result in short stature. GH replacement therapy has been used for more than three decades in children. Initially, however, only moderate success was observed in terms of ultimate height, mainly because of the limited supply of GH. With the unlimited availability of recombinant human GH (rhGH), treatment became more efficient, with most GH-deficient children reaching their target height. Pediatric endocrinologists targeted GH therapy on final height and stopped treatment when this goal was reached. However, clinical observations revealed that GH deficiency (GHD) in adults was associated with nonspecific signs and symptoms, including increased fat mass, reduced lean body mass and muscle strength, decreased extracellular water, decreased bone density, poor physical performance, and impaired sense of well-being. The first recorded GH substitution therapy in the adult human was reported by Raben (1). About 25 years later, the first results in groups of adults with GHD were published (2,3). Both studies provided strong evidence that GH treatment in adults with GHD has beneficial effects on body composition, quality of life, and bone mass.

A large number of studies on GH therapy in adults with GHD have since been published. Large interindividual responses to GH treatment have been found, leading to the recommendation that the dose of rhGH therapy should be individualized (4). In this chapter we discuss differences in response to rhGH between males and females with versus without estrogen replacement therapy.

Physiological GH Secretion in Adults

Multiple studies addressing physiological GH secretion have been performed in pubertal children, young adults, middle-aged subjects, and elderly subjects (5–8). GH secretion is maximal at the time of the growth spurt, subsequently decreasing by the end of growth. In the third decade a substantial amount of GH is still produced, but production falls considerably thereafter, although it is still detectable even in the elderly. Mean 24-h GH production is about twice as high in (middle-aged) females compared to males, and the female GH secretion pattern is less regular (5,6,9).

Available studies support the hypothesis that estrogen alone, or testosterone after aromatization to estrogen, can influence the GH–IGF axis in men and women. Oral administration of estrogen decreases circulating IGF-I concentration and increases mean 24-h GH concentration in postmenopausal women, whereas low doses of transdermal estrogen do not influence GH–IGF axis secretion (10–12). In a crossover study, O'Sullivan and coworkers found that oral estrogens increased fat mass and decreased lean mass, changes that were not observed during transdermal substitution (13). Lipid oxidation was also lower and carbohydrate oxidation higher during oral compared to transdermal estrogen replacement (13). These observations strongly point to route-dependent effects of estrogen on the GH–IGF-I axis and on substrate oxidation. In addition, recent results also suggest that (oral) progestagens can influence serum IGF-I concentration (14). Given concomitant differences in physiological GH secretion, it is thus likely that factors such as age, gender, and estrogen therapy may influence the daily amount of rhGH required to reverse GHD in adults.

Causes and Diagnosis of GHD in Adults

In adults, GHD is caused most frequently by (treatment of) pituitary adenomas. Other less common causes of GHD include epithelial tumors, craniopharyngioma, trauma, late effects of cranial radiation, and Sheehan syndrome. Most clinicians agree that the insulin tolerance test (ITT), which is a safe procedure in the hands of experienced medical staff, is the test of choice for establishing the diagnosis of GHD. It is generally agreed that a minimum serum GH value of 5.0 µg/l (13 mU/l) should be reached to exclude GHD (15).

Circulating IGF-I concentration cannot be used as a diagnostic tool for GH deficiency because many patients with a defective GH response to ITT have normal age- and sex-adjusted circulating IGF-I concentrations (16,17). This clinical observation introduces, however, a treatment paradox, because in the absence of other convenient treatment parameters, normalization of circulating IGF-I concentration is currently used as a target for treating GHD in most adults.

GH Substitution in Adults

In the first studies of rhGH therapy in adults, GH doses were based on earlier experience in children (0.07 IU, kg^{-1} day^{-1}; ~3 IU m^{-2} day^{-1}; ~5 IU/day). A high incidence of side effects (mainly related to fluid retention) and supranormal serum IGF-I levels was reported, suggesting that the dose used may have been too large in the adult patients with GHD. Reduced doses were used in subsequent studies (0.035 IU/kg^{-1} day^{-1}; ~1.5 IU m^{-2}day^{-1}; ~2.5 IU/day), resulting in smaller increases in serum levels of IGF-I (although supranormal levels were still reported), and a decrease in the incidence of side effects (18).

The difference in physiological GH secretion between men and women suggests that lower doses of rhGH are probably needed by men than women with GHD. Two studies indeed reported a greater response in males than in females with GHD during GH therapy (19,20). These studies used target doses based on body weight or body surface however, such that the target dose in male patients (generally heavier and with larger body surface) was higher than that in females, which could have introduced a bias. In the present study, GH doses were based on GH production rates in healthy middle-aged subjects. Our group reported daily GH production of about 200µg/day and 100 µg/day in females and males, respectively. Assuming an availability of sc-administered rhGH of 60%, a dose between 0.6 and 1.8 IU/day should be more nearly in agreement with physiological GH production in adults than the doses previously advocated. Moreover, the dose used in the present study was based on neither body weight nor body surface area.

In males and females with GHD, a mean GH dose of 0.6 and 1.2 IU/day, respectively, normalized serum IGF-I concentrations after short-term treatment (12 weeks) (17). After long-term treatment with individualized doses of rhGH for 2 years, we found that males and estrogen-deplete (ED) women needed lower doses of rhGH to normalize serum IGF-I concentrations than estrogen-replete (ER) women. Based on these findings, we have now analyzed short-term effects of GH therapy on selected biochemical parameters in men, ED women, and ER women.

Patients and Design of the Study

Sixty Caucasian patients (30 men and 30 women) with severe GHD, as evidenced by a peak serum GH response of less than 7 mU/l during ITT, were studied. None of the patients had a BMI greater than 32 kg/m^2. Patients were randomized to three groups to receive one of three treatment regimens: 0.6 IU/day (~0.2 mg/day) for 24 weeks; 0.6 IU/day for 4 weeks, followed by 1.2 IU/day (~0.4 mg/day) for 20 weeks; and 0.6 IU/day for 4 weeks, followed by 1.2 IU/day for a further 4 weeks and 1.8 IU/day for the rest of the study. Doses were thus independent of gender and body weight. For the present analysis we excluded 9 patients with childhood-onset GHD. Male and female patients were comparable in terms of mean age, causes of pituitary failure, history of irradiation, estimated number of years of GH deficiency, and other pituitary deficiencies. All women with GHD were also estrogen deficient, and 19 of them consequently received estrogen substitution therapy (3 with transdermal and 16 with oral estrogen therapy). Mean age was 61 years (range, 50–68 years) in ED and 40 years (range, 26–58 years) in ER women.

Results

Plasma IGF-I and IGF-Binding Protein (IGF-BP-3) Concentrations

The time course of plasma IGF-I concentrations is shown in Figure 20.1. Before treatment, the mean IGF-I concentration was significantly higher in men than in either ED or ER women ($p = 0.005$). After 24 weeks of rhGH therapy, serum IGF-I concentration increased in all three groups, but significantly more so in men than in women ($p = 0.007$). No difference in response was noted, however, between men and ED women ($p = 0.903$). The difference in IGF-I response between ER and ED women was significant ($p = 0.039$), although no dose–response relationship could be demonstrated ($p = 0.098$). The increase in IGF-BP-3 concentration is also depicted. In contrast to the IGF-I data, neither an influence of gender and estrogen status on response, nor a dose–response effect, was demonstrable.

Plasma Thyroxine and Triodothyronine Concentrations

During GH therapy, the serum thyroxine concentration decreased, while that of triodothyronine increased. All patients, except for two, were receiving thyroxine replacement therapy. There was no influence of gender or estrogen status on response to therapy, but a significant dose effect was present for thyroxine ($p = 0.033$). No change in serum thyroxine concentration was seen in the 0.6 IU/day dose group, and the largest decrease was found in the higher-dose group (i.e., 1.8 IU/day).

Plasma Lipids

There was no gender difference or therapy influence on serum triglyceride concentrations. Total plasma cholesterol decreased significantly during GH treatment ($p = 0.029$), but no gender difference or dose effect could be shown. HDL cholesterol was lowest in men and highest in ED women ($p = 0.035$). During therapy, the HDL concentration increased only in men ($p = 0.001$). LDL cholesterol decreased during therapy ($p = 0.007$), especially in men ($p < 0.005$).

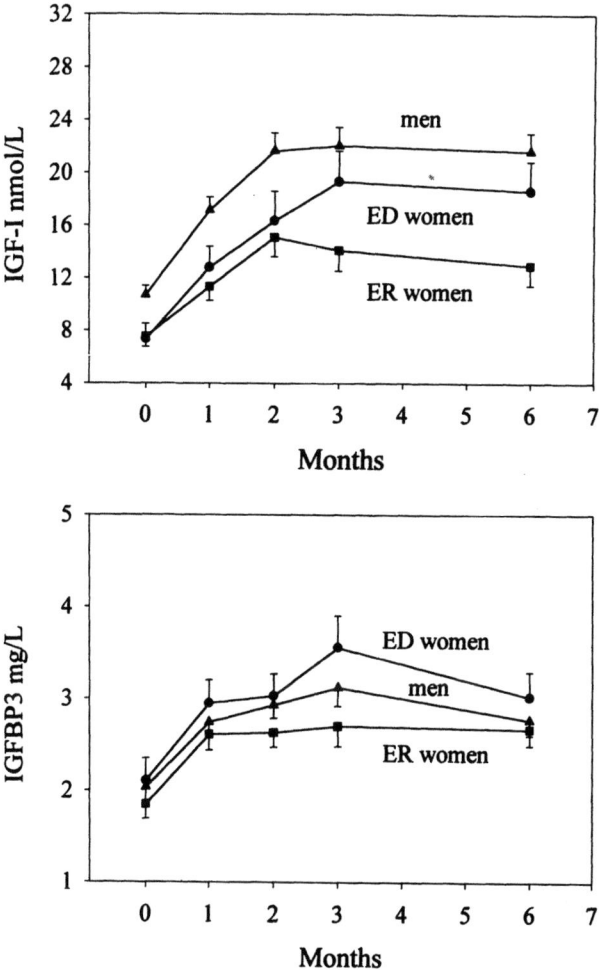

FIGURE 20.1. Serum concentrations of IGF-I (*upper panel*) and IGFBP-3 (*lower panel*) in GHD patients during substitution therapy with rhGH. Data are shown as mean and SEM. ED, estrogen-deplete state; ER, estrogen replete condition.

Hematology

Baseline hemoglobin, hematocrit, and erythrocyte counts were significantly higher in men than women with GHD. No changes were found during 6 months of GH therapy. Baseline numbers of leucocytes and thrombocytes were not significantly different among the three groups and did not change during rhGH treatment.

Glucose Homeostasis

Fasting glucose concentrations and HbA1c values were not different between sexes or among groups and remained unchanged during therapy.

Serum Proteins

The total serum protein concentration was lowest in ER women and highest in men ($p = 0.015$), and the concentration remained unchanged during GH substitution. Similar results were obtained for plasma albumin concentration. The sex hormone-binding globulin (SHBG) concentration was, as expected, lowest in men and highest in ER women. No significant changes in circulating levels were found during GH therapy. Cortisol-binding globulin (CBG) concentration also differed between sexes, being highest in ER women and lowest in men ($p < 0.005$). During therapy, CBG concentration decreased in both sexes ($p = 0.001$), independently of gender, substitution with estrogens, or GH dose.

Electrolytes and Kidney Function

Plasma potassium concentration was lowest in ED women and highest in men before treatment ($p < 0.005$). During substitution, plasma levels increased independently of gender and estrogen status, but were weakly dose dependent ($p = 0.047$). Plasma sodium concentration increased in all groups independent of the dose used ($p = 0.006$). Serum creatinine concentration was highest in men and lowest in ER women ($p = 0.001$). During therapy, creatinine concentration decreased in all patient groups, without a dose effect. Comparable results were obtained by estimating the 24-h urinary creatinine clearance, with the three groups of patients showing an increase of about 10%.

Biochemical Parameters of Bone Remodeling

Serum calcium concentration was highest at the start of the study in ER women and lowest in ED women ($p = 0.001$). During therapy, the calcium concentration increased in all groups ($p = 0.001$), with a tendency for dose-dependency ($p = 0.064$). As expected, fasting serum phosphate concentration increased in all three groups during therapy, with a significant dose–response effect ($p = 0.025$). Serum alkaline phosphatase activity was highest in ED women (Fig. 20.2), and lowest in men ($p = 0.016$). During treatment, an increase in alkaline phosphatase activity occurred only in ED women and male patients, but not in ER women. Comparable results were evident for serum osteocalcin concentration, although a significant dose effect was also present ($p = 0.002$). Urinary calcium excretion increased in all groups without gender or estrogen effect. Urinary hydroxyproline excretion was highest in ED women, and increased dose-dependently in all three groups during therapy, with a significant

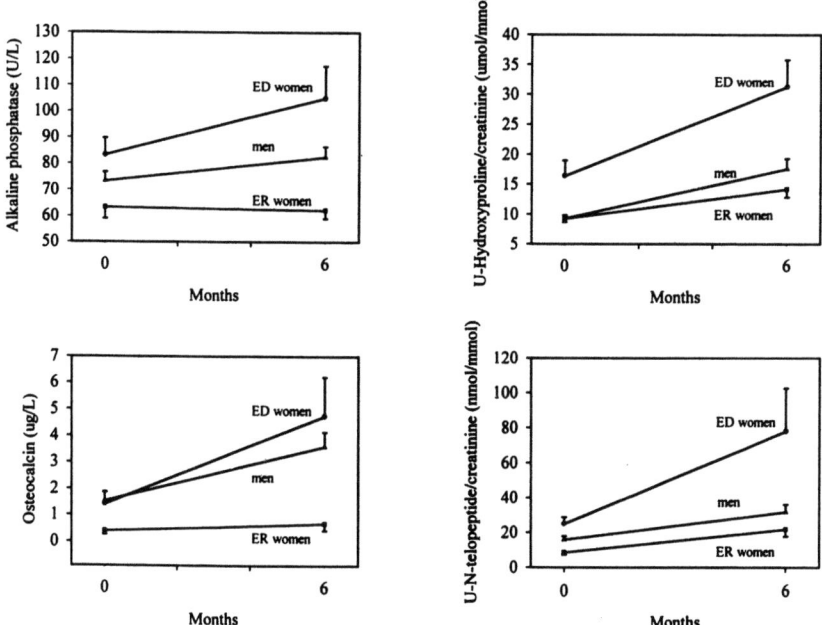

FIGURE 20.2. Serum parameters of bone turnover in GHD patients before and during GH substitution therapy.

dose effect ($p = 0.000$) and a smaller estrogen effect ($p = 0.017$), mainly the result of the large increase observed in ED women. Comparable differences and changes were found for the urinary excretion of collagen N-telopeptide (21).

Discussion

Several studies have reported a gender difference in responsiveness to GH substitution therapy in adult GHD (19,20). For example, larger increases in serum IGF-I concentration in men compared to women are reported. However, these results could have been biased by the target doses used in these studies, which were based on body weight or body surface area. In the present study, the gender difference in serum IGF-I response was confirmed by using both lower doses and doses of rhGH independent of weight. The present study also indicates that the gender difference results mainly from differences between ER women and men, as no significant differences were found between ED women and men and a larger response was found in ED women compared to ER women.

Baseline IGF-I concentrations were significantly higher in men compared to women with GHD, which is in agreement with the results reported by others (20,22). There was no difference in the maximum GH concentration during ITT, which also suggests that male patients are more sensitive to GH than females. This is consistent with the observation of higher GH production in healthy females compared with age- and BMI-matched males with similar serum IGF-I concentrations (6). No difference was found in baseline serum IGF-I between ER and ED women, which is in agreement with the results of Burman et al (20). In contrast, we observed a greater IGF-I response in ED than ER women with GH therapy and a similar response between ED women and men. Similar results were obtained in the age-adjusted IGF-I SD scores, thereby eliminating the potential bias of age in these patient groups.

Excessive bone loss caused by estrogen deficiency is an important risk factor for osteoporosis in postmenopausal women (23). As expected in the postmenopausal state in patients with no pituitary pathology (24) and in agreement with the results of Holloway et al. (25), older women with GHD and estrogen deficiency had a significantly greater bone turnover than their counterparts who were estrogen replete, both at baseline and during rhGH therapy. No differences were found in changes in the biochemical parameters of bone turnover in response to rhGH therapy between males and ED females. Increases during therapy were, however, less in ER females compared to ED females and males, suggesting that one potential influence of gender is through the relative contribution of estrogen repletion versus depletion.

Although men and ED women appear to respond more to rhGH treatment than do ER women, large intragroup differences are still present. We recently reported a large variability in the availability of sc-injected GH, which could explain some of the differences. Mean bioavailability of sc-injected GH in women was 63%, with a wide range of 40% to 80%. Other factors influencing responsiveness to treatment could be baseline age, plasma GHBP concentration, or BMI of the patient. In the study by Johannsson (19), the greatest responses were reported in the younger patients with low GHBP levels, the latter being positively correlated with body fat. Side effects are more likely to occur in older patients and in patients with larger weight and BMI, although these results could be biased by using weight-dependent doses (26).

In conclusion, the results of the present study indicate that changes in serum IGF-I during rhGH therapy of GHD adults are different in ER women compared to both ED women and men. Baseline differences and unequal alterations in biochemical parameters of bone turnover were also apparent between groups during rhGH therapy.

References

1. Raben MS. Clinical use of human growth hormone. N Engl J Med 1962;266:82–6.
2. Jørgensen JO, Pedersen SA, Thuesen L, et al. Beneficial effects of growth hormone treatment in GH-deficient adults. Lancet 1989;1:1221–5.

3. Salomon F, Cuneo RC, Hesp R, Sönksen PH. The effects of treatment with recombinant human growth hormone on body composition and metabolism in adults with growth hormone deficiency. N Engl J Med 1989;321:1797–803.
4. Johannsson G, Rosen T, Bengtsson BA. Individualized dose titration of growth hormone (GH) during GH replacement in hypopituitary adults. Clin Endocrinol (Oxf) 1997;47:571–81.
5. Ho KY, Evans WS, Blizzard RM, et al. Effects of sex and age on the 24-hour profile of growth hormone secretion in man: importance of endogenous estradiol concentrations. J Clin Endocrinol Metab 1987;64:51–8.
6. Van den Berg G, Veldhuis JD, Frölich M, Roelfsema F. An amplitude-specific divergence in the pulsatile mode of GH secretion underlies the gender difference in mean GH concentrations in men and premenopausal women. J Clin Endocrinol Metab 1996;81:2460–7.
7. Veldhuis JD, Liem AY, South S, et al. Differential impact of age, sex steroid hormones, and obesity on basal versus pulsatile growth hormone secretion in men as assessed in an ultrasensitive chemiluminescence assay. J Clin Endocrinol Metab 1995;80:3209–22.
8. Iranmanesh A, South S, Liem AY, et al. Unequal impact of age, percentage body fat, and serum testosterone concentrations on the somatotrophic, IGF-I, and IGF-binding protein responses to a three-day intravenous growth hormone-releasing hormone pulsatile infusion in men. Eur J Endocrinol 1998;139:59–71.
9. Pincus SM, Gevers EF, Robinson IC, et al. Females secrete growth hormone with more process irregularity than males in both humans and rats. Am J Physiol 1996;270:E107–15.
10. Weissberger AJ, Ho KK, Lazarus L. Contrasting effects of oral and transdermal routes of estrogen replacement therapy on 24-hour growth hormone (GH) secretion, insulin-like growth factor I, and GH-binding protein in postmenopausal women. J Clin Endocrinol Metab 1991;72:374–81.
11. Dawson-Hughes B, Stern D, Goldman J, Reichlin S. Regulation of growth hormone and somatomedin-C secretion in postmenopausal women: effect of physiological estrogen replacement. J Clin Endocrinol Metab 1986;63:424–32.
12. Bellantoni MF, Vittone J, Campfield AT, Bass KM, Harman SM, Blackman MR. Effects of oral versus transdermal estrogen on the growth hormone/insulin-like growth factor I axis in younger and older postmenopausal women: a clinical research center study. J Clin Endocrinol Metab 1996;81:2848–53.
13. O'Sullivan AJ, Crampton LJ, Freund J, Ho KY. The route of estrogen replacement therapy confers divergent effects on substrate oxidation and body composition in postmenopausal women. J Clin Invest 1998;102:1035–40.
14. Ho KKY, Crampton L, Nguyen T, Reutens AT, Leung KC. Progesterone modulates the effects of estrogen on IGFI in post-menopausal women (abstract). Growth Horm IGF Res 1998;8:316.
15. Consensus guidelines for the diagnosis and treatment of adults with growth hormone deficiency: summary statement of the Growth Hormone Research Society Workshop on Adult Growth Hormone Deficiency. J Clin Endocrinol Metab 1998;83:379–81.
16. Hoffman DM, O'Sullivan AJ, Baxter RC, Ho KK. Diagnosis of growth-hormone deficiency in adults. Lancet 1994;343:1064–8.
17. Janssen YJH, Frölich M, Roelfsema F. A low starting dose of genotropin in growth hormone-deficient adults. J Clin Endocrinol Metab 1997;82:129–35.

18. Mårdh G, Lindeberg A (on behalf of the investigators). Growth hormone replacement therapy in adult hypopituitary patients with growth hormone deficiency: combined clinical safety data from clinical trials in 665 patients. Endocrinol Metab 1995;2:11–6.
19. Johannsson G, Bjarnason R, Bramnert M, et al. The individual responsiveness to growth hormone (GH) treatment in GH-deficient adults is dependent on the level of GH-binding protein, body mass index, age, and gender. J Clin Endocrinol Metab 1996;81:1575–81.
20. Burman P, Johansson AG, Siegbahn A, Vessby B, Karlsson FA. Growth hormone (GH)-deficient men are more responsive to GH replacement therapy than women. J Clin Endocrinol Metab 1997;82:550–5.
21. Janssen YJH, Hamdy NA, Frolich M, Roelfsema F. Skeletal effects of two years of treatment with low physiological doses of recombinant human growth hormone (GH) in patients with adult-onset GH deficiency. J Clin Endocrinol Metab 1998;83:2143–8.
22. Attanasio AF, Lamberts SW, Matranga AM, et al. Adult growth hormone (GH)-deficient patients demonstrate heterogeneity between childhood onset and adult onset before and during human GH treatment. Adult Growth Hormone Deficiency Study Group. J Clin Endocrinol Metab 1997;82:82–8.
23. Richelson LS, Wahner HW, Melton LJ III, Riggs BL. Relative contributions of aging and estrogen deficiency to postmenopausal bone loss. N Engl J Med 1984;311:1273–5.
24. Steiniche T, Hasling C, Charles P, Eriksen EF, Mosekilde L, Melsen F. A randomized study on the effects of estrogen/gestagen or high dose oral calcium on trabecular bone remodeling in postmenopausal osteoporosis. Bone (NY) 1989;10:313–20.
25. Holloway L, Butterfield G, Hintz RL, Gesundheit N, Marcus R. Effects of recombinant human growth hormone on metabolic indices, body composition, and bone turnover in healthy elderly women. J Clin Endocrinol Metab 1994;79:470–9.
26. Holmes SJ, Shalet SM. Which adults develop side-effects of growth hormone replacement? Clin Endocrinol (Oxf) 1995;43:143–9.

21

How Sex Steroids Modulate GH Action on Target Tissues: Long-Term Follow-up in GH-Deficient Adults

GUDMUNDUR JOHANNSSON AND BENGT-ÅKE BENGTSSON

In GH-deficient children, the importance of puberty and gonadal steroids on the growth plate during growth hormone (GH) treatment has long been known. Knowledge about the impact and possible modulating effects of gonadal steroids on various aspects of GH deficiency and GH treatment in adults is, however, scarce.

Well-controlled clinical studies have shown that GH deficiency in adults is associated with multiple cardiovascular risk factors (1), which may help explain the doubling in cardiovascular mortality observed in these patients (2,3). In addition, quality of life is impaired and the bone mass and density are reduced (4). Whether gender, androgens, or estrogens may have attenuating effects on these aberrations has been addressed in only a few studies. The purpose of this review is to summarize the limited data available on the possible modulating effects of gonadal steroids and gender on GH treatment of adults with GH deficiency.

Serum IGF-I

Based on an IGF-I generation test, age and oral estrogen therapy were found to be important short-term determinants of the serum IGF-I response to a single dose of GH (5). The serum IGF-I response was more marked in young men than in older men, and postmenopausal women with oral estrogen therapy responded less than postmenopausal women not receiving oral estrogen. In a 9-month trial with 21 men and 13 estrogen-replaced women, with similar age, duration of hypopituitarism, and severity of GH deficiency, the serum IGF-I level was almost twofold higher in the men than the women after 9 months of GH treatment with a similar dose of GH per unit body surface area (6). Oral estrogen therapy therefore influences the serum IGF-I response to GH administration.

As the starting dose of GH has been based mostly on body weight or body surface area, women have received an initially lower total daily dose of GH than men. However, after more prolonged treatment periods, after the adjustment of doses because of side effects and high serum IGF-I concentrations, the daily (total) dose of GH has been similar in women and men. Despite this, the increment in and the final level of serum IGF-I were lower in women than men (7–9), further demonstrating the influence of sex and estrogen on IGF-I responses during GH replacement in adults.

Body Composition

In terms of body composition, GH-deficient adults have large gender differences at baseline, similar to those seen in healthy adults (7,9), thereby demonstrating the importance of gonadal steroids and gender for body composition. For example, the mean alternations in total body fat and extracellular water compared with healthy controls are similar in GH-deficient men and women (10), indicating that hypopituitarism and untreated GH deficiency are important to both men and women in altering normal body composition.

The ability of estrogen to antagonize the nitrogen-retaining effects of GH (11) was an early observation that gonadal steroids could influence or modulate the action of GH. An indication that gonadal steroids could influence the impact of GH on body composition came from a study in acromegalic patients before and after transphenoidal adenomectomy (12). As determined by computed tomography, total subcutaneous adipose tissue increased by 53% ($p < 0.01$) in the men and 20% ($p < 0.05$) in the women 1 year after treatment. In addition, visceral adipose tissue increased by 84% ($p < 0.01$) in the men but only by 25% (NS) in the women, and adipose tissue distribution changed significantly only in the men. These results indicate that men may be more responsive to the lipolytic action of GH than women.

The response to GH treatment in GH-deficient adults has recently been shown to be highly variable among individuals. This large variability was explained to some extent by differences in baseline body mass index (BMI), serum levels of GH-binding protein (GHBP), age, and gender (7). Large baseline gender differences in body composition were accounted for in these comparisons. Other studies have shown that the increase in fat-free mass, total body water, and the decrease in body fat are more marked in men than in women, as measured using bioelectrical impedance analysis (BIA) (7,13). Moreover, in response to 9 months of GH treatment, the reduction in percentage body fat (Fig. 21.1) and abdominal fat mass was more marked in men than in women (6). In human adipose tissue, testosterone inhibits lipid-accumulating pathways and stimulates lipid mobilization. These androgenic effects are more powerful in the presence of GH (14). A more marked lipolytic action

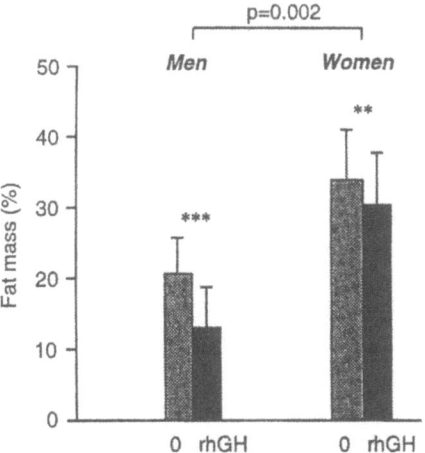

FIGURE 21.1. The percentage of total body fat in 21 men and 15 women with GH deficiency before and after 9 months of treatment with GH. Estimates were made using DEXA. Results are shown as the mean ± SD. **, $p \leq 0.01$; ***, $p \leq 0.001$, compared with pretreatment within-gender values. The difference in response to GH treatment between men and women was significant at $p = 0.002$. Reproduced with permission from Burman et al. (6).

might therefore be expected in men than in women in response to GH treatment.

The explanation for the more marked anabolic response in men than in women, as measured using total body nitrogen analysis from in vivo neutron activation and total body potassium by using a whole-body counter (9), is not so obvious, but could be an effect of the lower increment and the lower absolute level of serum IGF-I concentration attained in response to GH treatment in women compared with men. The differential response to GH in GH-deficient men and women is therefore probably explained by different interactions between GH and estrogens and androgens, respectively, in controlling body composition.

Bone

Gonadal steroids and GH exert important effects on both the growth plate and mature bone. The interaction between sex steroids and GH on linear growth is well demonstrated by the importance of both gonadal steroids and GH to the pubertal growth spurt. Postmenopausal bone loss and osteoporosis in women are clear examples of the importance of estrogen for bone mass. Analogously androgen deficiency is known to be an important risk factor for osteoporosis and fractures in men (15). Estrogen and testosterone therapy in both these conditions can increase bone density (16,17). The relevance of both GH and gonadal steroids to retaining bone mass and bone mineral density is also evident in previous studies in hypopituitary adults. Rosén et al. (18) demonstrated that both men and women, with and without gonadal steroid replacement therapy, had low bone mineral content. The importance of estrogen was inferred by the finding that female patients with estrogen replacement therapy had higher bone mineral content than those not receiving such treatment. Beshyah et al. (19) demonstrated that bone mineral density in the total body and the lumbar spine was reduced, compared with healthy control subjects, in both men and women with GH deficiency.

Gender differences in bone metabolism during GH treatment were studied by comparing 24 men and 20 women during 2 years of GH treatment (8). Men and women had similar baseline concentrations of serum markers reflecting bone formation and bone resorption (osteocalcin, PICP, and ICTP). The initial increment in these markers in response to GH treatment was more pronounced in men than in women. However, after 2 years of treatment, the serum concentrations of the bone markers were similar in men and women (8). A similar gender difference in bone markers in initial response to GH treatment has been reported during 9 months of GH treatment (6). The 2-year study demonstrated a more marked overall increment in total body bone mineral density in women, while the bone mineral density in lumbar spine and hip responded similarly to GH treatment in men and women (8). These results may suggest that the interaction between GH and estrogen induces a more positive remodeling balance with less increment in the bone-remodeling activity than the interaction between GH and androgen.

Lipoproteins

Androgens and estrogen have nearly opposite effects on very low density lipoprotein (VLDL) metabolism. Estrogen increases the hepatic production of triglyceride-rich VLDL and its plasma levels while androgens appears to lower VLDL by increasing the rate of its catabolism. GH affects both production and secretion of lipoproteins from the liver and their clearance from the circulation (2). Interactions between gonadal steroids and GH are therefore likely to occur, particularly in terms of VLDL metabolism.

Gender differences in the responses of lipoproteins to GH treatment have emerged in two previous trials, one lasting 9 months (6) and the other 12 months (13). The 12-month study showed that men increased their lipoprotein a (Lp)(a) concentration significantly more than women (Fig. 21.2). It has been suggested that both androgens and estrogens decrease serum Lp(a) levels. However, it is conceivable that estrogens are more potent in lowering Lp(a) serum concentrations than androgens (21), thereby explaining this gender difference in response to GH treatment. The same study showed that women had unchanged serum HDL cholesterol concentration and increased serum

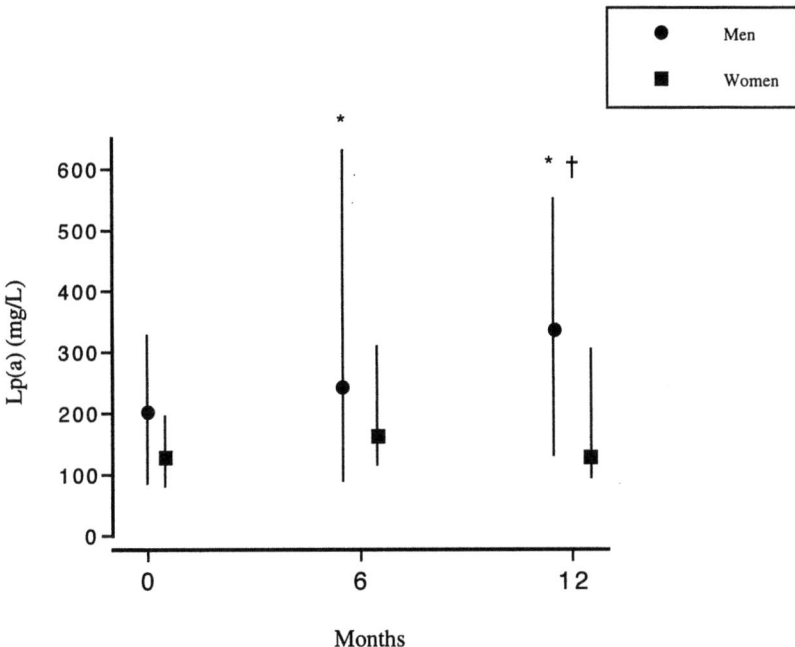

FIGURE 21.2. Median and 25th and 75th percentiles for serum lipoprotein (Lp)(a) concentrations during 12 months of GH administration in 30 men and 14 women with GH deficiency. *, $p < 0.05$ compared with baseline values for men; †, $p < 0.05$ compared with the percentage response in women. Adapted from Johannsson et al. (13).

triglyceride concentration after 12 months of GH treatment. In men, HDL cholesterol concentration increased and triglyceride concentration did not change. In both trials, the apoB concentration decreased in men but not in women, and in the 9-month study total cholesterol and LDL cholesterol concentrations also decreased only in the men. It should be stressed that in neither study were these changes significant in a group comparison, except for Lp(a). The suggested gender differences in triglycerides and HDL cholesterol might be explained by replacement gonadal steroids and their effects on VLDL metabolism, while the differential effects on total, LDL cholesterol, and apoB are more difficult to explain.

GH Doses

Trials employing doses of GH based on body weight or surface area have reported a high frequency of side effects, mainly associated with water retention (22–24). This dose regimen will yield higher daily doses of GH for men than women. This result is not in accordance with our knowledge of physiological GH secretion in adults, which is usually higher in women (25,26). The unphysiological approach of determining GH dosage according to body weight is further highlighted by the finding that dose reductions made in response to side effects and high serum concentration of IGF-I are less marked in women than men (8). By using an individualized dose schedule for GH treatment based on clinical response and normalization of serum IGF-I and body composition, it was found that women were given more GH per kilogram of body weight than men (27). Moreover, to achieve normalization of the serum IGF-I level in patients with severe GH deficiency (mean age, 47 years), a daily dose of 0.2 mg per day was sufficient in men whereas 0.4–0.6 mg per day was required in women (28). These results agree with our knowledge about individual responsiveness to GH (7) and physiological GH secretion in healthy men and women (25,26).

Conclusion

Both GH-deficient men and women have clinically significant aberrations in mortality, body composition, and metabolic indices, indicating that estrogen or androgen replacement alone does not have fully protective effects on the consequences of GH deficiency in adults. The few studies that have addressed the question of whether gonadal steroids influence the effects of GH treatment in GH-deficient adults have revealed significant gender differences in responsiveness to GH therapy in terms of serum IGF-I concentrations, body compositional changes, and bone metabolism. Important gender differences in the treatment response to GH in lipoprotein metabolism are also apparent in previous studies. In contrast, the improvement in quality of life and psychological well-being seems to be of similar magnitude in men and women (29).

The choice of GH dosage based on body weight and large baseline gender differences are important confounders when evaluating gender differences in response to GH therapy. These factors should be considered in future studies evaluating the influence of gender and gonadal steroids on GH action in adults. Further knowledge of gender differences in responses to GH treatment should help us to optimize hormone replacement therapy in hypopituitary adults.

References

1. Bengtsson B-Å, Rosén T. Johansson J-O, Johannsson G, Oscarsson J, Landin-Wilhelmsen K. Cardiovascular risk factors in adults with growth hormone deficiency. Endocrinol Metab 1995;2(suppl B):29–35.
2. Rosén T, Bengtsson B-Å. Premature mortality due to cardiovascular diseases in hypopituitarism. Lancet 1990;336:285–8.
3. Bülow B, Hagmar L, Mikozy Z, Nordström C-H, Erfurth EM. Increased cerebrovascular mortality in patients with hypopituitarism. Clin Endocrinol 1996;46:75–81.

4. De Boer H, Blok G-J, Van der Veen EA. Clinical aspects of growth hormone deficiency in adults. Endocr Rev 1995;16:63–86.
5. Lieberman SA, Holloway L, Marcus R, Hoffman AR. Interactions of growth hormone and parathyroid hormone in renal phosphate, calcium, and calcitriol metabolism and bone remodeling in postmenopausal women. J Bone Miner Res 1994;9:1723–8.
6. Burman P, Johansson AG, Siegbahn A, Vessby B, Karlsson FA, Growth hormone (GH)-deficient men are more responsive to GH replacement therapy than women. J Clin Endocrinol Metab 1997;82:550–5.
7. Johannsson G, Bjarnason R, Bramnert M, et al. The individual responsiveness to growth hormone (GH) treatment in GH-deficient adults is dependent on the level of GH binding protein, body mass index, age and gender. J Clin Endocrinol Metab 1996;81:1575–81.
8. Johannsson G, Rosén T, Bosaeus I, Sjöström L, Bengtsson B-Å. Two years of growth hormone (GH) treatment increases bone mineral content and density in hypopituitary patients with adult-onset GH deficiency. J Clin Endocrinol Metab 1996;81:2865–73.
9. Johannsson G, Grimby G, Stibrant Sunnerhagen K, Bengtsson B-Å. Two years of growth hormone (GH) treatment increases isometric and isokinetic muscle strength in GH-deficient adults. J Clin Endocrinol Metab 1997;82:2877–84.
10. Rosén T, Bosaeus I, Tölli J, Lindstedt G, Bengtsson B-Å. Increased body fat mass and decreased extracellular fluid volume in adults with growth hormone deficiency. Clin Endocrinol 1993;38:63–71.
11. Schwartz E, Wiedemann E, Simon S, Schiffer M. Estrogenic antagonism of metabolic effects of administered growth hormone. J Clin Endocrinol Metab 1969;26:1176–81.
12. Brummer R-JM, Lönn L, Kvist H, Grangård U, Bengtsson B-Å, Sjöström L. Adipose tissue and muscle volume determination by computed tomography in acromegaly, before and one year after adenomectomy. Eur J Clin Invest 1993;23:199–205.
13. Johannsson G, Oscarsson J, Roén T, et al. Effects of 1 year of growth hormone therapy on serum lipoprotein levels in growth hormone-deficient adults: influence of gender and apo(a) and apoE phenotypes. Arterioscler Thromb Vasc Biol 1995;15:2142–50.
14. Yang S, Xu X, Bjöorntorp P, Edén S. Additive effects of growth hormone and testosterone on lipolysis in adipocytes of hypophysectomized rats. J Endocrinol 1995;147:147–52.
15. Seeman E, Melton LK, O'Fallon WM, Riggs BL. Risk factos for spinal osteoporosis in men. Am J Med 1983;75:977–83.
16. Riggs BL, Melton LJ. The prevention and treatment of osteoporosis. N Engl J Med 1992;327:620–7.
17. Katznelson L, Finkelstein JS, Schoenfeld DA, Rosenthal DI, Anderson EJ, Klibanski A. Increase in bone density and lean body mass during testosterone administration in men with acquired hypogonadism. J Clin Endocrinol Metab 1996;81:4358–65.
18. Rosén T, Hansson T, Granhed H, Szücs J, Bengtsson B-Å. Reduced bone mineral content in adult patients with growth hormone deficiency. Acta Endocrinol 1993;129:201–8.
19. Beshyah SA, Freemantle C, Thomas E, et al. Abnormal body composition and reduced bone mass in growth hormone deficient hypopituitary adults. Clin Endocrinol 1995;42:179–89.

20. Angelin B, Rudling M. Growth hormone and hepatic lipoprotein metabolism. Curr Opin Lipidol 1994;5:160–5.
21. Henricksson P, Angelin B, Berglund L. Hormonal regulation of secrum Lp(a) levels. Opposite effects after estrogen treatment and orchidectomy in males with prostatic carcinoma. J Clin Invest 1992;89:1166–71.
22. Salomon F, Cuneo RC, Hesp R, Sönksen PH. The effects of treatment with recombinant human growth hormone on body composition and metabolism in adults with growth hormone deficiency. N Engl J Med 1989;321:1797–803.
23. Bengtsson B-Å, Edén S, Lönn L, et al. Treatment of adults with growth hormone (GH) deficiency with recombinant human GH. J Clin Endocrinol Metab 1993;76:309–17.
24. De Boer H, Blok GJ, Popp-Snijders C, Stuurman L, Baxter RC, Van der Veen E. Monitoring of growth hormone replacement therapy in adults based on measurements of serum markers. J Clin Endocrinol Metab 1996;81:1371–7.
25. Ho KKY, Evans WS, Blizzard RM, et al. Effects of sex and age on the 24-hour profile of growth hormone secretion in man. Importance of endogenous estradiol concentrations. J Clin Endocrinol Metab 1987;64:51–8.
26. Van den Berg G, Veldhuis JD, Frölich M, Roelfsema F. An amplitude-specific divergence in the pulsatile mode of GH secretion underlies the gender difference in mean GH concentration in men and premenopausal women. J Clin Endocrinol Metab 1996;81:2460–7.
27. Johannsson G, Rosén T, Bengtsson B-Å. Individualized dose titration of growth hormone (GH) during GH replacement in hypopituitary adults. Clin Endocrinol 1997;47:571–81.
28. Janssen YJH, Frölich M, Roelfsema F. A low starting dose of genotropin in growth hormone-deficient adults. J Clin Endocrinol Metab 1997;82:129–35.
29. Wirén L, Bengtsson B-Å, Johannsson G. Beneficial effects of long-term GH replacement therapy on quality of life in adults with GH deficiency. Clin Endocrinol 1997;48:613–20.

22

Growth Hormone Treatment in Turner Syndrome: Rationale for Therapy

E. KIRK NEELY

The Efficacy of Growth Hormone Therapy in Turner Syndrome

Human growth hormone (hGH) is now an established therapy for the growth failure of primary gonadal dysgenesis, or Turner syndrome (TS), yet the magnitude of response expected from hGH is much debated. In clinical trials of hGH treatment, gains over initial projected heights have varied from 0 to 10 cm (1). Naturally, opponents of therapy cite the former figure, and proponents the latter. Clinical trials have varied widely in their use of estrogen replacement therapy, which is now known to be detrimental to final height attainment if initiated too early, concomitant androgen therapy, and the age at initiation of hGH treatment. Remarkably, the earliest mean reported starting age is 9 years (2), and that study has reported one of the best responses to hGH, namely a gain of 7–8 cm. Investigators are prone to extrapolate from the association of early start with better outcome to conclude that an earlier treatment will result in greater gains in final height, but little in the literature directly supports this assumption. Optimal age of initiation of hGH is perhaps a major unresolved question in medical management of TS in childhood.

One basis of uncertainty regarding the expected response to hGH is the final height of untreated patients, the "historical controls" inferred in most clinical trials. (No published hGH trials have utilized randomized untreated control groups after the first year.) Lyon et al. (3) constructed a standardized growth curve utilizing cross-sectional heights from untreated European girls with TS in the absence of hormonal therapy (4). This curve is widely used in clinical practice, and individual TS patients, with reasonable reliability, do follow a particular percentile throughout childhood and adolescence. Mean final height was 142.9 ± 7.3 cm in the Lyon data, a figure confirmed by a survey of European and U.S. studies between 1970 and 1990 (5). The accuracy of this historic mean adult height in TS is a focus of residual contention among investigators, as several TS series have reported higher mean final heights of 147–148 cm in the absence of growth hormone therapy (6–8). The difference between 143 and 148 cm is a major discrepancy, because the average for reported gains is only about 5–6 cm, with mean final heights in the range of 146–154 cm after hGH therapy in North American and European populations.

In my opinion, the preponderance of evidence supports an untreated mean adult height in the lower portion of this disputed range; accordingly, the reported height gains from hGH treatment are clinically substantial. Preliminary results of a large randomized controlled Canadian study, presented at the FDA hearings before approval of hGH for use in TS, confirmed a final height for controls close to the 143-cm figure, and a mean 6–7 cm difference in final heights between the treatment group and controls. In the largest uncontrolled hGH treatment population reported to date, 622 subjects had a mean pretreatment projected height of 141.9 cm (from the Lyon curve) and a mean final height of 148.3 cm, for an average gain of 6.4 cm when hGH was administered for 3.7 years starting at 12.9 years old (9). This is a reasonable current expectation of height response to hGH therapy.

Final height in TS patients is influenced by karyotype, ethnicity, parental height, and hormonal therapies. The effect of karyotype generally is minimal. Subjects with mosaicisms for common structural abnormalities of the second X chromosome, such as ring X and long arm isochromosome, are just as short as 45,X subjects. Those with 45,X/46,XX and 45,X/46,XY mosaicisms may be on average slightly taller. Effects of ethnicity and parental height can be summarized by the expectation that adults with TS are approximately 20 cm shorter than predicted from maternal or midparental height. Untreated Japanese TS adults are about 138 cm tall and Northern European TS adults are about 148 cm tall; thus, there are discrepancies in untreated final heights in different populations. The 50th percentile of untreated North American TS heights is 20 cm below the 50th percentile North American adult female height. The single best predictor of final height in TS is parental height.

Estrogens and androgens were widely tried as growth therapies in TS before the availability of recombinant hGH, and thereafter continued as adjunctive therapies. Both result in short-term growth stimulation. However, estrogens deliver no benefit in final height and are detrimental in many circumstances. In contrast, several studies support a long-term growth benefit from nonaromatizable androgens, either alone (10,11) or in combination with hGH (2,12).

Recently published final heights from the early Swedish and U.S. studies are instructive about the relative benefits of hormonal therapies. The Swedish study (12) began at a fairly late mean age of 12.2 years (range, 9–16) and had no hGH-alone arm. Final height after combined hGH and oxandrolone therapy was 154.2 cm, which is 8.5 cm greater than the original projected final height; addition of ethinyl estradiol (100 ng^{-1} kg d^{-1}) to this regimen resulted in a mean height of 151.1 cm, or only a 3.2-cm gain. Other studies have confirmed the reduction in final height resulting from initiation of estrogen, even low doses, before age 13 years (13,14; see also Chapter 23). The U.S. study (2) did have an hGH-alone group and a larger hGH plus oxandrolone group. Historical controls had a mean final height of 144.2 cm, matching their initial projected height. Final heights from hGH and hGH plus oxandrolone were 152.1 and 150.4 cm, respectively, which reflect gains over pretreatment projected height of 10.3 and 8.4 cm, respectively. Both studies had earlier reported much greater gains in projected adult height, which now can be properly attributed to a more rapid tempo of growth.

Effects of the various hormonal therapies can be summarized in the following manner. Although the height deficit in untreated TS is approximately 20 cm, hGH treatment alone can add up to 10 cm to final height, although the average from studies is about 5–6 cm. Height gain is greater if hGH precedes estrogen therapy by several years (i.e., hGH started by 9–10 years of age). Conversely, introduction of estrogen before 13 years of age may reduce final height by 3 cm or more. Oxandrolone increases height by about 3 cm, alone or in combination with hGH. Addition of oxandrolone to hGH allows for earlier termination of therapy and for lower costs of therapy, but it is currently unavailable in the United States. Therefore, from the reports to date, the optimal strategy for growth therapy in TS is use of combined hGH and oxandrolone beginning by 9 years of age, followed by estrogen at age 13. Nevertheless, only half the growth deficit appears to be overcome by such hormonal therapy. Additional possible statural benefit from earlier hGH is strictly conjectural, although some investigators are optimistic that more of the height deficit can be ameliorated by longer duration of therapy.

The Nature of Growth Failure in Turner Syndrome

After decades of clinical trials in TS, we know that estrogens fail to augment final height, androgens can marginally improve it, and hGH can make up almost half of the total growth deficit. What does this tell us about the etiology of short stature in Turner syndrome?

Manifestation of growth failure in TS begins with mild intrauterine growth retardation (IUGR). Retrospective evaluations of birth weight and length reveal values approximately 1 SD deviation below the mean. Clinically, this intrauterine growth deficit is usually overlooked unless the diagnosis is discerned prenatally or is made as a consequence of features such as neonatal lymphedema, neck webbing, or coarctation of the aorta. Despite this deficit, growth velocity in the first 2 or 3 years of life is relatively normal, then declines progressively during the remainder of childhood. Growth failure in TS becomes even more obvious in adolescence because of the absence of a pubertal growth spurt. In the absence of hormonal therapies, girls with TS continue to grow slowly during late adolescence, with epiphyseal fusion occurring as late as 18–20 years because of the estrogen deficiency. This persistent slow growth provides some catch-up growth.

Any explanation of short stature in TS must account for at least three distinct phases of poor growth: intrauterine growth retardation, subnormal growth velocity before the age of puberty, and growth failure at the age of puberty. Plausible mechanisms of growth failure are listed in Table 22.1. The etiology of growth failure in TS is probably multifactorial, involving skeletal dysplasia, a degree of growth hormone secretory dysfunction, and estrogen deficiency. All three phases of subnormal growth could be caused by the intrinsic chromosomal imbalance; delayed cell replication is a putative mechanism common to a multitude of chromosomal abnormalities with associated growth failure.

Patients with Turner syndrome are not profoundly GH deficient as classically defined by provocative GH testing, although any girl whose growth velocity is below the TS norm should be screened for GH deficiency and hypothyroidism. The dogma is that GH secretion and insulin-like growth factor I (IGF-I) levels in TS are normal in childhood and partially deficient in adolescence (15–21). It is certainly convincing that growth hormone (GH) secretion in adolescence is less than that of controls matched for chronological age or skeletal age. The failure in TS of the usual pubertal increase in GH and IGF-I is at least partially caused by estrogen deficiency, as confirmed by the increase in GH and IGF-I concentrations in response to low or replacement doses of estrogens (21–23). However, some factor other than estrogen deficiency must be invoked to explain the absolute decline in GH secretion at adolescence (17,18). Some of this phenomenon is attributable to the increase in body mass (19).

Estrogen deficiency is probably only a small factor in the overall growth deficit in TS. One study (24) compared 75 adult TS patients, who had received estrogen replacement therapy at 13–17 years of age but no hGH, with a control group with equally severe estrogen-unreplaced primary ovarian failure (46,XX pure gonadal dysgenesis). The median height of all Turner groups was 141.8 cm, in contrast with 156.0 cm for pure gonadal dysgenesis. Thus, ovarian failure alone accounts for little of the TS height deficit. Furthermore, it is still unclear clinically whether TS patients with functional ovaries are taller than those without. In the minority of patients in whom

TABLE 22.1. Possible etiologies of growth failure in Turner syndrome.

Chromosome imbalance, delayed cell replication.

Hormonal
  Estrogen deficiency
  Diminished GH secretion
  GH or IGF-I unresponsiveness?

Skeletal dysplasia
  Secondary to intrauterine lymphedema
  Primary defect of X chromosome haploinsufficiency—SHOX?

spontaneous puberty occurs, adult height is associated with only a slight pubertal growth spurt. Massa et al. (25) have reported a modest increase in adult height in patients with spontaneous puberty, but other studies have found no significant difference or even reduced height associated with endogenous puberty (26). Thus, endogenous estrogen production does not markedly affect the growth failure of TS.

What then causes the growth failure apparent in most girls with TS during childhood, before deficiencies in the estrogen–GH–IGF axes are demonstrable? One consideration is mild but clinically pertinent prepubertal hormonal deficiency. Close analysis of the GH and IGF-I concentrations in Turner children reveals that most fall at the low end of the normal range, equivalent to a partial GH deficiency. It is also tempting to speculate about deleterious effects of possible relative deficiency of the extremely low endogenous estrogen levels detectable in normal prepubertal girls (27). Could this prepubertal deficiency cause the subnormal growth, progressive skeletal age delay, and slightly low GH secretion in prepubertal TS? Several lines of reasoning weigh against this thesis. The normal physiological consequences of prepubertal estrogen secretion have yet to be determined; the source, gonadal versus adrenal, has not been established; and TS patients with spontaneous puberty, and presumably normal estrogen levels in childhood, do not escape growth retardation. Direct testing of the proposition is needed.

In my opinion, an intrinsic bone defect, not an endocrine abnormality, is likely responsible for much of the short stature of TS. Pathological bone development could represent either an intrinsic defect in ossification secondary to loss of critical genes on the X chromosome, or alternatively might be one of the protean manifestations of intrauterine edema. The underlying skeletal dysplasia is manifested both as short stature and in the variety of skeletal abnormalities seen in TS. Increased bone resorption and defective osteoblast function have both been reported in TS, and cartilage from TS patients has been noted to have narrowed growth zones (28–30). Many patients with Turner syndrome have a short, square appearance, and some investigators, but not all, have reported short-leggedness and abnormal body segment ratios (31,32). Nonetheless, there are no unique histological findings recognized as yet in the bones of patients with TS.

What are the genetic correlates of short stature and a putative bone dysplasia in TS? Phenotype–karyotype correlations in TS are notoriously difficult, but deletions of the short arm of the X chromosome are crudely associated with short stature, whereas long arm deletions are associated with ovarian dysgenesis (33). In a literature review of short (Xp) or long (Xq) arm terminal X-chromosome deletions, short stature occurred in 88% and 43% of the Xp- and Xq- cases, respectively (34). In the paper by Cohen et al. (24), TS patients with deletions of the long arm had a median height of 148.9 cm, significantly taller than the other TS groups (141.8 cm). Evidence from terminal deletions of the X or Y chromosome have led to the conclusion that a genetic locus affecting height resides on the distal short arm of the X chromosome.

Rao et al. (35) recently narrowed the search for the X chromosome skeletal growth gene to a distal 170-kb interval of the pseudoautosomal region, and then isolated and sequenced a short stature homeobox-containing gene (SHOX). It is highly conserved across species and has two transcripts derived from alternative splicing: SHOXa is widely expressed, whereas expression of SHOXb is highest in bone marrow fibroblasts. Importantly, this group demonstrated deletion of the locus in 36 of 36 short individuals with Xp22 or Yp11.3 breakpoints, and absence of a deletion in relatives, normal controls, or subjects with X or Y rearrangements and normal height. A SHOX mutation was found in 1 of 91 subjects with otherwise unexplained short stature.

Ellison's group had previously described the same gene, which they named the pseudoautosomal homeobox-containing osteogenic gene (PHOG) (36). They recognized that the location meets gene dosage requirements for a Turner growth gene, in that a pseudoautosomal locus is not X-inactivated and a homolog exists on Yp. They also determined that postnatal expression is confined to osteogenic cells, such as trabecular cells and bone marrow fibroblasts, consistent with a putative role for the gene in bone physiology.

In light of prior reports of XY translocations in patients with a bone dysplasia called Leri–Weill dyschondrosteosis (LWD), families with this syndrome were evaluated for deletions and mutations of SHOX (PHOG). Belin et al. (37) reported large deletions of SHOX in seven LWD families and a nonsense mutation in an additional family. In one family, a fetus with Langer mesomelic dysplasia (severe shortening of distal extremities, long considered the homozygous form of LWD) had loss of both SHOX alleles. Shears et al. (38) identified SHOX deletions in five LWD families, including a six-generation pedigree. In one family these investigators found a point mutation truncating both SHOXa and -b transcripts. Thus, loss of one SHOX allele results in Leri–Weill dyschondrosteosis and loss of both results in Langer mesomelic dysplasia.

LWD was first described in 1929 and is thought to be autosomal dominant, but more severe in females. It is characterized by a short forearm with Madelung deformity and limited mobility, variable tibiofibular shortening, variable metacarpal and metatarsal involvement, and mild short stature (39). Although the Turner phenotype is broader and more variable than that of LWD, it seems likely that some of the skeletal features of TS result from deletion of SHOX, including Madelung deformity, other bony abnormalities, and some degree of the short stature. It remains to be determined whether the component of growth failure in TS presumptively caused by SHOX haploinsufficiency is remediable with human growth hormone therapy.

References

1. Albertsson-Wikland K, Ranke MB, eds: Turner syndrome in a life span perspective: research and clinical aspects. Amsterdam: Elsevier, 1995.
2. Rosenfeld RG, Attie KM, Frane J, et al. Growth hormone therapy of Turner's syndrome: beneficial effect on adult height. J Pediatr 1998;132:319–24.

3. Lyon AJ, Preece MA, Grant DB. Growth curve for girls with Turner syndrome. Arch Dis Child 1985;60:932–5.
4. Ranke MB, Pfluger H, Rosendahl W, et al. Turner syndrome: spontaneous growth in 150 cases and review of the literature. Eur J Pediatr 1983;141:81–8.
5. Frane JW, Sherman BM, and the Genentech Collaborative Group. Predicted adult height in Turner syndrome. In: Rosenfeld RG, Grumbach MM, eds. Turner syndrome. New York: Marcel Dekker, 1990:405–19.
6. Sybert VP. Adult height in Turner syndrome with and without androgen therapy. J Pediatr 1984;104:365–9.
7. Naeraa RW, Nielsen J. Standards for growth and final height in Turner's syndrome. Acta Paediatr Scand 1990;79:182–90.
8. Taback SP, Collu R, Deal CL, et al. Does growth-hormone supplementation affect adult height in Turner's syndrome? Lancet 1996;348:25–7.
9. Plotnick L, Attie KM, Blethen SL, Sy JP. Growth hormone treatment of girls with Turner syndrome: the National Cooperative Growth Study Experience. Pediatrics 1998;102:479–81.
10. Naeraa RW, Nielsen J, Pedersen IL, Sorensen K. Effect of oxandrolone on growth and final height in Turner's syndrome. Acta Paediatr Scand 1990;79:784–9.
11. Crock P, Werther GA, Wettenhall HNB. Oxandrolone increases final height in Turner syndrome. J Paediatr Child Health 1990;26:221–4.
12. Nilsson KO, Albersson-Wikland K, Alm J, et al. Improved final height in girls with Turner's syndrome treated with growth hormone and oxandrolone. J Clin Endocrinol Metab 1996;81:635–40.
13. Werther GA, Dietsch S (for the Australasian Paediatric Endocrine Group): Multicentre trial of synthetic growth hormone and low dose oestrogen in Turner syndrome: analysis of final height. In: Albertsson-Wikland K, Ranke MB, eds. Turner syndrome in a life span perspective: research and clinical aspects. Amsterdam: Elsevier, 1995:105–12.
14. Attie KM, Chernausek S, Frane J, Rosenfeld RG: Growth hormone use in Turner syndrome. In: Albertsson-Wikland K, Ranke MB, eds. Turner syndrome in a life span perspective: research and clinical aspects. Amsterdam: Elsevier, 1995:285–95.
15. Laczi F, Julesz J, Janaky T, Laszlo FA. Growth hormone reserve capacity in Turner's syndrome. Horm Metab Res 1979;11:664–6.
16. Ross JL, Long LM, Loriaux DL, Cutler GB Jr. Growth hormone secretory dynamics in Turner syndrome. J Pediatr 1985;106:202–6.
17. Massarano AA, Brook CGD, Hindmarsh PC, et al. Growth hormone secretion in Turner's syndrome and influence of oxandrolone and ethinyl estradiol. Arch Dis Child 1989;64:587–92.
18. Ranke MB, Blum WF, Bierich JR. Growth hormone secretion and somatomedin C/IGF-I levels in Turner syndrome and in patients with idiopathic growth hormone deficiency. In: Rosenfeld RG, Grumbach MM, eds. Turner syndrome. New York: Marcel Dekker, 1990;281–300.
19. Lu PW, Cowell CT, Jimenez M, Simpson JM, Silink M. Effect of obesity on endogenous secretion of growth hormone in Turner's syndrome. Arch Dis Child 1991;66:1184–90.
20. Rosenfeld RG, Hintz RL, Johanson AJ, et al. Methionyl human growth hormone and oxandrolone in Turner syndrome: preliminary results of a prospective randomized trial. J Pediatr 1986;109:936–43.

21. Cuttler L, Van Vliet G, Conte FA, Kaplan SL, Grumbach MM. Somatomedin-C levels in children and adolescents with gonadal dysgenesis: differences from age-matched normal females and effect of chronic estrogen replacement. J Clin Endocrinol Metab 1985;60:1087–92.
22. Schober E, Frisch H, Waldhauser F, Bieglmayr CH. Influence of estrogen administration on growth hormone response to GHRH and L-Dopa in patients with Turner's syndrome. Acta Endocrinol 1989;120:442–6.
23. Mauras N, Rogol AD, Veldhuis JD. Increased hGH production rate after low-dose estrogen therapy in prepubertal girls with Turner's syndrome. Pediatr Res 1990;28:626–30.
24. Cohen A, Kauii R, Pertzelan A, et al. Final height of girls with Turner's syndrome: correlation with karyotype and parental height. Acta Paediatr 1995;84:550–4.
25. Massa G, Vanderschueren-Lodeweyckx M, Malvaux P. Linear growth in patients with Turner syndrome: influence of spontaneous puberty and parental height. Eur J Pediatr 1990;149:246–50.
26. Page LA. Final heights in 45,X Turner's syndrome with spontaneous sexual development. Review of European and American reports. J Pediatr Endocrinol 1993;6:153–8.
27. Klein KO, Baron J, Colli MJ, McDonnell DP, Cutler GB Jr. Estrogen levels in childhood determined by an ultrasensitive recombinant cell bioassay. J Clin Invest 1994;94:2475–80.
28. Lubin MB, Gruber HE, Rimoin DL, Lachman RS. Skeletal abnormalities in the Turner syndrome. In: Rosenfeld RG, Grumbach MM, eds. Turner syndrome. New York: Marcel Dekker, 1990:281–300.
29. Rubin KR. Osteoporosis. In: Rosenfeld RG, Grumbach MM, eds. Turner syndrome. New York: Marcel Dekker, 1990.
30. Horton WA. Growth plate biology and the Turner syndrome. In: Rosenfeld RG, Grumbach MM, eds. Turner syndrome. New York: Marcel Dekker, 1990.
31. Rongen-Westerlaken C, Rikken B, Vastrick P, et al. Body proportions in individuals with Turner syndrome. Eur J Pediatr 1993;152:813–7.
32. Hughes PCR, Ribeiro J, Hughes IA. Body proportions in Turner syndrome. Arch Dis Child 1986;61:506–7.
33. Ferguson-Smith MA. Karyotype-phenotype correlations in gonadal dysgenesis and their bearing on the pathogenesis of malformations. J Med Genet 1965;2:142–5.
34. Therman E, Susman B. The similarity of phenotypic effects caused by Xp and Xq deletions in the human female: a hypothesis. Hum Genet 1990;85:175–83.
35. Rao E, Weiss B, Fukami M, et al. Pseudoautosomal deletions encompassing a novel homeobox gene cause growth failure in idiopathic short stature and Turner syndrome. Nat Genet 1997;16:54–63.
36. Ellison JW, Wardak Z, Young MF, et al. PHOG, a candidate gene for involvement in the short stature of Turner syndrome. Hum Mol Genet 1997;6:1341–7.
37. Belin V, Cusin V, Viot G, et al. SHOX mutations in dyschondrosteosis (Leri-Weill syndrome). Nat Genet 1998;19:67–9.
38. Shears DJ, Vassal HJ, Goodman FR, et al. Mutation and deletion of the pseudoautosomal gene SHOX cause Leri-Weill dyschondrosteosis. Nat Genet 1998;19:70–3.
39. Dawe, C, Wynne-Davies R, Fulford GE. Clinical variation in dyschondrosteosis. J Bone Joint Surg 1982;64B:377–81.

23

Practical Issues of Clinical Growth Hormone Therapy in Turner Syndrome

PAUL M. MARTHA, JR., AND KENNETH M. ATTIE

Short stature and delayed sexual development are cardinal features of the syndrome described by Henry Turner in 1938 (1). On average, adult stature of women with Turner syndrome is 20 cm less than the population mean for normal adult women, and at least some degree of short stature is almost universally present (2). Because of extremely short stature, this condition has long been considered a natural target for growth-promoting therapies. Early attempts at growth promotion using androgens generally yielded disappointing effects on adult stature.

Treatment of Turner syndrome with human pituitary growth hormone (GH) began in the early 1960s (3–6). By the early 1970s, it was clear that growth hormone could increase short-term growth rate in these patients (7,8). However, because supplies of human pituitary-derived GH were limited, experience with this therapy in Turner syndrome remained limited. In the early 1980s, the situation changed dramatically with the availability of somatrem (recombinant methionyl human growth hormone) and somatropin (recombinant human growth hormone). With this increase in the availability of human growth hormone (hGH), clinical trials were soon initiated to address the safety and efficacy of hGH for treating the short stature of girls with Turner syndrome. In addition to their short stature, girls with Turner syndrome experience a lifelong deficiency of endogenous estrogen production as a consequence of their primary ovarian failure. The need for estrogen replacement therapy presents special clinical challenges, particularly with regard to growth hormone therapy. However, this need also provides a unique opportunity to gain insight into the role played by estrogen in modulating the effects of growth hormone on adult height.

Estrogen, whether derived from normal ovarian production or exogenously administered, impacts both the rate of statural growth and the rate of skeletal maturation. Therefore, an important decision in the medical care of any girl with Turner syndrome concerns the timing of estrogen treatment to promote development of female secondary sexual characteristics. At the heart of this issue, for each patient, is the relative importance of adult height and the age at which she undergoes physical feminization. This is particularly true for girls with Turner syndrome receiving GH therapy, because augmentation of adult height is a primary treatment goal. The results of several studies now provide strong evidence that treatment with GH during childhood increases adult height in Turner syndrome, but it is also clear that the age at which estrogen treatment begins is an important variable influencing treatment outcome.

To understand better the impact of growth hormone and gonadal steroids on childhood growth in girls with Turner syndrome, two long-term multicenter clinical trials were initiated by U.S. investigators in the mid-1980s. The first, begun in 1983, examined the safety and efficacy of recombinant human growth hormone (hGH), alone or in combination with the nonaromatizable androgen oxandrolone, for the treatment of short stature associated with Turner syndrome. Previously published reports have detailed the results of 1 (9), 3 (10), and 6 years (11) of therapy and of final height (12) from this trial. The second trial, initiated in 1987, was designed to assess the impact of age of induction of feminization and of duration of hGH therapy on the final height of individuals with Turner syndrome (13,14). A summary of the salient findings from these two important trials is provided next.

Growth Hormone Therapy with or Without Oxandrolone

Seventy girls with Turner syndrome previously confirmed by karyotype enrolled in this trial (9–12). After 6 months of observation to document a pretreatment growth velocity of less than 6 cm/year, the girls were randomized to one of the following four treatment groups for phase I of the study:

- Control group receiving no treatment
- Oxandrolone, 0.125 mg/kg per day orally
- HGH, 0.375 mg kg^{-1}week^{-1} given three times weekly parenterally
- Combination, hGH 0.375 mg kg^{-1}week^{-1} plus oxandrolone 0.125 mg kg^{-1}day^{-1}

Following 1–2 years in the phase I groups, the subjects were reassigned to the following phase II groups:

- hGH alone (the same individuals as those receiving hGH alone in phase I)
- hGH plus oxandrolone at 0.0625 mg kg^{-1}day^{-1} (all other subjects)

Estrogen therapy was withheld in this trial until a girl had reached a minimum age of 14 years and had participated in the study for at least 3 years. Treatment with hGH, with or without oxandrolone, continued until the subject had either attained a skeletal age of greater than 14 years and was growing less than 2 cm per year, or chose to discontinue participation in the trial.

Results

Sixty of the original 70 girls completed the study and maintained an adequate level of compliance. Results were compared to a historical control group of 25 U.S. women with Turner syndrome, who were followed to final height but did not receive treatment with hGH.

The treatment groups were well matched for baseline age (mean), height, Turner syndrome height SDS, and midparental target height. Seventeen girls received hGH alone and reached a final height of 150.4 ± 5.5 cm (mean ± SD), which is 8.4 ± 4.5 cm taller than their mean projected adult height at enrollment. The girls who received a hGH plus oxandrolone reached a mean height of 152.1 ± 5.9 cm, which was 10.3 ± 4.7 cm taller than their projected adult height. Similar gains in height were apparent when the heights of the subjects were compared to those of the matched historical control group (15).

The Impact of Age of Estrogen Replacement upon Final Height

One hundred and seventeen girls with Turner syndrome were enrolled in this trial to evaluate the impact upon final height of the age at which estrogen replacement therapy is introduced during hGH treatment. During the first 12 months of the study, the subjects where randomized to untreated control or hGH treatment (0.375 mg kg^{-1}week^{-1}). At month 12, the controls were switched to hGH therapy. Subjects who began hGH treatment before age 11 were randomized to begin estrogen replacement therapy at either age 12 or age 15. All subjects who initiated hGH therapy after age 11 began estrogen replacement after 12 months of hGH therapy. The three treatment groups and number of subjects in each group were as follows:

Timing of hGH initiation	Timing of estrogen	N
hGH before 11 years	Age 12 years	30
hGH before 11 years	Age 15 years	30
hGH after 11 years	At month 12 of hGH	57

Estrogen replacement therapy comprised conjugated estrogens (Premarin, Wyeth-Ayerst) at a dosage of 0.3 mg daily for 6 months, after which the dosage was increased to 0.625 mg daily. After a year of estrogen therapy, subjects received 10 mg of medroxyprogesterone acetate (Provera, Upjohn) for 10 days each calendar month.

Resulting heights of the study subjects were compared with the heights of 14 historical control patients matched by age with the treatment group receiving hGH before age 11 and of 55 historical control subjects matched to the group first receiving hGH at a later age. These control subjects were selected from a database of more than 1300 U.S. Turner syndrome patients never treated with hGH or estrogens (15). The mean age for initiation of estrogen replacement for these matched historical control subjects was similar to or later than that of the delayed estrogen group (16.2 ± 1.9 years for historical controls vs. 15.0 ± 0.5 years for the group receiving estrogen after age 15).

Results

For subjects younger than 11 years of age at initiation of hGH therapy, the two groups (estrogen at age 12 and estrogen at age 15) were well balanced with respect to baseline age (mean, 9.5 years), baseline height, midparental height, proportion of patients with monosomy 45x, and bone age. During the first year following introduction of estrogen therapy, the group receiving estrogen at age 12 grew an average of 1 cm/year faster than those continuing on hGH alone. However, during the second year the growth rates of the two groups were no different. In subsequent years, the average growth rate of the group beginning estrogen at age 12 declined to and remained below the levels expected for untreated girls with Turner syndrome. In contrast, the mean growth rate for those subjects beginning estrogen at age 15 remained above the same predicted rate. Furthermore, the rate of skeletal maturation (as judged by radiologic bone age) advanced at a significantly more rapid rate in the group receiving estrogen at age 12 years.

When compared to pretreatment predicted adult height, the gain in final or near-final height was 5.1 cm for girls receiving estrogen at age 12 years and 8.4 cm for those in whom estrogen treatment was delayed until age 15. The results were similar when height gain was computed using the matched historical control group as a reference. The mean near-final height of the subjects who were older than age 11 when hGH therapy was begun was approximately 5 cm, similar to the gain of the group beginning estrogen at age 12.

To determine the factors that most affected ultimate height gain, the investigators performed multiple regression analyses using height gain as the dependent variable and age, height, and bone age at baseline, as well as karyotype, midparental height, age at last measured height, and duration of hGH therapy before introduction of estrogen as independent variables. When all subjects with a height measured after age 13.5 were considered in the analysis ($n = 58$), the significant predictive factors were baseline age, baseline bone age, age at last measured height, and hGH therapy duration before estrogen ($p < 0.01$ for each). Because of the apparent impact of the age at last measured height, an additional analysis was conducted including those subjects most likely to have achieved true final height (those with a height measured after age 16; $n = 29$). In this second analysis the only factor identified as having a significant effect on treatment outcome was the duration of hGH therapy before initiation of estrogen replacement ($p < 0.0001$, $r^2 = 0.41$).

Conclusions

When considered together with the larger body of published studies, these prospective clinical trials add substantially to the already strong evidence that adequate therapy with hGH can increase the final adult height of most individuals with Turner syndrome. In addition, these reports provide important insight into the critical role served by the specific gonadal hormone milieu in modulating, and ultimately determining, the final treatment outcome. Perhaps the most notable finding to emerge from these trials is the importance of the number of years of hGH treatment before feminization is induced in determining final adult height.

In selected patients, low-dose oxandrolone treatment may be useful during the childhood years to maximize the height achieved by the time the appropriate age for feminization is reached (see Chapter 22). Based on the results of these and other studies, early diagnosis and treatment of Turner syndrome with hGH should allow patterns of growth and external physical development that approach the normal range in most girls.

References

1. Turner HH. A syndrome of infantilism, congenital webbed neck, and cubitus valgus. Endocrinology 1938;23:566–74.
2. Ranke MB. Growth in Turner's syndrome. Acta Pediatr 1994;83:343–4.
3. Tzagouris M. Response to long-term administration of human growth hormone in Turner's syndrome. JAMA 1969;210:2373.
4. Wright JC, Brasel JA, Aceto TJ. Studies with human growth hormones in Turner's syndrome. Amer J Med 1965;38:499–516.
5. Hutchings JJ, Escanilla RF, Li CH, Forsham PF. Human growth hormone administration in gonadal dysgenesis. Am J Dis Child 1965;109:318.
6. Soyka LF, Ziskind A, Crawford JD. Treatment of short stature in children and adolescents with human pituitary growth hormone (Raben). N Engl J Med 1964;271:754.
7. Escamilla R. Nonhypopituitary dwarfs and human growth hormone therapy. In: Ratti S, ed. Advances in human growth hormone research. Washington, D.C. U.S. Department of Health, Education and Welfare, 1973;766–85.
8. Tanner JM, Whitehouse RH, Hughes PCR, Vince FP. Effect of human growth hormone treatment for 1–7 years on growth of 100 children with growth hormone deficiency, low birth weight, inherited smallness, Turner's syndrome and other complaints. Arch Dis Child 1971;46:745.
9. Rosenfeld RG, Hintz RL, Johanson AJ, et al. Methionyl human growth hormone and oxandrolone in Turner syndrome: preliminary results of a prospective randomized trial. J Pediatr 1986;109:936–43.
10. Rosenfeld RG, Hintz RL, Johanson AJ, et al. Three-year results of a randomized prospective trial of methionyl human growth hormone and oxandrolone in Turner syndrome. J Pediatr 1988;113:393–400.
11. Rosenfeld RG, Frane J, Attie KM, et al. Six-year results of a randomized, prospec-

tive trial of human growth hormone and oxandolone in Turner syndrome. J Pediatr 1992;121:49–55.
12. Rosenfeld RG, Attie KM, Frane J, et al. Growth hormone therapy of Turner's syndrome: beneficial effect on final height. J Pediatr 1998;132:319–24.
13. Attie KM, Chernausek S, Frane J, Rosenfeld RG, The Genentech study group. Growth hormone use in Turner syndrome: a preliminary report on the effect of early vs. delayed estrogen. In: Albertsson-Wikland K, Ranke M, eds. Turner syndrome in a Life-Span perspective. Amsterdam: Elsevier Science, 1995;175–81.
14. Chernausek SD, Attie KM, Frane J, Genentech Collaborative Group. Predictors of adult height in girls with Turner syndrome treated with growth hormone. In: Program of the Annual Meeting of the Society for Pediatric Research, 1998 (abstract 408).
15. Lippe BM, Plotnick LP, Attie KM, Frane J. Growth in Turner syndrome: updating the United States experience. In: Hibi I, Takano K, eds. Basic and clinical approach to Turner syndrome. Amsterdam: Elsevier, 1995:77–82.

Part IV

GH Secretagogues: Differential Effects in Men and Women

24
Influence of Gonadal Function on GH Secretion

Francisca Lago, Angela Peñalva, Eva Carro,
Rosa Maria Señaris, Vera Popovic, Manuel Pombo,
Felipe F. Casanueva, and Carlos Dieguez

In contrast to the relatively clear-cut sexually dimorphic secretory pattern in rodents, episodic growth hormone (GH) secretion in humans is unpredictable, with the exception of the first cycle of slow-wave sleep, thus making their clinical assessment particularly difficult. Nevertheless, it has been reported that spontaneous GH secretion is increased around puberty in both male and female subjects, thus suggesting a facilitative role of gonadal steroids (1,2). Increased serum testosterone levels during normal puberty in boys is associated with increased GH secretory burst mass and amplitude (3). Similarly, testosterone replacement in hypogonadal boys increased the number of GH pulses and their amplitude (4). Finally, in normal aging there is an association between reduction of spontaneous GH secretion and decreased androgenic activity (5). Taken together these findings suggest that androgens play a facilitative role in spontaneous GH secretion in humans.

A stimulatory role for estrogens in spontaneous GH secretion is supported by the finding that the age-related fall in GH secretion is greater in women than in men, being correlated in females with circulating free-estradiol concentrations (2,6,7). In keeping with this hypothesis, ethinyl estradiol administration to postmenopausal women also increases pulse amplitude and 24-h GH concentrations but not GH pulse frequency. However, because a decrease in circulating IGF-I levels was also found after ethinyl-estradiol treatment, it is unclear whether these changes are related to the effects of estrogens at a central or peripheral (feedback) level (2). In support of the first possibility are the findings that the serum estradiol concentration is a strongly positive determinant of the mean serum GH level and predicts higher-amplitude nocturnal GH peaks in pubertal girls and adult women (2,8). Also, GH pulse mass and amplitude are correlated with serum estradiol concentrations in women at different phases of the menstrual cycle. Finally, middle-aged women exhibit a higher daily GH secretion than age-matched men (2).

Data regarding sex-related differences in exogenously stimulated GH secretion in humans failed to show clear-cut differences to most GH secretagogues, with the exception of arginine or galanin-induced GH release. Although GH responses to arginine or galanin are greater in female than in male subjects, such differences are not observed with many other stimuli such as growth hormone-releasing factor (GHRH), clonidine, or hypoglycemia (1,9; and the authors' unpublished observations). For example, most workers failed to find any differences in GH responses to GHRH between male and female subjects, or in relation to the different phases of the menstrual cycle in females (10,11). Similarly, GH responses to GHRP-6 are similar in male and female subjects, and across different stages of the menstrual cycle in young women (12) (Fig. 24.1).

GH responses to some of the most powerful GH stimuli in humans also failed to demonstrate any sexual dimorphism in this response. Pyridostigmine (PD), an indirect cholinergic agonist that presumably acts by inhibiting hypothalamic somatostatin release, when administered in combination with GHRH led to a marked and similar increase in plasma GH levels in both male and female subjects (13). Also, combined administration of GHRH plus GHRP-6 led to a massive and similar GH secretory response in both groups (14). Taken together, these data show that GH responses to different GH secreta-

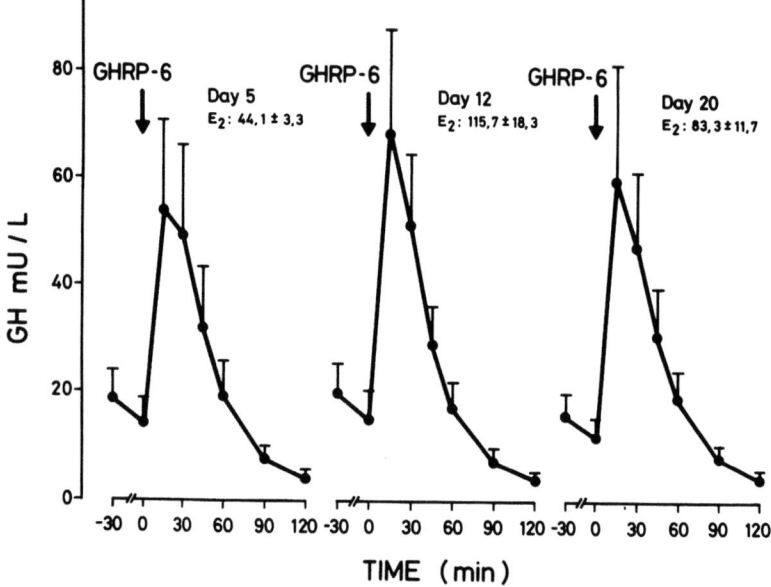

FIGURE 24.1. Mean ± SEM serum GH concentration responses to GHRP-6 (1 µg/kg i.v. at 0 min) (*arrows*) in three groups of women on different days (as indicated) in the menstrual cycle. E_2 values are (mean ± SEM) serum estradiol levels.

gogues are very similar in male and female subjects and that the effects of gonadal function on GH secretion appear to be exerted mostly at the hypothalamic level. Whether this gender difference could be related to changes in neuropeptide gene expression or to different patterns of pulsatile GH release driven by the different neuropeptides involved in the genesis of GH secretory patterns is unknown.

Nonandrogenic Testicular Factors and GH Secretion

There is clear evidence that the sexual dimorphism observed in GH secretion is at least in part mediated by gonadal steroids. Nevertheless, recent indirect evidence has also suggested that, at least in the male rat, nonsteroidal testicular factors secreted by cell types other than Leydig cells may well be involved in the regulation of the GH axis. In vivo GH responses to GHRH were reduced in gonadectomized male rats but remained unchanged after testosterone suppression by administration of a specific toxin Leydig cell (ethylene-1,2-dimethane sulfonate, EDS) (15). A similar experimental model has recently also suggested that nonandrogenic testicular factors are involved in the regulation of hypothalamic GHRH and somatotropin release-inhibiting factor (SRIF) mRNA levels, as well as in pituitary mRNA content of somatostatin (SRIF) receptor subtypes (16,17). Selective removal of Leydig cells with EDS greatly reduced the SRIF mRNA content in the periventricular nucleus of the hypothalamus. These levels were significantly lower than those found in gonadectomized rats. Furthermore, replacement treatment with dihydrotestosterone (DHT) did not completely restore SRIF mRNA levels in EDS-treated rats, contrasting with the complete recovery of SRIF mRNA levels in gonadectomized rats (18,19). On the other hand, gonadectomy and EDS treatment produced a significant reduction in GHRH mRNA levels in the arcuate nucleus. DHT administration reversed the action of gonadectomy, but did not restore GHRH mRNA content in EDS-treated rats (20). Using the same experimental paradigm, sexual dimorphism in SRIF subtype receptor (SSTR), SSTR-1, SSTR-2, and SSTR-3, mRNA levels in the rat anterior pituitary has been assessed (16). SSTR-1 mRNA levels were higher in female than male rats, but SSTR-3 mRNA levels display the opposite pattern (Fig. 24.2). Moreover, SSTR-1 mRNA levels appear to be regulated by testosterone, while SRIF mRNA content is regulated by nonandrogenic testicular factors. In contrast, SSTR-2 levels in the anterior pituitary are not influenced by gonadal function. These findings indicate that testicular factors other than testosterone, secreted by cell types different from Leydig cells, may well be involved in the regulation of the GH axis, as they are in the gonadotropin axis (17).

Assuming that nonandrogenic testicular factors synthesized by the Sertoli cells can also influence the GHRH–somatostatin–GH axis, then inhibin and activins are some of the obvious candidates. Regarding inhibin, recent data suggest that it may well be involved in the regulation of GH secretion (21).

246   F. Lago et al.

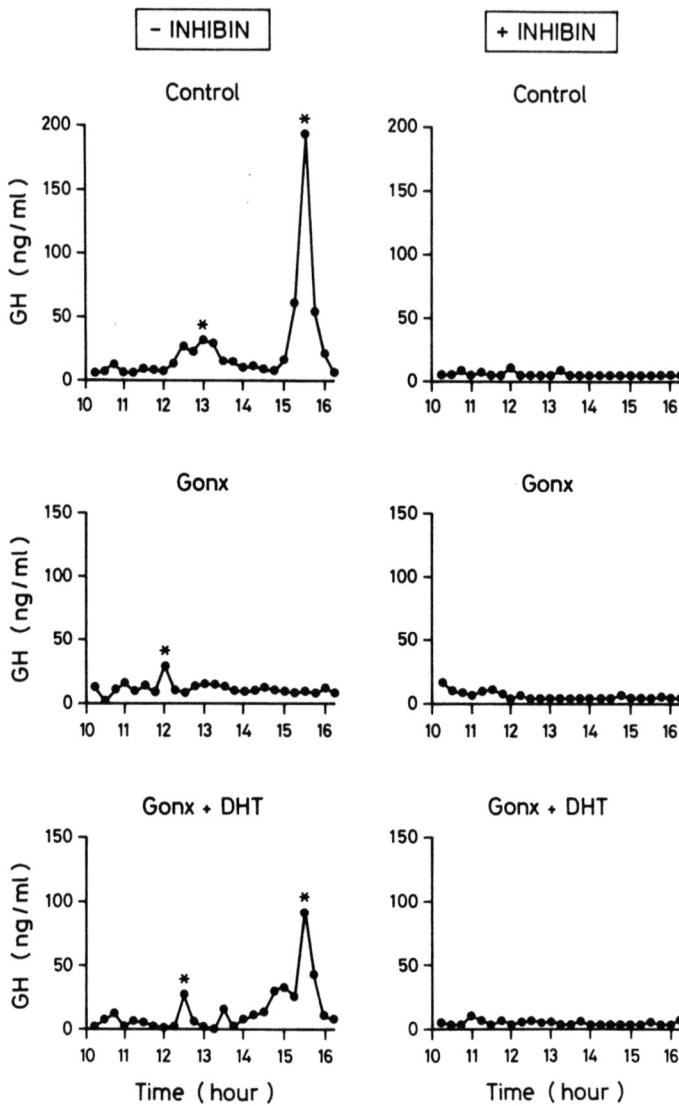

FIGURE 24.2. Representative plasma GH profiles in individual male rats after injection of inhibin (100 µg/kg, i.v.) or vehicle (saline solution) in intact (control), gonadectomized (GonX), and DHT-treated gonadectomized (GonX+DHT) animals. The compounds were administered at 1015 (n = 6–8 rats per group).

Thus, inhibin administration to intact, gonadectomized, or DHT-treated gonadectomized rats led to a marked supression in spontaneous GH secretion in all groups. Furthermore, inhibin markedly reduced in vivo GH response to GHRH in both intact and gonadectomized rats. The possibility that inhibin could be acting at the hypothalamic level is supported by data showing that inhibin administration to intact male rats led to an increase in somatostatin mRNA levels in the periventricular nuclei and a decrease in GHRH mRNA levels in the arcuate nucleus. Nevertheless, in the physiological setting, the relative roles played by inhibin (synthesized in the gonads), the pituitary gland and the CNS in the integrated control of GH secretion need clarification. Also, the effects of other nonandrogenic testicular factors synthesized by the Sertoli cells on GH secretion await elucidation.

Acknowledgments. This work was supported with grants from the Fondo de Investigaciones Sanitarias, Spanish Ministry of Health, and the Xunta de Galicia.

References

1. Dieguez C, Page MD, Scanlon MF. Growth hormone neuroregulation and its alterations in disease states. Clin Endocrinol 1988;28:109–43.
2. Veldhuis JD. Gender differences in secretory activity of the human somatotropic (growth hormone) axis. Eur J Endocrinol 1996;134:287–95.
3. Martha PM Jr, Gooman KM, Blizzard RM, Rogol AD, Veldhuis JD. Endogenous growth hormone secretion and clearance rates in normal boys as determined by deconvolution analysis: relationship to age, pubertal status and body mass. J Clin Endocrinol Metab 1992;74:336–44.
4. Giustina A, Scalvini T, Tassi C, Desenzani P, Poiesi C, Wehenberg W, et al. Maturation of the regulation of growth hormone secretion in young males with hypogonadotropic hypogonadism pharmacologically exposed to progressive increments in serum testosterone. J Clin Endocrinol Metab 1997;82:1210–9.
5. Veldhuis JD, Liem AY, South S, Weltman A, Weltman J, Clemmons DA, et al. Differential impact of age, sex-steroid hormones, and obesity on basal versus pulsarile growth hormone secretion in men as assessed in an ultrasensitive chemiluminescence assay. J Clin Endocrinol Metab 1995;80:3209–22.
6. Veldhuis JD, Metzger DL, Martha PM Jr, Mauras N, Kerrigan JR, Keenan B, et al. Estrogen and testosterone, but not a non-aromatizable androgen, direct network integration of the hypothalamo-somatotrope (growth hormone)-insulin-like growth factor-I axis in the human: evidence from pubertal pathophysiology and sex-steroid hormone replacement. J Clin Endocrinol Metab 1997;34:3414–20.
7. Ovesen P, Vahl N, Fisker S, Veldhuis JD, Christiansen JS, Jorgensen JOL. Increased pulsatile, but not basal, growth hormone secretion rates and plasma insulin-like growth factor I levels during the preovulatory interval in normal women. J Clin Endocrinol Metab 1998;83:1662–7.
8. Wennink JM, Delemarre-van de Waal HA, Schoemaker R, Blaauw G, van den Brakem C, Schoemaker J. Growth hormone secretion patterns in relation to LH and estradiol secretion throughout normal female puberty. Acta Endocrinol 1991;124:129–35.
9. Giustina A, Gastaldi C, Bugari G, Chiesa L, Loda G, Tironi C, et al. Role of galanin in the regulation of somatotrope and gonadotrope function in young ovulatory women. Metabolism 1995;44:1028–32.

10. Gelato MC, Pescovitz OH, Cassorla F, Loriaux DL, Merriam GR. Dose-response relationships for the effects of growth hormone releasing factor-(1-44)-NH2 in young adult men and women. J Clin Endocrinol Metab 1984;59:197–201.
11. Lima L, Arce V, Lois N, Fraga C, Lechuga MJ, Tresguerres JA, et al. Growth hormone (GH) responsiveness to GHRH in normal adults is not affected by short-term gonadal blockade. Acta Endocrinol 1989;120:31–6.
12. Peñalva A, Pombo M, Carballo A, Barreiro J, Casanueva FF, Dieguez C. Influence of sex, age, and adrenergic pathways on the growth hormone response to GHRP-6. Clin Endocrinol 1993;38:87–91.
13. Arvat A, Cappa M, Casanueva FF, Dieguez C, Ghigo E, Nicolosi M, et al. Pyridostigmine potentiates growth hormone (GH)-releasing hormone-induced GH release in both men and women. J Clin Endocrinol Metab 1993;76:374–7.
14. Peñalva A, Pombo M, Mallo F, Barreiro J, Carballo A, Alvarez C, et al. Mechanism of action of growth hormone releasing hexapeptide (GHRP-6) on in vivo growth hormone secretion. In: Pombo M, Rosenfeld RG, eds. Two decades of experience in growth. New York: Raven Press, 1993:253–300.
15. Aguilar E, Pinilla L, Tena-Sempere M. Growth hormone-releasing hormone-induced growth hormone secretion in adults rats orchidectomized or injected with ethylene dimethane sulphonate. Neuroendocrinology 1993;57:132–4.
16. Lago F, Señaris RM, Emson PC, Dominguez F, Dieguez C. Evidence for the involvement of non-androgenic testicular factors in the regulation of hypothalamic somatostatin and GHRH mRNA levels. Mol Brain Res 1996;35:220–6.
17. Señaris RM, Lago F, Dieguez C. Gonadal regulation of somatostatin receptor 1, 2, and 3 mRNA levels in the rat anterior pituitary. Mol Brain Res 1996;38:171–5.
18. Chowen-Breed JA, Steiner RA, Clifton DK. Sexual dimorphism and testosterone-dependent regulation of somatostatin gene expression in the periventricular nucleus of the rat brain. Endocrinology 1989;125:357–62.
19. Argente J, Chowen-Breed JA, Steiner RA, Clifton DK. Somatostatin messenger RNA in hypothalamic neurons is increased by testosterone through activation of androgen receptors and not by aromatization to estradiol. Neuroendocrinology 1990;52:342–9.
20. Zeitler P, Argente J, Chowen-Breed JA, Clifton DK, Steiner RA. Growth hormone-releasing hormone messenger ribonucleic acid in the hypothalamus of the adult male rat is increased by testosterone. Endocrinology 1990;127:1362–8.
21. Carro E, Señaris RM, Mallo F, Dieguez C. Inhibin suppresses in vivo growth hormone secretion. Neuroendocrinology 1998;68:293–6.

25

Endocrine Responses to GH Secretagogues in Relation to Sex and Age in Humans

EMANUELA ARVAT, ROBERTA GIORDANO, LAURA GIANOTTI, FABIO BROGLIO, GIAMPIERO MUCCIOLI, FRANCO CAMANNI, AND EZIO GHIGO

Growth hormone secretagogues (GHS) are synthetic peptidyl and nonpeptidyl molecules that possess strong, dose-dependent, and reproducible GH-releasing activity in vivo in several species and in humans after intravenous, subcutaneous, intranasal, and oral administration (1–4). In addition to GHRP-6, other members of the GHS family studied in humans include peptidyl analogs, such as GHRP-1 (a heptapeptide), GHRP-2 and Hexarelin (both hexapeptides), and nonpeptidyl GHRP mimetics, such as MK-0677, a spiroindoline, which shows high bioavailability and long-lasting effects after oral administration (1–4).

The releasing activity of GHS compounds, however, is not fully specific for GH. In fact, they also possess significant prolactin- (PRL-) and ACTH/cortisol-releasing effects (2,5). Moreover, GHS stimulate food intake, influence sleep patterns, and exert cardiotropic effects (6–8). The activities of GHS are mediated by specific receptors subtypes, which are present mainly in the pituitary and hypothalamus but also in other CNS areas and in peripheral tissues (9–13). The existence of specific human GHS receptors suggested the existence of a natural GHS-like ligand, the identity of which remains unknown.

In this chapter, we focus on the endocrine activities of GHS in relation to sex and age in humans. The majority of the data herein presented mainly refer to the effects of Hexarelin, a hexapeptide, the activities of which are believed to be common to other peptidyl and nonpeptidyl GHS, and which overlap those of GHRP-2, another GHRP-6 superanalog (1–4).

Human GHS Receptors

Specific animal and human GHS receptors have recently been cloned. The GHS receptor is encoded by a rare mRNA with a predicted open reading frame of 366 amino acids, resulting in a transmembrane-spanning topography typified by the G-protein-coupled receptor family. The receptor sequence does not show significant homology with other known receptors, and receptor transcripts are expressed in the pituitary and the hypothalamus (3,9). The existence of GHS-receptor subtypes has already been shown (11).

The human and animal (1) hypothalamus and pituitary gland show the highest specific GHS binding, which is also present in other CNS areas, such as the cerebral cortex, hippocampus, medulla oblongata, and choroid plexuses, but not in the cerebellum, thalamus, striatum, substantia nigra, and corpus callosum (10). GHS receptors have also been demonstrated in fetal human pituitary, and GHS stimulate GH release from human fetal pituitary in vitro as well as in newborns (14,15). The existence of specific GHS-binding sites within the pituitary gland and the CNS may explain the neuroendocrine and also some of the extraneuroendocrine activities of GHS, such as the control of sleep and food intake (6,7).

We studied the effects of sex and age on specific ^{125}I-Tyr-Ala-Hexarelin-binding sites in the human pituitary gland, hypothalamus, and other CNS areas from subjects of both sexes (age ranging from 18 to 93 years). As illustrated in Figure 25.1, GHS-receptor density did not vary as a function of sex in the pituitary, hypothalamus, other human brain areas. These data agree with evidence that the GH response to a maximal dose of GHS in men and women is similar (1–3,16,17). As shown in Figure 25.2, age did not affect the binding of ^{125}I-Tyr-Ala-Hexarelin to membranes from the pituitary gland of middle-aged and elderly subjects. However, an age-related decrease of GHS-receptor density was observed in the hypothalamus of both middle-aged and elderly subjects. As the stimulatory effect of GHS on GH secretion is mainly mediated by hypothalamic actions, these findings could explicate observations that the

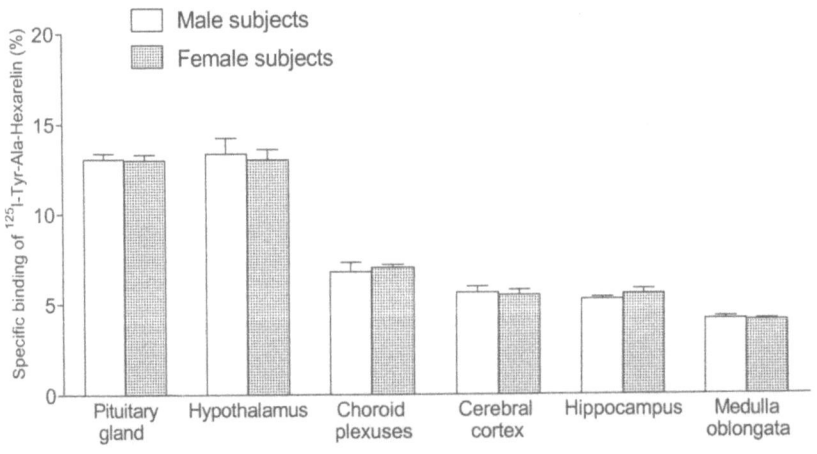

FIGURE 25.1. Specific binding of ^{125}I-Tyr-Ala-Hexarelin to membranes from the pituitary gland and different brain regions of young adult (18–38 years-old) male and female subjects. Values are mean ± SEM of six to eight subjects per group.

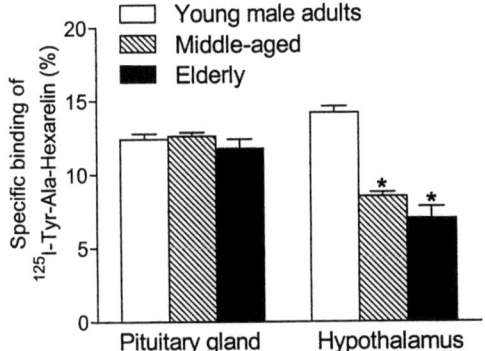

FIGURE 25.2. Specific binding of ^{125}I-Tyr-Ala-Hexarelin to membranes from the pituitary gland and hypothalamus of 21 male subjects of different ages (6 young adults, 18–34 years; 7 middle-aged subjects, 50–65 years; 8 elderly subjects, 75–93 years). *, $p < 0.01$ vs. young adults.

GH-releasing effect of GHS undergoes an age-related reduction from adulthood to aging (1–3,18,19).

Our recent studies demonstrate that GHS receptors are also present in peripheral tissues. In fact, specific binding for Tyr-Ala-Hexarelin was found in the adrenal, heart, ovary, testis, lung, and skeletal muscle, which was even more remarkable than, or at least overlapping with, that found in the pituitary and the hypothalamus (12,13). Significant binding was also evident in kidney, bone epiphyses, and the thyroid gland, but not in smooth muscle, pancreas, parotid gland, and spleen (13). Noteworthy was the ability of Hexarelin, but not MK677, to displace the radioligand from heart membranes (12,20), while there is converse evidence showing that MK677, but not peptidyl GHS, binds specifically in the pancreas (3). Such findings indicate the probable existence of various GHS-receptor subtypes; some could be specific for peptidyl and others for nonpeptidyl GHS.

The functional significance of peripheral GHS receptors is largely unknown. However, there is evidence showing that GHS have GH-independent cardiotropic activities in both animals and humans. In fact, prolonged treatment with Hexarelin and other GHS protects against cardiovascular damage in aged rats as well as in GH-deficient (GHD) rats with postischemic ventricular dysfunction (8). Moreover, Hexarelin, but not rhGH, has inotropic effects in normal young volunteers as well as in hypopituitary patients with severe GHD in the absence of any variations of mean blood pressure, heart rate, or catecholamine levels (21).

Endocrine Activities of GHS in Humans

GH-Releasing Activity

The GH response to GHS is dose related and shows good intraindividual reproducibility, unlike the variability of GHRH (1–4). The GH-releasing activity of GHS is higher in vivo than in vitro, and it is also higher in humans than in animals (1–4). In humans as well as in animals, the GH response to GHS is strongly reduced, although not abolished, by hypothalamopituitary disconnection (22,23); these findings agree with the assumption that a major action of GHS takes place at the hypothalamic level (1–4).

GHS and GHRH have a synergistic effect, and even a very low GHS dosage is able to strikingly potentiate the GHRH-induced GH rise, indicating that these peptides act, at least partially, via different mechanims of action, in agreement with other in vitro and in vivo data in animals (1–3). Nevertheless, there is evidence that GHS need GHRH activity to fully express their GH-releasing effect and that GHS probably trigger the activity of GHRH-secreting neurons. In fact, in humans the GH response to GHS is strongly inhibited, although not abolished, by a GHRH antagonist (24) as well as by hypothalamopituitary disconnection (23).

Among mechanisms underlying the GH-releasing activity of GHS, some evidence favors the hypothesis that these agents act as functional somatostatin antagonists at both the pituitary and the hypothalamic level (1–4). In agreement with this postulate, in humans the GH response to GHS is not modified by substances acting via somatostatin inhibition (such as indirect cholinergic agonists, L–arginine) which, however, potentiate the GHRH-induced GH rise (2,25,26). Moreover, the GH-releasing activity of GHS generally is partially refractory to the inhibitory effect of substances acting via stimulation of hypothalamic somatostatin (such as cholinergic antagonists, beta-adrenergic agonists, and glucose), which almost abolish somatotrope responsiveness to GHRH (2,25–28). GHS are also partially refractory to the inhibitory effects of substances acting directly on somatotrope cells such as free fatty acids and exogenous somatostatin (26,28,29).

GHS are also partially refractory to the negative GH autofeedback (30,31) while showing peculiar sensitivity to IGF-I negative feedback actions. In fact, rhIGF-I blunts the GH response to GHRH, but more markedly inhibits that to HEX, in young volunteers (32).

In adulthood, the GH-releasing effect of GHS is gender independent (16,17). In our experience, the peak GH responses to Hexarelin in a population of normal young men and women are similar (Fig. 25.3). On the other hand, the GH response to GHS undergoes marked age-related variations (1–3,18,19), different from those recorded after GHRH. The GH response to GHRH is maximal in newborns, and then progressively decreases with aging without any change at puberty, while that to HEX is low at birth, strikingly increases at puberty, persists in adulthood, and decreases thereafter in middle-aged individuals in whom activity is similar to that in elderly subjects (Fig. 25.4). We found similar age-related variations also in the synergistic effect of HEX plus GHRH with a clear increase at puberty and decrease in aging (Fig. 25.4). The age-associated reduction of the GH response to GHS alone or combined with GHRH has been found by some authors (18,19,33) but not by others (34).

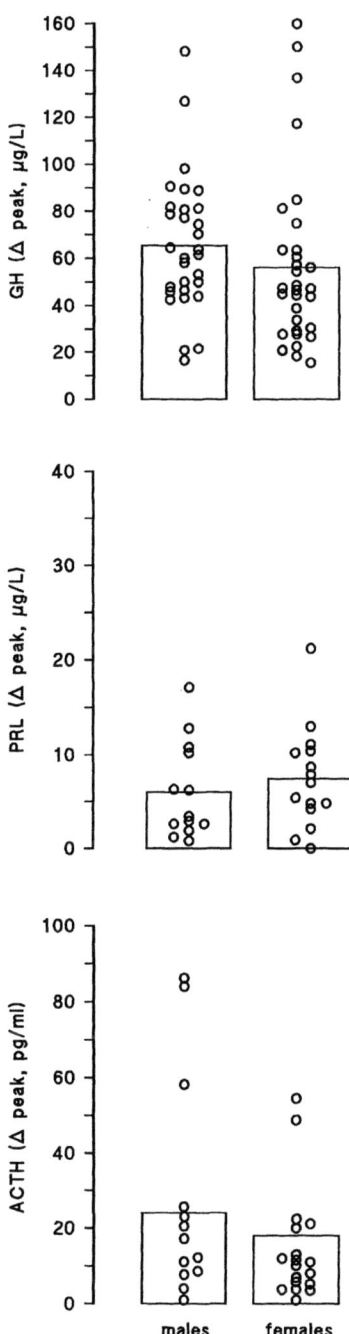

FIGURE 25.3. Group mean (±SEM) and individual responses to Hexarelin (HEX, 2.0 µg/kg i.v.) in normal young men and women.

254　E. Arvat et al.

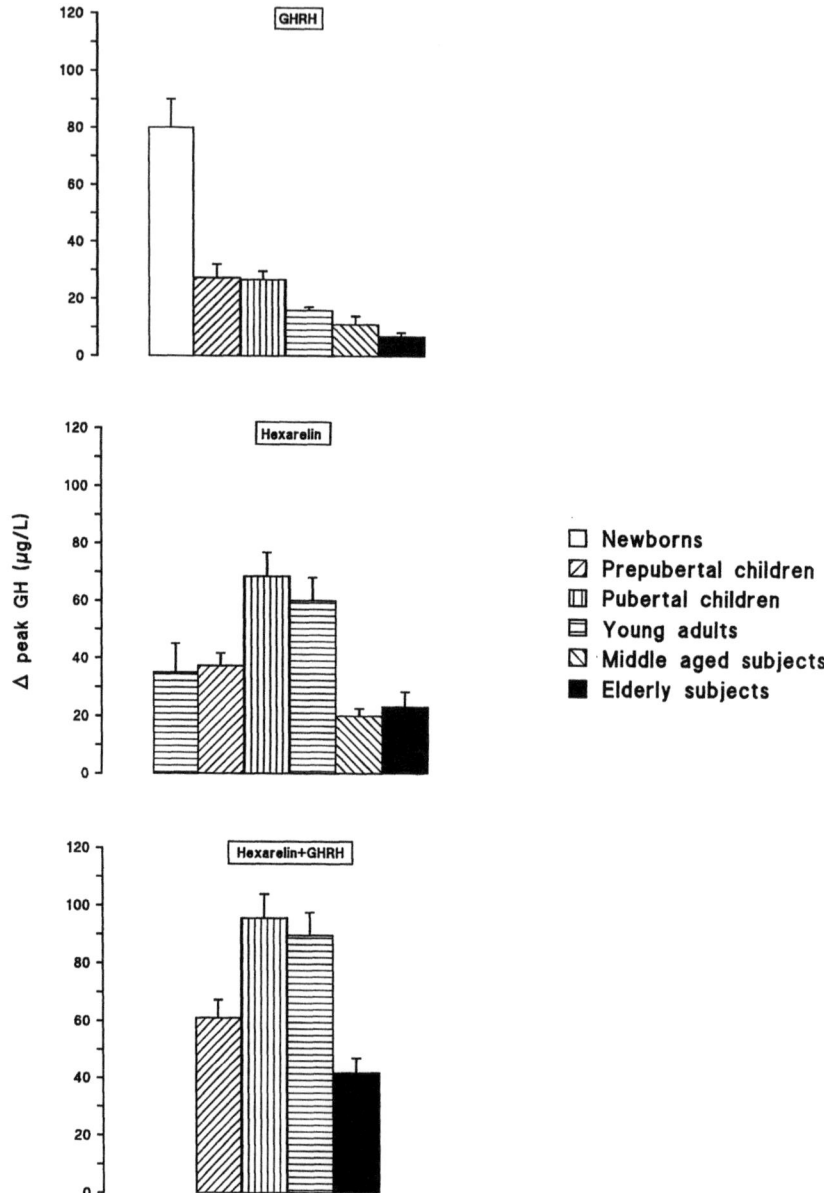

FIGURE 25.4. Age-related variations in the GH response to GHRH, Hexarelin, and Hexarelin plus GHRH in humans. Data are means ± SEM.

The mechanisms underlying the age-related variations in the GH-releasing activity of GHS differ by age. In childhood, the GH response to Hexarelin in prepubertal girls and boys is similar. Interestingly, the somatotrope responsiveness to Hexarelin increases at puberty much more in girls than in boys

(35). Moreover, we have also shown that in prepubertal children the GH response to HEX is increased to pubertal levels by short-term estradiol and testosterone pretreatment, but not by oxandrolone, a nonaromatizable androgen (36). Such findings indicate that estrogens may play a critical role in enhancing the somatotrope sensitivity to the stimulatory activity of GHS, in agreement with some data in animals (1).

The positive influence of estrogens on GHS activity in humans is, however, limited to puberty. In fact, as alluded to earlier, the GHS effect in adult and elderly men and women is similar. Moreover, the low GH response near-maximal to Hexarelin in postmenopausal women is not modified by 3 months of treatment with transdermal estradiol (37), although this intervention is able to restore spontaneous GH secretion (38). Thus, the reduced activity of GHS in aging women does not depend on the fall in gonadal steroid levels. In agreement with data showing a significant reduction in hypothalamic GHRP receptors in the aging human brain, we found that the GH response to Hexarelin in elderly subjects can be increased by higher Hexarelin dosage but remains clearly lower than that elicited by the maximally effective dose in young adults (33).

The most important mechanism accounting for the reduced GH-releasing activity of GHS in aging is probably represented by age-related variations in the neural control of somatotrope function, including GHRH hypoactivity and somatostatinergic hyperactivity (38). In agreement with this assumption, in aged humans the GH response to Hexarelin is increased, but not restored, by GHRH. Somatotrope responsiveness to Hexarelin alone, as well as to Hexarelin plus GHRH, is fully restored to young levels by L-arginine, which probably acts via inhibition of hypothalamic somatostatin release. These effects of L-arginine in aging are more impressive considering that it does not modify the GH response to Hexarelin alone and combined with GHRH in young adults (26,33).

Prolactin-Releasing Activity (PRL)

The stimulatory effect of GHS on PRL secretion in humans is small and dose dependent, with PRL levels remaining within the normal range of basal values and markedly lower than that recorded after thyrotropin-releasing hormone (TRH), metoclopramide, or L-arginine administration (1–3). The lactotroph responsiveness to GHS is not dependent on gender (Fig. 25.3) or age (39), in contrast to that observed for the somatotrope responsiveness. The mechanism underlying the PRL-releasing activity of GHS is not mediated by opioidergic, serotoninergic, or histaminergic pathways (40–42), and could be mediated by direct stimulation of somatomammotrope cells (43,44).

ACTH- and Cortisol-Releasing Activity

The stimulatory effect of GHS on the activity of the HPA axis in humans is not negligible. In fact, the ACTH and cortisol response to GHS overlaps with that after arginine vasopressin or naloxone and is even similar to that after corticotropin-releasing hormone (CRH) in young adults (40,45,46). The stimulatory effect of GHS, however, seems to be an acute neuroendocrine effect, being lost during prolonged treatment (19). The ACTH-releasing activity of GHS is not dependent on gender (Fig. 25.3) but shows peculiar age-related variations. It increases in puberty, then falls in adulthood and, again, tends to increase in aging (39). The increased effect of GHS on ACTH release in puberty could depend on estrogens whereas the later rise in aging would agree with evidence showing HPA hyperactivity caused by neuroendocrine changes in the older brain (47).

Interestingly, the age-related pattern of ACTH response to GHS is different from that to CRH, which effect does not show any increase in puberty but tends to increase in aging (48). On the other hand, the age-related pattern of adrenocorticotroph responsiveness to GHS is dissociated from that of somatotrope cells in aging. Taking also into account the age-independent effect of GHS on lactotroph secretion, these findings indicate that GHS can act at different levels or on different receptors to induce a variety of neuroendocrine responses, in agreement with the existence of different GHS-receptor subtypes both in the pituitary and the CNS (3,11).

The stimulatory effects of GHS on cortisol secretion are the result of their ACTH-releasing activity, which, in turn, totally depends on CNS-mediated mechanisms. In fact, the stimulatory effect of GHS on HPA is abolished by hypothalamopituitary disconnection, and GHS do not stimulate ACTH release from rat pituitary (3,22). Studies in animals have suggested a CRH-mediated action for GHS (49), although the possibility that GHS act via AVP, neuropeptide Y or the putative endogenous ligand for GHS receptors has also been hypothesized (18,50). Interestingly, in young adults the coadministration of Hexarelin and AVP or CRH or naloxone has no interaction or an effect less than additive (40,45), in spite of the well-known synergistic interaction between CRH and AVP on ACTH secretion (48). This result suggests that the ACTH-releasing activity of GHS could be, at least partially, independent of CRH or AVP. It is unlikely that GHS action on HPA is mediated by serotoninergic or histaminergic pathways (41,42), although GABAergic pathways could be involved; in fact, the ACTH-releasing activity of Hexarelin is abolished by pretreatment with alprazolam (51).

The ACTH response to GHS remains sensitive to negative cortisol feedback control under physiological conditions. In normal subjects, the ACTH response to Hexarelin is inhibited by dexamethasone and enhanced by metyrapone (18,51), while it is exaggerated, and even exceeds that to the hCRH stimulus, in patients with pituitary ACTH-dependent Cushing's disease, in spite of their hypercortisolism (46). As specific GHS receptors have been demonstrated in ACTH-secreting tumors (52), the stimulatory effect of GHS in Cushing's disease could take place, at least partially, through direct pituitary actions.

Conclusions

The neuroendocrine actions of GHS are markedly influenced by age in several respects. The GH-releasing activity of GHS increases at puberty, and then declines from adulthood to aging. The PRL-releasing activity does not change throughout the lifespan, while the ACTH-releasing activity increases at puberty, decreases in mid adulthood, and rises slightly in aging. Actions at different loci or on different GHS-receptor subtypes may underly these specific age-related variations. Estradiol probably plays a major role in increasing the GH- and the ACTH-releasing activity of GHS at puberty. The influence of estrogens on GHS activity is, however, limited to puberty, and no gender-related differences to single-dose GHS are present in adulthood. On the other hand, age-related variations in neuroendocrine control underlie the variations in the GH- and ACTH-releasing activities of GHS in aging. The reduced GH response to GHS in elderly subjects likely reflects hypothalamic hypoactivity of GHRH-secreting neurons and concomitant somatostatinergic hyperactivity, although impairment of GHS receptor activity or of putative endogenous GHS-like tone also should be considered.

Acknowledgments. Personal studies reported in this paper were supported by CNR, MURST (9706151106), FSMEM, and Europeptides. The authors thank Dr. M.F. Boghen and R. Deghenghi as well as Dr. F. Catapasso, L. Di Vito, C. Ghè, B. Maccagno, and J. Ramunni for their collaboration to our studies.

References

1. Bowers CY, Veeraragavan K, Sethumadhavan K. Atypical growth hormone releasing peptides. In: Bercu BB, Walker RF, eds. Growth hormone II: Basic and clinical aspects. New York: Springer-Verlag, 1993:203–22.
2. Ghigo E, Arvat E, Muccioli G, Camanni F. Growth hormone-releasing peptides. Eur J Endocrinol 1997;136:445–60.
3. Smith RG, Van der Ploeg LXT, Howard AD, Feighner SD, Cheng K, Hickey GJ, et al. Peptidomimetic regulation of growth hormone secretion. Endocr Rev 1997;18(5): 621–45.
4. Deghenghi R. Growth hormone releasing peptides. In: Bercu B, Walker R, eds. Growth hormone secretagogues. New York: Springer-Verlag, 1993:85–102.
5. Ghigo E, Arvat E, Camanni F. Growth hormone secretagogue as corticotrophin-releasing factors. Growth Horm IGF Res 1998;8:145–8.
6. Locke W, Kirgis HD, Bowers CY, Abdo AA. Intracerebroventricular growth-hormone-releasing peptide-6 stimulates eating without affecting plasma growth hormone responses in rats. Life Sci 1995;56:1347–52.
7. Frieboes RM, Murck H, Maier P, Schier T, Holsboer F, Steiger A. Growth hormone-releasing peptide-6 stimulates sleep, growth hormone, ACTH and cortisol release in normal man. Neuroendocrinology 1995;61:584–9.

8. Berti F, Muller E, De Gennaro Colonna V, Rossoni G. Hexarelin exhibits protective activity against cardiac ischaemia in hearts from growth hormone-deficient rats. Growth Hormone & IGF Res 1998;8:149–52.
9. Howard AD, Feighner SD, Cully DF, Arena JP, Liberator PL, Rosenblum CI, et al. A receptor in pituitary and hypothalamus that functions in growth hormone release. Science 1996;273:974–7.
10. Muccioli G, Ghè C, Ghigo MC, Papotti M, Arvat E, Boghen MF, et al. Specific receptors for synthetic GH secretagogues in the human brain and pituitary gland. J Endocrinol 1998;157:99–106.
11. Ong H, McNicoll N, Escher E, Collu R, Deghenghi R, Locatelli V, et al. Identification of a pituitary GHRP receptor subtype by the photoaffinity labeling approach using ^{125}I p-benzoyl-phenylalanine Hexarelin derivative. Endocrinology 1998;139:432–5.
12. Muccioli G, Ghe' C, Ghigo MC, Arvat E, Papotti M, Boghen MF, et al. GHRP receptors in pituitary, central nervous system and peripheral human tissues. J Endocrinol Invest 1997;20(Suppl)4:52.
13. Muccioli G, Ghe' C, Catapano F, Papotti M, Ghigo E, Ong H, et al. Specific receptors for synthetic growth hormone secretagogue in human tissues. In: Proceedings, XIII. International Congress of Pharmacology, Munchen, Germany, 1998, p 548.
14. Shimon I, Yan X, Melmed S. Human fetal pituitary expresses functional growth hormone-releasing peptide receptors. J Clin Endocrinol Metab 1998;83:174–8.
15. Bartolotta E, Bellone J, Aimaretti G, Arvat E, Benso L, Deghenghi R, et al. The GH releasing effect of Hexarelin, a new synthetic hexapeptide, in newborns is lower than in young adults. J Pediatr Endocrinol Metab 1997;10:491–7.
16. Penalva A, Pombo M, Carballo A, Barreiro J, Casanueva FF, Dieguez C. Influence of sex, age and adrenergic pathways on the growth hormone response to GHRP-6. Clin Endocrinol (Oxf) 1993;38:87–91.
17. Ghigo E, Arvat E, Gianotti L, Imbimbo BP, Lenaerts V, Deghenghi R, et al. Growth hormone-releasing activity of Hexarelin, a new synthetic hexapeptide, after intravenous, subcutaneous, intranasal and oral administration in man. J Clin Endocrinol Metab 1994;78:693–8.
18. Arvat E, Camanni F, Ghigo E. Age-related growth hormone-releasing activity of growth hormone secretagogues in humans. Acta Paediatr Suppl 1997;423:92–6.
19. Chapman IM, Bach MA, Van Cauter E, Farmer M, Krupa D, Taylor AM, et al. Stimulation of the growth hormone (GH)-insulin-like growth factor I axis by daily oral administration of a GH secretagogue (MK-677) in healthy elderly subjects. J Clin Endocrinol Metab 1996;81:4249–57.
20. Bodard V, McNicoll N, Carriere P, Bouçhard JF, Lamontagne D, Sejlitz T, et al Identification and characterization of a new GHRP receptor in the heart. In: Proceedings, 80th annual meeting of the Endocrine Society, New Orleans, LA, 1998, p 302.
21. Valetto MR, Podio V, Broglio F, Bertuccio G, Arvat E, et al. The acute administration of Hexarelin, a peptidyl GH secretagogues, has GH-independent, positive inotropic effect in humans (abstract) In: Hormones and the heart, Naples, 1998, p 54.
22. Hickey GJ, Drisko J, Faidley T, Chang C, Anderson L, Nicolich S, et al. Mediation by the central nervous system is critical to the in vivo activity of the GH secretagogue L-692,585. J Endocrinol 1996;148:371–80.
23. Popovic V, Damjanovic S, Micic D, Djurovic M, Dieguez C, et al. Blocked growth hormone-releasing peptide (GHRP-6)-induced GH secretion and absence of the synergic action of GHRP-6 plus GH-releasing hormone in patients with hypothalamopituitary disconnection: evidence that GHRP-6 main action is exerted at the hypothalamic level. J Clin Endocrinol Metab 1995;80:942–7.

24. Pandya N, De Mott-Friberg R, Bowers CY, Barkan AL, Jaffe CA. Growth hormone (GH)-releasing peptide-6 requires endogenous hypothalamic GH-releasing hormone for maximal GH stimulation. J Clin Endocrinol Metab 1998;83:1186–9.
25. Penalva A, Carballo A, Pombo M, Casanueva FF, Dieguez C. Effect of growth hormone (GH)-releasing hormone (GHRH), atropine, pyridostigmine, or hypoglycemia on GHRP-6-induced GH secretion in man. J Clin Endocrinol Metab 1993;76:168–71.
26. Arvat E, Gianotti L, Di Vito L, Imbimbo BP, Lenaerts V, Deghenghi R, et al. Modulation of growth hormone-releasing activity of hexarelin in man. Neuroendocrinology 1995;61:51–6.
27. Arvat E, Gianotti L, Ramunni J, Di Vito L, Deghenghi R, Camanni F, et al. Influence of beta-adrenergic agonists and antagonists on the GH-releasing effect of Hexarelin in man. J Endocrinol Invest 1996;19:25–9.
28. Maccario M, Arvat E, Procopio M, Gianotti L, Grottoli S, Imbimbo BP, et al. Metabolic modulation of the growth-hormone-releasing activity of hexarelin in man. Metabolism 1995;44(1):134–8.
29. Massoud AF, Hindmrsh PC, Brook CGD. Interaction of the growth hormone releasing peptide hexarelin with somatostatin. Clin Endocrinol 1997;47:537–47.
30. Massoud AF, Hindmrsh PC, Brook CG. Hexarelin induced growth hormone release is influenced by exogenous growth hormone. Clin Endocrinol 1995;43:617–21.
31. Arvat E, Di Vito L, Gianotti L, Ramunni J, Boghen MF, Deghenghi R, et al. Mechanisms underlying the negative growth hormone (GH) autofeedback on the GH-releasing effect of Hexarelin in man. Metabolism 1997;46:83–8.
32. Ghigo E, Gianotti L, Arvat E, Ramunni J, Valetto MR, Broglio F, et al. Effects of rhIGF-I administration on GH secretion both spontaneous and stimulated by GHRH or Hexarelin, a peptidyl GH secretagogue, in humans. J Clin Endocrinol Metab 1999;84:285–90.
33. Arvat E, Ceda GP, Di Vito L, Ramunni J, Gianotti L, Broglio F, et al. Age-related variations in the neuroendocrine control, more than impaired receptor sensitivity, cause the reduction in the GH-releasing activity of GHRPs in human aging. Pituitary 1998;1:51–8.
34. Micic D, Popovich V, Doknic M, Macut D, Dieguez C, et al. Preserved growth hormone (GH) secretion in aged and very old subjects after testing with the combined stimulus GH-releasing hormone plus GH-releasing Hexapeptide-6. J Clin Endocrinol Metab 1998;83:2569–72.
35. Bellone J, Aimaretti G, Bartolotta E, Benso L, Imbimbo BP, Lenhaerts V, et al. Growth hormone-releasing activity of Hexarelin, a new synthetic hexapeptide, before and during puberty. J Clin Endocrinol Metab 1995;80:1090–4.
36. Loche S, Colao A, Cappa M, Bellone J, Aimaretti G, Farello G, Faedda A, et al. The growth hormone response to Hexarelin in children: reproducibility and effect of sex steroids. J Clin Endocrinol Metab 1997;82:861–4.
37. Arvat E, Gianotti L, Broglio F, Maccagno B, Bertagna A, Deghenghi R, et al. Oestrogen replacement does not restore the reduced GH-releasing activity of hexarelin, a synthetic hexapeptide, in postmenopausal women. Eur J Endocrinol 1997;136:483–7.
38. Ghigo E, Arvat E, Gianotti L, Ramunni J, Di Vito L, Maccagno B. Human aging and GH-IGF-I axis. J Pediatr Endocrinol Metab 1996;9:271–8.
39. Arvat E, Ramunni J, Bellone J, Di Vito L, Baffoni C, Broglio F, et al. The GH, prolactin, ACTH and cortisol responses to Hexarelin, a synthetic hexapeptide, undergo different age-related variations. Eur J Endocrinol 1997;137:635–42.

40. Korbonits M, Trainer PJ, Besser GM. The effect of an opiate antagonist on the hormonal changes induced by hexarelin. Clin Endocrinol 1995;43:365–71.
41. Arvat E, Maccagno B, Ramunni J, Gianotti L, Di Vito L, Deghenghi R, et al. Effects of histaminergic antagonists on the GH-releasing activity of GHRH or hexarelin, a synthetic hexapeptide, in man. J Endocrinol Invest 1997;20:122–7.
42. Arvat E, Maccagno B, Ramunni J, Broglio F, Lanfranco F, Giordano R, et al. Influence of galanin and serotonin on the endocrine response to Hexarelin, a synthetic peptidyl GH-secretagogue, in normal women. J Endocrinol Invest 1998;21:673–9.
43. Ciccarelli E, Grottoli S, Razzore P, Gianotti L, Arvat E, Deghenghi R, et al. Hexarelin, a synthetic growth hormone releasing peptide, stimulates prolactin secretion in acromegalic but not in hyperprolactinemic patients. Clin Endocrinol 1996;44:67–71.
44. Renner U, Brockmeier S, Strasburger CJ, Lange M, Schopohl J, Muller OA, et al. Growth hormone (GH)-releasing peptide stimulation of GH release from human somatotroph adenoma cells: interaction with GH-releasing hormone, thyrotropin-releasing hormone, and octreotide. J Clin Endocrinol Metab 1994;78:1090–6.
45. Arvat E, Maccagno B, Ramunni J, Di Vito L, Broglio F, Deghenghi R, et al. Hexarelin, a synthetic growth-hormone releasing peptide, shows no interaction with corticotropin-releasing hormone and vasopressin on adrenocorticotropin and cortisol secretion in humans. Neuroendocrinology 1997;66:432–8.
46. Ghigo E, Arvat E, Ramunni J, Colao A, Gianotti L, Deghenghi R, et al. Adrenocorticotropin- and cortisol-releasing effect of hexarelin, a synthetic growth hormone-releasing peptide, in normal subjects and patients with Cushing's syndrome. J Clin Endocrinol Metab 1997;82:2439–44.
47. Sapolsky RM, Krey LC, McEwen B. The neuroendocrinology of stress and aging: the glucocorticoid cascade hypothesis. Endocr Rev 1986;7:284–301.
48. Orth DN. Corticotropin-releasing hormone in humans. Endocr Rev 1992;13:164–91.
49. Thomas GB, Fairhall KM, Robinson ICAF. Activation of the hypothalamo-pituitary-adrenal axis by the growth hormone (GH) secretagogue, GH-releasing peptide-6, in rats. Endocrinology 1997;138:1585–91.
50. Dickson SL, Luckman SM. Induction of c-*fos* messenger ribonucleic acid in neuropeptide Y and growth hormone (GH)-releasing factor neurons in the rat arcuate nucleus following systemic injection of the GH segretagogue, GH-releasing peptide-6. Endocrinology 1997;138:771–7.
51. Arvat E, Maccagno B, Ramunni J, Di Vito L, Gianotti L, Broglio F, et al. Effects of dexamethasone and alprazolam, a benzodiazepine, on the stimulatory effect of Hexarelin, a synthetic GHRP, on ACTH, cortisol and GH secretion in humans. Neuroendocrinology 1998;67:310–6.
52. De Keyzer Y, Lenne F, Bertagna X. Widespread transcription of the growth hormone-releasing peptide receptor gene in neuroendocrine human tumors. Eur J Endocrinol 1997;137:715–8.

26

Gender Impact on the GH Response to Exercise

Laurie Wideman, Judy Y. Weltman, Nikhita Shah,
E. Shannon Story, Cyril Y. Bowers, Johannes D. Veldhuis,
and Arthur L. Weltman

Aerobic exercise is a powerful, physiological stimulus for GH release (1–7). Although the mechanisms underlying exercise-induced GH release have not been elucidated fully, they likely include growth hormone-releasing hormone (GHRH) release, somatostatin (SMS) withdrawal, putative natural ligand release (i.e., GH-releasing peptide, GHRP), or a combination of these hypothalamic responses. If the mechanism entails solely somatostatin withdrawal, than the combination of L-arginine plus exercise should result in a GH response that is similar to that of either given alone. Likewise, if the mechanism involves solely GHRH or GHRP release, then stimulation of GHRH or GHRP plus exercise should result in a GH response that is similar to either given alone. Clearly, the issue of mechanistic control of exercise-induced GH release becomes more difficult if a combination of hypothalamic responses is involved.

GH-Releasing Peptides and Arginine

GH-releasing peptides (GHRP) are a new class of synthetic oligopeptides that are potent stimulators of GH release in animals as well as humans (8). The GHRPs may be related to an endogenous peptide that is involved in the regulation of GH secretion along with GHRH or SMS (8) and activates the same receptor as synthetic GHRPs. However, a natural GHRP-like ligand has not yet been isolated. It is now well accepted that GHRP-induced GH release occurs via a novel nonopiate, non-GHRH receptor that uses a calcium-dependent, cAMP-independent mechanism (9). Expression of the GHRP receptor has been localized to both the hypothalamus (10) and the pituitary (11), and has recently been cloned (12,13). The GHRPs release GH in a dose-related fashion (14). Several studies (8,15–17) have demonstrated that GHRP and

GHRH have synergistic effects on GH release in humans, which indicates that these substances are working through nonidentical mechanisms (8,16,18–20). GHRP and GHRH have been shown to stimulate GH release in a synergistic manner in both adults and children (21,22). In contrast to these results, Tiulpakov et al. (23) did not find a synergistic reaction between GHRH and GHRP-2. The actions of the GHRPs are independent of and complementary to native GHRH, but it appears that GHRP and GHRH also interact, with GHRH acting in a permissive fashion (8). Currently, there are no published studies investigating the effects of GHRP-2 during exercise. The possible role of an endogenous GHRP-2-like molecule in the mechanistic control of exercise-induced GH release is plausible and warrants investigation.

Arginine (A) is a secretagogue for GH that exerts its effects presumptively by inhibiting endogenous SMS release (24,25). Arginine potentiates GHRH-induced GH release (25–27). Arginine potentiation of GHRH-induced GH release is similar to the interaction of pyridostigmine and propranolol (both affect endogenous somatostatinergic tone) (28,29). Ghigo et al. (26) showed that pyridostigmine and A given alone or together resulted in a similar increase in GH. These results further support the hypothesis that arginine exerts its effects via inhibition of endogenous SMS release. Currently, it is accepted that the β_2-adrenergic inhibition of GH release occurs via increased SMS release, because β_2-agonists block L-arginine- and pyridostigmine-stimulated GH secretion. β_2-adrenergic stimulation of SMS release may be more potent than the inhibitory effects of A or pyridostigmine on SMS release (27,30). This may be an important consideration when interpreting data from studies where exercise and arginine are used together as a feed-forward system for α-2-adrenergic stimulation of GH release has been hypothesized (4).

Gender Differences in Basal GH Release

The correlations between GH release and characteristics such as body fat, age, and fitness level are stronger in males than premenopausal females (31). For example, 24-h GH area under the curve (AUC) in women is less negatively influenced by the effects of increased body fat than this same parameter in men. The 24-h calculated GH AUC is higher in young females than young males and the magnitude of the effect ranges between 1.5- and 3 fold, dependent upon the investigation reported (32–34). By employing deconvolution analysis to assess gender differences in mean serum GH concentration, van den Berg et al. (33) reported that GH secretory pulse frequency and endogenous half-life were not different in males and females. The authors noted that females had higher maximal and incremental serum GH peak amplitudes and that these differences were caused by increased GH-secretory burst mass. In addition, elevated interpulse GH basal secretion has been observed in females compared to males (33,34). These gender differences may be the result of estradiol, as estrogen clearly has an amplifying effect on GH secretion (32).

Veldhuis et al. (35) suggested from studies in men that estradiol may have a larger effect on basal GH secretion, while regulation of pulsatile secretion may be more dependent on testosterone. Although gender differences in GH secretion have been noted during rest, the gender-based control of GH secretion during exercise has not been fully investigated in the human. Gender differences in the orderliness of the GH release pattern in humans were noted by Pincus et al. (36). Men were observed to have more orderly GH release patterns compared to women (36), and this difference may be the result of increased GHRH responsiveness or reduced somatostatin inhibitory tone in women (33,37).

In some exercise studies, males and females responded similarly (2,38). Lassarre et al. (2) noted that females tended to have unspecified GH peaks before exercise that did not return to baseline within 1 h, but the response pattern to exercise was similar to that of males regardless of the baseline differences. Data from our laboratory indicate that the response pattern to exercise was similar in men and women, with the exception that women attained peak GH concentration earlier than men (24 vs. 32 min) (unpublished observations). Despite the temporal disparity, the incremental GH AUC was not different, indicating that the magnitude of the added GH release in response to exercise was similar in men and women. Lang et al. (39) showed a greater GHRH-induced incremental release of GH in premenopausal women than age-matched males. The GHRH dose needed to produce a half-maximal effect (ED_{50}) was 0.2 µg/kg for women and 0.4 µg/kg for men. In a more recent study, Benito et al. (40) reported a higher basal GH and greater maximal GH response to GHRH in women versus men. Bowers (8) reported that there was no significant difference in GH release in men and women when maximal GHRP was used to stimulate GH release, which suggests that gonadal hormones may have little influence on GHRP-induced release of GH. Similar results for GHRP-6-induced GH release were reported by Penalva et al. (14). Recently, data from Dr. Bowers' laboratory have shown that females have a lower ED_{50} for GHRP-2 compared to men, at submaximal doses. Our studies revealed that in postmenopausal women, estrogen treatment increased GHRP-2 potency (heightened pituitary sensitivity) (41). With maximal GHRP doses, there was no gender difference in GHRP-2-induced GH release. The GHRP-2 dose used in the current study (1 µg/kg) has been shown to cause maximal GH release in both young males and females (unpublished observations), although it may not be maximal in the presence of exogenous estradiol (41).

Exercise-Induced GH Release

Acute Exercise

Exercise of appropriate intensity and duration is a potent physiological stimulus for GH release (2–4,6,38,42,43). Exercise-induced GH release is delayed until approximately 10–15 min into exercise (2,3) and peaks toward the end of exercise (2,3) or slightly after its termination (42). Felsing et al. (43) reported that high-intensity exercise must last at least 10 min to reliably stimulate GH release. Intersubject variability in peak exercise-induced GH concentration is large (2–4,6,42,43). Similar variability in GH response has been noted with various other stimuli (44). At least some of this variability can be attributed to differences in the age and body mass index of subjects used in different studies (31). Exercise-induced GH release appears to be intensity dependent (42,45). Some reports suggested that an intensity threshold must be reached before exercise-induced GH release is activated (46). Viru (45) suggested that the activation threshold is 50% of VO_2max. However, recent data from our laboratory show a linear response between GH release and exercise intensity (47). Lassarre et al. (2) and Thompson et al. (4) reported that the exercise-associated half-life of GH decreased slightly from that reported at rest (decreased from ~19 min to ~16 min).

Hartley et al. (48) observed that extremely high intensity exercise actually decreased exercise-induced GH release compared to moderate-intensity exercise. However, the high-intensity exercise bout was completed on the same day as the moderate- and low-intensity bouts and was always completed last. Another study suggested that multiple bouts of exercise in a single day will decrease the exercise-induced GH release after several sessions (49). This decrease in exercise-induced GH release was alleviated when longer periods of rest (2–3 h) were allowed between exercise sessions (49). Conversely, Felsing et al. (43) showed that three bouts of exercise each separated by 1 h caused increasing amounts of GH release. However, the exercise bouts also increased in duration each time. We recently examined the effects of repeated bouts of exercise on GH release and concluded that regardless of whether the three constant-load exercise sessions (equal intensity and duration) were separated by 1 h or 4 h, there was no difference in the magnitude of the exercise-induced GH release (1). Indeed, the GH response to repeated bouts of exercise appeared to be augmented during subsequent exercise.

Training Studies

Hartley et al. (48) reported that 7 weeks of training (short-term) decreased exercise-induced GH release after subsequent mild, moderate, and high-intensity work. Similarly, Weltman et al. (7) noted that the GH response to constant load exercise was reduced after only 3 weeks of training and did not decrease further after 6 weeks of training. Chronic exercise training has been shown to increase the 24-h integrated AUC for GH and pulse amplitude in women who trained for 1 year (long-term), if at least some of the training was done at greater than lactate threshold (LT). This effect may be physiologically important because the increase in pulsatile GH release may result in increased tissue trophic actions (6).

Potential Mechanisms of Exercise-Induced GH Release

Exercise is a potent physiological stimulus of GH release, but the actual mechanism(s) by which this occurs remain(s) to be elucidated. Many neurotransmitters, including dopamine, norepinephrine, acetylcholine, and opioids, are likely involved in exercise-induced GH release. Investigators have attempted to discern the contribution of these various neurotransmitters to the exercise-induced release of GH, but no one pathway has been established as the primary control mechanism. It is likely that several of these mechanisms are active concomitantly during exercise-induced GH release. Regardless of the neurotransmitter(s) involved, the final common effect or pathway must still involve natural ligand release, increases in GHRH, and/or decreases in SMS.

The adrenergic system has been shown to influence the exercise-induced release of GH. Studies using metoprolol (cardioselective β-blocking agent) and propranolol (nonselective β-antagonist) have unequivocally indicated that β-receptor blockade increases the exercise-induced release of GH (50–53). The results for α-blockade are less conclusive. Hansen (50) demonstrated that phentolamine nearly totally suppressed GH release during exercise, while Sutton and Lazarus (51) found that phentolamine had little effect. The experimental conditions and the exercise bouts in the two studies were similar, and the disparity in results is difficult to explain. However, the conflicting results might be caused by the great variability in GH responsiveness among individuals and within individuals tested at different times (54).

Cappa et al. (55) found that pyridostigmine (cholinergic agonist) administered with exercise caused an increase in GH that was not significantly different from the response obtained when each was used separately and the results added. Pyridostigmine likely decreases SMS release, and because the combination of pyridostigmine plus exercise is additive, one could conclude that the effect of exercise involves increased release of GHRH (55). Thompson et al. (4), however, reported that pyridostigmine alone or pyridostigmine plus naltrexone potentiated exercise-induced GH release compared to placebo. The authors suggest the possibility of a feed-forward system with cholinergic innervation modulating α_2-adrenergic neurons, thereby, elevating GH release. According to this schema, increased cholinergic tone could potentiate the GH response to adrenergic stimulation (via norepinephrine), and the former response may be mediated by decreased SMS release from the hypothalamus (4). Atropine (a muscarinic antagonist, thought to release SMS) inhibits the GH response to exercise (56), which would suggest that atropine and exercise work through a similar (final) mechanism. However, enhanced somatostatinergic tone caused by cholinergic antagonists inhibits GH responsiveness to several stimuli including GHRH. Casanueva et al. (56) suggested that their results indicate that exercise-induced GH release works through the cholinergic system. However, whether exercise acts specifically via GHRH stimulation, SMS inhibition, or both could not be distinguished.

Previous experiments investigating the effects of naloxone (opiate antagonist) on exercise-induced GH release have been equivocal. Moretti et al. (57) used high doses of naloxone with 20-min of high-intensity exercise in highly trained competitive athletes and found that naloxone completely blocked exercise-induced GH release. Unlike these results, Coiro et al. (58) found that naloxone did not change exercise-induced GH release. However, subjects in the Coiro et al. (58) study were not trained athletes, which allows the speculation that training can actually induce the development of opioid pathways for the regulation of GH release during exercise. This may explain some of the disparity in studies investigating the effects of endogenous opioids on the exercise-induced release of GH. Coiro et al. (58) reported that the combination of sodium vaplroate (a GABAergic agonist) and naloxone completely abolished the exercise-induced release of GH. This result suggests that there may be a GABAergic mechanism involved in the GH response to exercise in humans and that an opioid pathway can counter the inhibitory effects of GABA in this situation (58). The ability of cholinergic and opioid pathways to release GH during exercise also was investigated by Thompson et al. (4). Neither resting GH concentration nor its half-life was altered by naltrexone (opioid antagonist) or pyridostigmine (cholinergic agonist). However, both the mean and integrated serum GH concentrations during exercise were significantly increased with pyridostigmine alone or pyridostigmine plus naltrexone (4). The fact that naltrexone did not alter GH release tends to indicate that endogenous opioids are not important for the exercise-induced GH release (4).

In summary, the precise mechanism(s) of exercise-induced GH release have not been elucidated fully, although the final pathway likely involves GHRH stimulation or decreased SMS release. The ability of GHRP-2 to synergize with GHRH is an intriguing phenomenon, although the in vivo mechanism(s) of GHRP-2 action and interactions with the exercise stimulus are not yet understood, which makes it difficult to predict the outcome of an exercise plus GHRP-2 infusion trial. In the remainder of this chapter, we present preliminary data from a clinical study that examined further the mechanism underlying exercise-induced GH release. The differences in GH release between genders have not been extensively investigated during exercise. Therefore, we also compared exercise-induced GH responses between genders.

Methods

Subjects between the ages of 18 and 35 years completed this study. All subjects provided written informed consent as approved by the Human Investigation Committee at the University of Virginia. Subjects were not taking any medications and were habitual exercisers (20–30 min of aerobic exercise, 3–4 times/week). Subjects completed a peak oxygen consumption/lactate threshold test on a cycle ergometer. Initial power output was 40 W for women and 60 W for men, and the power output was increased 15 W every 3 min until volitional fatigue. Metabolic measures were collected using standard open-circuit spirometric techniques (Sensormedics metabolic cart 2700Z; Yorba Linda, CA) and heart rate was determined electrocardiographically. Blood samples were taken at rest and during the last 15 s of each stage for lactate determination (YSI Instruments 2700; Yellow Springs, OH). The lactate threshold (LT) was determined from the blood lactate–power output (PO) relationship (25). The power output for the constant load aerobic exercise sessions (CLPO) was calculated as follows: CLPO = PO at LT + 0.50(PO at V_{o2} peak – PO at LT). After the initial exercise test, each subject was studied at the GCRC on a total of eight occasions, four with exercise (Ex) and four at rest (R). The following conditions were examined: (1) saline (S), (2) arginine (A), (3) GHRP-2 (G), and (4) arginine + GHRP-2 (AG). Admissions were randomly assigned, scheduled at least 2 days apart (i.e., no closer than 48 h between consecutive start times) and all females were tested in the early follicular phase (days 2 to 8) of the menstrual cycle. At 1700 on the evening before the study, volunteers received a standardized constant meal (55% carbohydrate, 30% fat, and 15% protein), based on body weight. After an overnight fast, blood samples were withdrawn at 10-minute intervals from 0600 to 1200. At 0730, an i.v. infusion of either saline or 30 g of L-arginine was given over 30 min. At 0800 a bolus i.v. injection of saline or 1 µg/kg of GHRP-2 was given. On the aerobic exercise admission, immediately after the i.v. bolus, subjects exercised for 30 min (0810–0840) at the predetermined power output. Subjects rested quietly in their rooms before and after aerobic exercise and on the nonexercise admission.

Serum GH concentrations for all samples (0600–1200, all admissions) were measured in the GCRC Core Laboratory using an ultra-sensitive (0.002 µg/l threshold) chemiluminescence assay (Nichols, San Juan Capistrano, CA). The mean intraassay coefficient of variation (CV) for the GH assay was 5.97% and the interassay CV was 9.9%. Incremental GH area under the curve (AUC) was measured using the trapezoidal rule (59). The prefitting procedure (PULSE2) and multiple-parameter deconvolution method followed a set fitting pathway and was employed to derive quantitative estimates of attributes of GH secretory events from the measured serum GH concentrations (60). A pulse of GH secretion was approximated by a Guassian distribution of secretory rates, and the subject-specific monoexponential half-life of apparent metabolic removal of endogenous GH was estimated (61). Basal secretion was estimated concurrently as previously described (35). GH secretory pulses were considered significant if the fitted amplitude could be distinguished from zero with 95% statistical confidence. The GH secretory pulse half-duration, defined as the duration in minutes of the calculated secretory burst event at half-maximal amplitude, GH half-life of elimination, and GH distribution volume were assumed to be constant throughout any one study period for an individual. The mass of GH secreted per pulse was estimated as the area of the calculated secretory pulse (µg/l distribution volume) (61). The endogenous pulsatile GH production rate was estimated as the product of the number of secretory pulses and the mean GH mass secreted per pulse.

Results

Representative serum GH concentration data from one male and one female are presented in Figure 26.1A, deconvolved and secretion curves for GH from the same two subjects are presented in Figure 26.1B. Visual inspection of the preliminary data reveals that regardless of gender or condition (R vs. Ex), GH release increased in each of the three stimulated conditions and the magnitude of the response was S<A<G<AG. In all treatments, the addition of exercise resulted in further increases in GH release above that observed at rest. Preliminary data reveal higher basal serum GH secretion in women compared to men (deconvolution parameters not shown). Males and females attained similar GH concentrations for G and AG (Fig. 26.1A).

Discussion

Inspection of Figure 26.1A and 26.1B reveals that GH responses to A, G, AG, and exercise are highly variable among individuals. Both men and women exhibit a wide range of GH response to the different stimuli. In addition, the variability of the GH response is not condition dependent (i.e., there is similar variability in the rest or exercise condition). Individual variability in GH release occurs following other stimuli (37,54), including exercise (2–4,6,42,43), and may reflect the varying endogenous hypothalamic-somatotroph rhythm (54). Similar to data reported previously (33,34), inspection of the deconvolution results (not shown) revealed that women maintain greater basal GH secretion than men.

Our preliminary data indicate that women attained higher GH concentrations in both the S and A treatments. Previously, Merimee et al. (62) demonstrated an increased GH response in women to L-arginine. Further, Merimee et al. (62) reported that the increase in GH in women was not actually caused by a higher L-arginine dose (when expressed per kg of body weight) in women. The four subjects presented here had similar body weights and therefore the average load of arginine at a fixed dose of 30 g was similar in men and women. Merimee et al. (62) also demonstrated that estrogen could increase the L-arginine effect and therefore concluded that the gender differences in response to arginine were caused by higher estrogen exposure in women versus men. The role of estrogen in the actual mechanism driving increased responsiveness of women to this amino acid has not been elucidated but may entail heightened GHRH release made evident by the withdrawal of SMS by L-arginine pretreatment. Because the increased GH release in response to arginine has been shown to be independent of the arginine load (62), it is possible that, as observed in the rat model, there may be differences in the human in the role(s)

26. Gender Impact on GH Response to Exercise   269

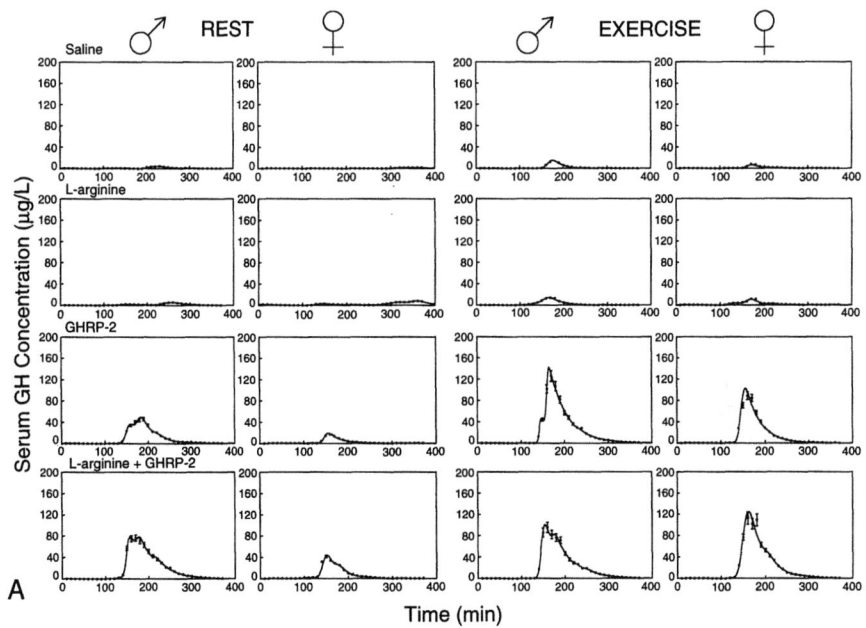

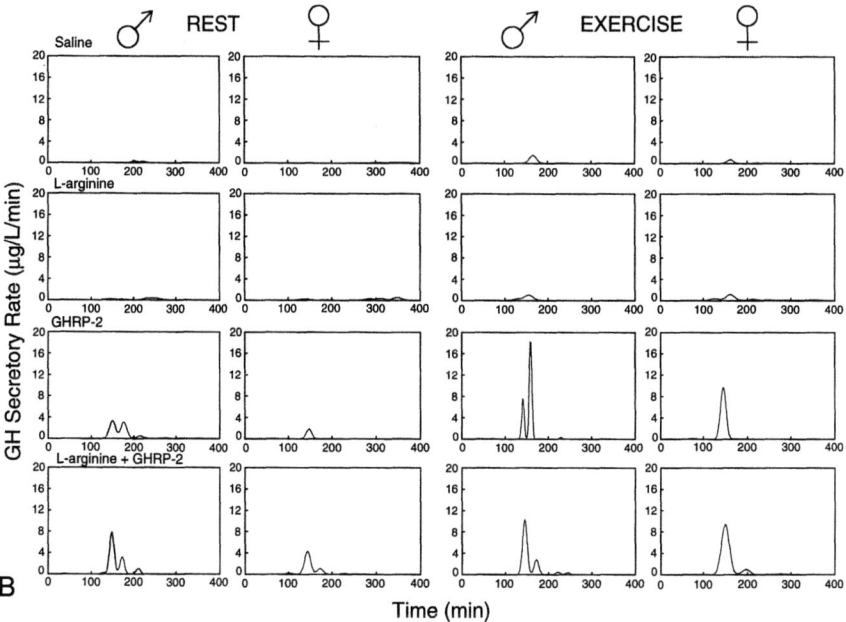

FIGURE 26.1. Representative serum GH concentration curves from one male and one female are shown in *A*; the corresponding GH secretion profiles given in *B*. Treatments or interventions are noted above each subpanel. L-Arginine infusion over 30 min was begun at 90 min (corresponding to 0730) and other stimuli at 120 min (0800).

of GHRH and SMS in the pulsatile control of GH release in the males and females and that these differences result in a differential GH response(s) to arginine. This idea is supported by the findings of Pincus et al. (36), namely that women have less orderly release of GH compared to men.

The peak serum GH concentration attained during the G and AG treatments appears to be similar for men and women. The lack of gender difference in the GHRP-2 treatment supports previous data suggesting that, for GHRP given alone, there is no gender difference in GH release (14). Although recent studies (33,34), have combined highly sensitive GH assay techniques with deconvolution analysis to delineate gender differences in GH secretion, elucidating the neural mechanisms subserving gender differences in humans is challenged by obvious limitations. Therefore, there is less consensus in the literature about gender differences in GH release in humans, specifically in stimulated states. Clearly, further investigation in this area is needed.

The preliminary data from the present study support earlier reports that exercise is a potent stimulus of GH release (2–4,6,38,42,43). The qualitative pattern of the GH response during rest and exercise in each of the treatments appears to be similar in men and women (Figure 26.1A,B). These data confirm observations reported by Lassarre et al. (2) that males and females have a similar exercise-induced GH pattern. Interestingly, we did not observe the extent of delay in GH release that has been reported with exercise (2,3), which may be due to the more frequent blood sampling and higher intensity of exercise employed in the current study. Some data suggest that the adrenergic, cholinergic, and GABAergic systems modulate exercise-induced GH release (4,52,53,55,58). Although this is not established, these systems may be differentially affected by GHRP-2 given alone, compared to GHRP-2 and arginine given together. Because catecholamines were not measured in the present study, the contribution of these neural pathways to the exercise-induced GH AUC cannot be ascertained. In addition, we recognize that the 1 µg/kg dose of GHRP-2 elicits maximal release of GH at rest (i.e., further increases in dose do not cause further increases in GH release) (unpublished observations), but not necessarily during exercise. Also, the impact of arginine and exercise on the near-maximal GH release otherwise caused by GHRP-2 alone might exert differential effects via other neural pathways, compared to only GHRP-2 and exercise.

The rank order of the intervention effects was AG>G>A>S. This was an expected trend because G has been shown previously to cause dramatic increases in GH release (21–23). Although L-arginine also consistently increases GH release (24,25,27), the maximal serum GH concentration attained following arginine stimulation was significantly lower than that after GHRP-2 stimulation (Fig. 26.1A). Current literature supports the notion that GHRP-2 works through a novel, non-GHRH receptor mechanism (9), which includes opposing the effects of SMS on somatotroph membranes and synergism with (released) GHRH. Therefore, the effect of AG on GH release would be expected to be greater than that of either A or G given alone. In fact, preliminary data from the current study revealed that the effect of AG given on the same day was nonadditive compared to either A or G given separately.

Possible interactions or mechanisms that control exercise-induced GH release are represented schematically in Figure 26.2. Exercise may effect GHRH, SMS, or putative GHRP release directly and thereby cause the increase in GH observed with exercise. If the sole mechanism of exercise-induced GH release were hypothalamic SMS withdrawal, then arginine combined with exercise should not significantly increase GH AUC (because the current clinical consensus is that arginine increases GH release via hypothalamic decreases of SMS) (24,25). If the sole mechanism of exercise-induced GH release were increased GHRH release from the hypothalamus, then arginine combined with exercise should significantly increase GH release. The comparison of the exercise and rest GH response within the A treatment demonstrates that the addition of exercise to arginine does not appear to result in so large an increase in GH release as the addition of exercise to GHRP-2. Therefore, data from the current investigation suggested that the mechanism of exercise-induced GH release is likely to involve, at least in part, hypothalamic SMS withdrawal. Because of intrahypothalamic connections, SMS withdrawal it-

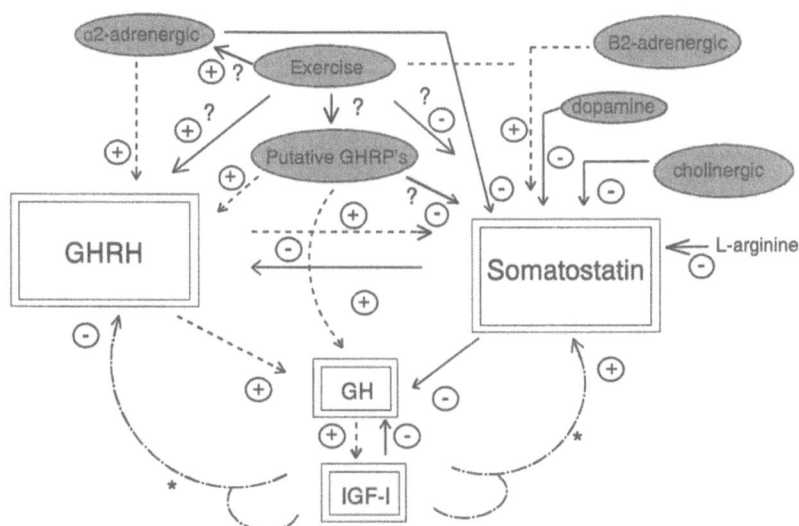

FIGURE 26.2. Schematic representation of the possible interactions or mechanisms that control exercise-induced GH release. Asterisk indicates the possibility that exercise modifies the normal autonegative feedback control of GH on GHRH and SMS. Plus and minus signs denote stimulation and inhibition, respectively. Adapted from Giustina and Veldhuis (37), with permission.

self could promote GHRH release. Because we did not use GHRH in the current investigation, we cannot rule out the possibility that exercise-induced GH release is also somewhat dependent on increased hypothalamic release of GHRH, which is suggested by the large response to GHRP-2. It is also possible that exercise-induced GH release may alter other neuromodulators, thereby increasing GH release indirectly. The α_2-adrenergic feed-forward system hypothesized by Thompson et al. (4) is such a possibly relevant collateral pathway for exercise-induced GH release. Beta-adrenergic systems should also be considered, as several studies have shown that β_2-adrenergic stimulation of SMS release is more potent than the inhibitory effect of arginine or pyridostigmine on SMS release (27,30). Exercise stimulates the β-adrenergic system (50,51,53), which may attenuate the expected response to arginine and, if not recognized, leads to false conclusions about the mechanism of exercise-induced GH release.

The current preliminary study indicates that exercise alone, arginine alone, or exercise with arginine results in a similar GH response, and the L-arginine stimulus appears to be more effective in women compared to men, which indicates a plausible role for SMS withdrawal in exercise-induced GH release. It appears that GHRP-2 infusion tends to elicit a higher GH response in men compared to women. The combination of AG appears to be supraadditive compared to either G or A given alone, and this supraadditivity appears to be gender independent. The addition of exercise to each treatment increased GH release compared to rest, although the magnitude of the increases differed among the four treatments and the smallest increment was observed in the AG treatment. Because the combination of exercise and GHRP-2 supraadditively released GH in both men and women, combined exercise and GHRP-2 administration could be valuable therapeutically. In addition, this synergy suggests that GHRH is involved in the GHRP-2–exercise interaction or that SMS opposition by GHRP-2 alone is not maximal. When AG was combined with exercise, the GH response was supraadditive in women, which raises the possibility of non-GHRH and non-SMS mechanisms.Regardless of concurrent exercise, the gradation of the treatment effects was AG>G>A>S, and this order was gender independent. Exercise and GHRP-2 probably activate different but complementary mechanisms, which cannot be determined fully by the present study. Accordingly, the individual and combined roles of GHRH, SMS, and putative endogenous GHRP in exercise-induced GH release need to be investigated further.

References

1. Kanaley JA, Weltman JY, Veldhuis JD, Rogol AD, Hartman ML, Weltman A. Human growth hormone response to repeated bouts of aerobic exercise. J Appl Physiol 1997;83:1756–61.
2. Lassarre C, Girard F, Durand J, Raynaud J. Kinetics of human growth hormone during submaximal exercise. J Appl Physiol 1974;37:826–30.

3. Raynaud J, Capderou A, Martineaud J-P, Bordachar B, Durand J. Intersubject variability in growth hormone time course during different types of work. J Appl Physiol 1983;55:1682–7.
4. Thompson DL, Weltman JY, Rogol AD, Metzger DL, Veldhuis JD, Weltman A. Cholinergic and opioid involvement in release of growth hormone during exercise and recovery. J Appl Physiol 1993;75:870–8.
5. Weltman A, Snead D, Stein P, Seip R, Schurrer R, Rutt R, et al. Reliability and validity of a continuous incremental treadmill protocol for the determination of lactate threshold, fixed blood lactate concentrations and VO_2max. Int J Sports Med 1990;11:26–32.
6. Weltman A, Weltman JY, Schurrer R, Evans WS, Veldhuis JD, Rogol AD. Endurance training amplifies the pulsatile release of growth hormone: effects of training intensity. J Appl Physiol 1992;72:2188–96.
7. Weltman A, Weltman JY, Womack CJ, Davis SE, Blummer JL, Gaesser GA, et al. Exercise training decreases the growth hormone (GH) response to acute constant-load exercise. Med Sci Sports Exercise 1997;29:669–76.
8. Bowers CY. GH releasing peptides—structure and kinetics. J Pediatr Endocrinol 1993;6:21–31.
9. Akman MS, Girard M, O'Brien LF, Ho AK, Chik CL. Mechanisms of action of a second generation growth hormone-releasing peptide (Ala-His-D-Nal-Ala-Trp-D-Phe-Lys-NH_2) in rat anterior pituitary cells. Endocrinology 1993;132:1286–91.
10. Sethumadhavan K, Veeraragavan K, Bowers CY. Demonstration and characterization of the specific binding of growth hormone-releasing peptide to rat anterior pituitary and hypothalamic membranes. Biochem Biophys Res Commun 1991;178:31–7.
11. Codd EE, Shu AYL, Walker RF. Binding of a growth hormone releasing hexapeptide to specific hypothalamic and pituitary binding sites. Neuropharmacology 1989;28:1139–44.
12. Howard AD, Feighner SD, Cully DF, Arena JP, Liberator PA, Rosenblum CI, et al. A receptor in pituitary and hypothalamus that functions in growth hormone release. Science 1996;273:974–7.
13. Pong S-S, Chaung L-YP, Dean DC, Nargund RP, Patchett AA, Smith RG. Identification of a new G-protein-linked receptor for growth hormone secretagogues. Mol Endocrinol 1996;10:57–61.
14. Penalva A, Pombo M, Carballo A, Barreiro J, Casanueva FF, Dieguez C. Influence of sex, age and adrenergic pathways on the growth hormone response to GHRP-6. Clin Endocrinol 1993;38:87–91.
15. Bowers CY, Maumenee FA, Reynolds GA, Hong A. On the in vitro and in vivo activity of a new synthetic hexapeptide that acts on the pituitary to specifically release growth hormone. Endocrinology 1984;114:1537–45.
16. Popovic V, Damjanovic S, Micic D, Djurovic M, Dieguez C, Casanueva FF. Blocked growth hormone-releasing peptide (GHRP-6)-induced GH secretion and absence of the synergic action of GHRP-6 plus GH-releasing hormone in patients with hypothalamopituitary disconnection: evidence that GHRP-6 main action is exerted at the hypothalamic level. J Clin Endocrinol Metab 1995;80:942–7.
17. Thorner MO, Vance ML, Rogol AD, Blizzard RM, Veldhuis JD, Cauter EV, et al. Growth hormone-releasing hormone and growth hormone-releasing peptide as potential therapeutic modalities. Acta Pediatr Scand 1990;367(Suppl):29–32.
18. Arvat E, Gianotti L, DiVito L, Imbimbo BP, Lenaers V, Deghenghi R, et al. Modulation of growth hormone-releasing activity of hexarelin in man. Neuroendocrinology 1995;61:51–6.

19. Blake AD, Smith RG. Desensitization studies using perifused rat pituitary cells show that growth hormone-releasing hormone and His-D-Trp-Ala-Trp-D-Phe-Lys-NH$_2$ stimulate growth hormone release through distinct receptor sites. J Endocrinol 1991;129:11–9.
20. Robinson BM, Friberg RD, Bowers CY, Barkan AL. Acute growth hormone (GH) response to GH-releasing hexapeptide in humans is independent of endogenous GH-releasing hormone. J Clin Endocrinol Metab 1992;75:1121–4.
21. Bowers CY, Reynolds GA, Durham D, Barrera CM, Pezzoli SS, Thorner MO. Growth hormone (GH)-releasing peptide stimulates GH release in normal men and acts synergistically with GH-releasing hormone. J Clin Endocrinol Metab 1990;70:975–82.
22. Pihoker SM, Middleton R, Reynolds GA, Bowers CY, Badger TM. Diagnostic studies with intravenous and intranasal growth hormone-releasing peptide-2 in children of short stature. J Clin Endocrinol Metab 1995;80:2987–92.
23. Tiulpakov AN, Brook CGD, Pringle PJ, Pererkova VA, Volevodz NN, Bowers CY. GH responses to intravenous bolus infusions of GH releasing hormone and GH releasing peptide-2 separately and in combination in adult volunteers. Clin Endocrinol 1995;43:347–50.
24. Alba-Roth J, Muller OA, Schopohl J, vonWerder K. Arginine stimulates growth hormone secretion by suppressing endogenous somatostatin secretion. J Clin Endocrinol Metab 1988;67:1186–9.
25. Ghigo E, Arvat L, Valente F, Nicolosi M, Boffano GM, Procopio M, et al. Arginine reinstates the somatotrope responsiveness to intermittent growth hormone-releasing hormone administration in normal adults. Neuroendocrinology 1991;54:291–4.
26. Ghigo E, Bellone J, Mazza E, Imperiale E, Procopio M, Valente F, et al. Arginine potentiates the GHRH- but not the pyridostigmine-induced GH secretion in normal short children. Further evidence for a somatostatin suppressing effect of arginine. Clin Endocrinol 1990;32:763–7.
27. Ghigo E, Arvat E, Gianotti L, Ranunni K, Maccario M, Camanni F. Interaction of salbutamol with pyridostigmine and arginine on both basal and GHRH-stimulated GH secretion in humans. Clin Endocrinol 1994;40:799–802.
28. Arosio M, Losa M, Bazzoni N, Bochicchio D, Palmieri E, Nava C, et al. Effects of propranolol on GH responsiveness to repeated GH-releasing hormone stimulations in normal subjects. Acta Endocrinol (Copenh) 1990;122:735–9.
29. Ross RJM, Tsagarakis S, Grossman A, Nhagafoong L, Touzel RJ, Rees LH, et al. GH feedback occurs through modulation of hypothalamic somatostatin under cholinergic control: studies with pyridostigmine and GHRH. Clin Endocrinol 1987;27:727–33.
30. Ghigo E, Arvat E, Bellone J, Ramunni J, Camanni F. Neurotransmitter control of growth hormone secretion in humans. J Pediatr Endocrinol 1993;6:263–6.
31. Weltman A, Weltman JY, Harman ML, Abbott RD, Rogol AD. Relationship between age, percentage body fat, fitness, and 24-hour growth hormone release in healthy young adults. J Clin Endocrinol Metab 1994;78:543–8.
32. Ho KY, Evans WS, Blizzard RM, Veldhuis JD, Merriam GR, Samojlik E, et al. Effects of sex and age on the 24-hour profile of growth hormone secretion in man: importance of endogenous estradiol concentrations. J Clin Endocrinol Metab 1987;64:51–8.
33. van den Berg G, Veldhuis JD, Frolich M, Roelfsema F. An amplitude-specific divergence in the pulsatile mode of growth hormone (GH) secretion underlies the gender difference in mean GH concentrations in men and premenopausal women. J Clin Endocrinol Metab 1996;81:2460–7.

34. Veldhuis JD. The neuroendocrine regulation and implications of pulsatile GH secretion: gender effects. Endocrinologist 1995;5:198–213.
35. Veldhuis JD, Liem AY, South S, Weltman A, Weltman JY, Clemmons DA, et al. Differential impact of age, sex steroid hormones, and obesity on basal versus pulsatile growth hormone secretion in men as assessed in an ultrasensitive chemiluminescence assay. J Clin Endocrinol Metab 1995;80:3209–22.
36. Pincus SM, Gevers EF, Robinson IC, van den Berg G, Roelfsema F, Hartman ML, et al. Females secrete growth hormone with more process irregularity than males in both humans and rats. Am J Physiol 1996;270:E107–15.
37. Giustina A, Veldhuis JD. Pathophysiology of the neuroregulation of GH secretion in experimental animals and the human. Endocr Rev 1998;19:717–97.
38. Cappon JP, Brasel J, Mohan S, Cooper DM. Effect of brief exercise on circulating insulin-like growth factor I. J Appl Physiol 1994;76:2490–6.
39. Lang I, Schernthaner G, Pietschmann R, Kurz R, Stephenson JM, Templ H. Effects of sex and age on growth hormone response to growth hormone-releasing hormone in healthy individuals. J Clin Endocrinol Metab 1987;65:535–40.
40. Benito P, Avila L, Corpas MS, Jimenez JA, Cacicedo L, Sanchez Franco F. Sex differences in growth hormone response to growth hormone-releasing hormone. J Endocrinol Invest 1991;14:265–8.
41. Shah N, Anderson SM, Azimi P, Evans WS, Bowers CY, Veldhuis JD. Estradiol augments basal GH pulse mass and amplifies GH-secretory responsiveness to low doses of GHRP-2. Presented at the Fifth International Pituitary Congress, Naples, Florida, June 28–30, 1998.
42. Sutton J, Lazarus L. Growth hormone in exercise: comparison of physiological and pharmacological stimuli. J Appl Physiol 1976;41:523–7.
43. Felsing N, Brasel J, Cooper DM. Effect of low and high intensity exercise on circulating growth hormone in men. J Clin Endocrinol Metab 1992;75:157–62.
44. Thorner MO, Rivier J, Spiess J, Borges JLC, Vance ML, Bloom SR, et al. Human pancreatic growth hormone-releasing factor selectively stimulates growth hormone secretion in men. Lancet 1983;1:24–8.
45. Viru A. Pancreatic hormones and the somatotropin-somatomedian system. In: Hormones in muscular activity, Vol 1. Boca Raton: CRC Press, 1985:61–75.
46. Chang FE, Dodds WG, Sullivan M, Kim MH, Malarkey WB. The acute effects of exercise on prolactin and growth hormone secretion: comparison between sedentary women and women runners with normal and abnormal menstrual cycle. J Clin Endocrinol Metab 1986;62:551–6.
47. Pritzlaff CJ, Wideman L, Weltman JY, Gutgesell ME, Hartman ML, Veldhuis JD, et al. Effects of exercise intensity on growth hormone (GH) release. Med Sci Sports Exercise 1998;30(Suppl)S48.
48. Hartley LH, Mason JW, Hogan RP, Jones LG, Kotchen TA, Mougey EH, et al. Multiple hormonal responses to graded exercise in relation to physical training. J Appl Physiol 1972;33:602–6.
49. Parkin JM. Exercise as a test of growth hormone secretion. Acta Endocrinol 1986;279(Suppl)47–50.
50. Hansen P. The effect of adrenergic receptor blockade on the exercise-induced serum growth hormone rise in normals and juvenile diabetics. J Clin Endocrinol Metab 1971;33:807–12.
51. Sutton J, Lazarus L. Effect of adrenergic blocking agents on growth hormone responses to physical exercise. Horm Metab Res 1974;6:428–9.

52. Maclaren NK, Taylor GE, Raiti S. Propranolol-augmented, exercise-induced human growth hormone release. Pediatrics 1975;56:804–7.
53. Uusitupa M, Siitonen O, Harkonen M, Gordin A, Aor A, Hersio K, et al. Modification of the metabolic and hormonal response to physical exercise by beta-blocking agents. Ann Clin Res 1982;14(Suppl)165–7.
54. Devesa J, Lima L, Lois N, Fraga C, Lechuga MJ, Arce V, et al. Reasons for the variability in growth hormone (GH) responses to GHRH challenge: the endogenous hypothalamic-somatotroph rhythm (HSR). Clin Endocrinol 1989;30:367–77.
55. Cappa M, Grossi A, Benedetti S, Drago F, Loche S, Ghigo E. Effect of the enhancement of the cholinergic tone by pyridostigmine on the exercise-induced growth hormone release in man. J Clin Endocrinol Invest 1993;16:421–4.
56. Casanueva FF, Villanueva L, Cabranes JA, Cerrato J, Cruz A. Cholinergic mediation of growth hormone secretion elicited by arginine, clonidine and physical exercise in man. J Clin Endocrinol Metab 1984;59:526–30.
57. Moretti C, Favvri A, Gnessi L, Cappa M, Calzolari A, Fraioli F, et al. Naloxone inhibits exercise-induced release of PRL and GH in athletes. Clin Endocrinol 1983;18:135–8.
58. Coiro V, Volpi R, Maffei ML, Caoazza A, Caffarri G, Capretti L, et al. Opioid modulation of the gamma-aminobutyric acid-controlled inhibition of exercise-stimulated growth hormone and prolactin secretion in men. Eur J Endocrinol 1994;131:50–5.
59. Veldhuis JD, Johnson ML. Cluster analysis: a simple, versatile and robust algorithm for endocrine pulse detection. Am J Physiol 1986;250:E486–93.
60. Friend K, Iranmanesh A, Veldhuis JD. The orderliness of the growth hormone (GH) release process and the mean mass of GH secreted per burst are highly conserved in individual men on successive days. J Clin Endocrinol Metab 1996;81:3746–53.
61. Veldhuis JD, Carlson ML, Johnson ML. The pituitary gland secretes in bursts: appraising the nature of glandular secretory impulses by simultaneous multiple-parameter deconvolution of plasma hormone concentrations. Proc Natl Acad Sci USA 1987;84:7686–90.
62. Merimee TJ, Rabinowitz D, Fineberg SE. Arginine-initiated release of human growth hormone: factors modifying the response in normal man. N Engl J Med 1969;280:1434–8.

27

Stimulated Release of GH in Normal Younger and Older Men and Women

CYRIL Y. BOWERS AND RAMONA GRANDA-AYALA

An earlier finding of the growth hormone-releasing peptides (GHRPs) was that they released GH more effectively in humans than in a number of different animal species, including rats, cows, pigs, sheep, rhesus monkeys, and chickens (1,2). To better appreciate the acute GH-releasing action of GHRP-2 and GHRH 1-44NH$_2$, studies have been performed in normal younger and older men and women. The hypothesis pursued was that new insight into the actions of sex steroids or GHRP on GH release can be revealed from the acute GH-releasing action of GHRP-2 and GHRH in diverse clinical contexts. The objective was to compare the GH-releasing action of GHRP-2 and GHRH alone and together to assess specific peptide, dosage, age, and sex dependencies.

The approach in the normal younger subjects consisted of administration of four doses of GHRP-2 (0.03, 0.1, 0.3, and 1 µg/kg) and three doses of GHRH (0.1, 0.3, and 1 µg/kg) by i.v. bolus alone and together in equal and unequal dosages to the same nine normal younger men and the same seven normal younger women. Also, studies were performed with 1 µg/kg GHRP-2 and GHRH alone and together in normal older men and women as well as in younger and older normal women on and off estrogen. Collectively, these results underscore some of the difficulties of comparing, interrelating, and analyzing the acute GH-releasing action of GHRP-2 and GHRH. The data demonstrate that GH secretion is not only peptide but also dosage, age, and sex dependent.

Age and Sex

Figures 27.1 and 27.2 record the acute GH responses in normal younger men and women to 1 µg/kg i.v. bolus GHRP-2, GHRH 1-44NH$_2$, and GHRP-2 + GHRH. The GH responses are much larger in younger men and women than in older individuals. At this dosage, there was no sex difference in the GH re-

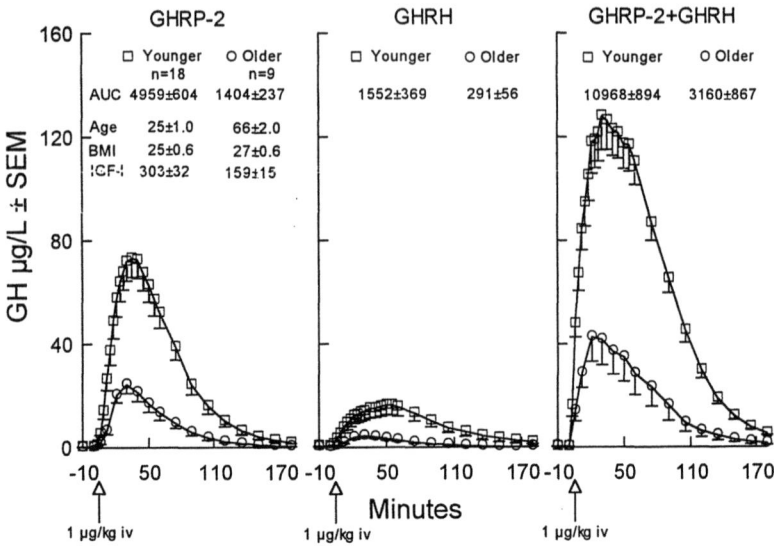

FIGURE 27.1. Serum concentration GH responses to GHRP-2, GHRH, and GHRP-2 + GHRH in normal younger and older men. Values are the mean ± SEM. IGF-I, µg/l;. AUC (area under the curve), µg/l•4 h; BMI, body mass index (kg/m²). Reproduced by permission from Marcel Dekker, Inc., New York (4).

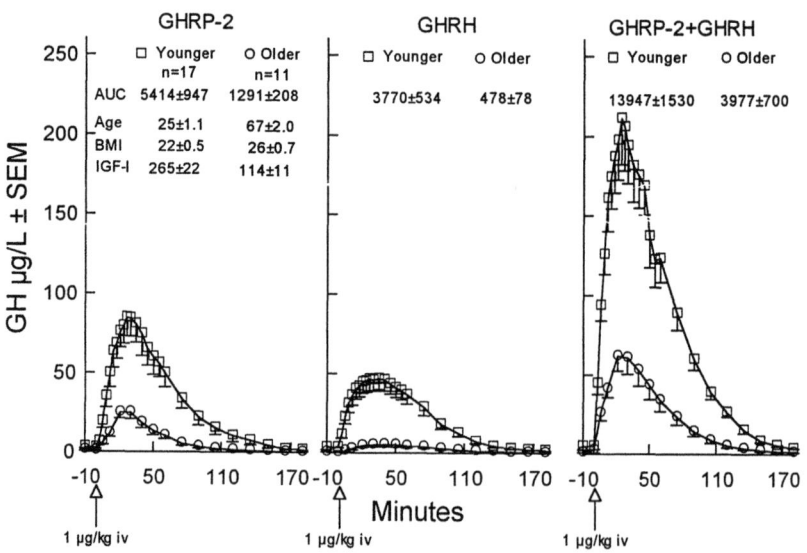

FIGURE 27.2. Serum GH responses to GHRP-2, GHRH, and GHRP-2 + GHRH in normal younger and older women. Values are the mean ± SEM. IGF-I, µg/l; AUC (area under the curve), µg/l•4 h. BMI, body mass index (kg/m²). Reproduced by permission from Marcel Dekker, Inc., New York (4).

sponse to GHRP-2 in either the younger or older subjects, but the GH response to GHRH was sex dependent. Although the amount of GH released by 1 µg/kg i.v. bolus GHRH was essentially the same in the older men and women, the GH response to GHRH in younger women was definitely greater than that in younger men. As the GH response to 1 µg/kg GHRH was decreased to the same degree in normal older men and women but was greater in normal younger women than younger men, this also indicates that stimulated GH secretion decreases more in older women than older men.

GHRP-2 and GHRH Dose Response Studies

Figures 27.3 and 27.4 reveal some sex differences in the GH responses to i.v. bolus GHRP-2 and GHRH 1-44NH$_2$ in normal men and women. The GH response to 1 µg/kg GHRP-2 is the same in men and women, while at lower GHRP-2 dosages of 0.1 and 0.3 µg/kg they are twofold higher in women (Fig. 27.3). At the lowest dose of 0.03 µg/kg, the GH response was not significantly increased in either men or women. A sex difference also was observed for i.v. bolus GHRH 1-44NH$_2$ (Fig. 27.4). At the high dose of 1 µg/kg, but not the

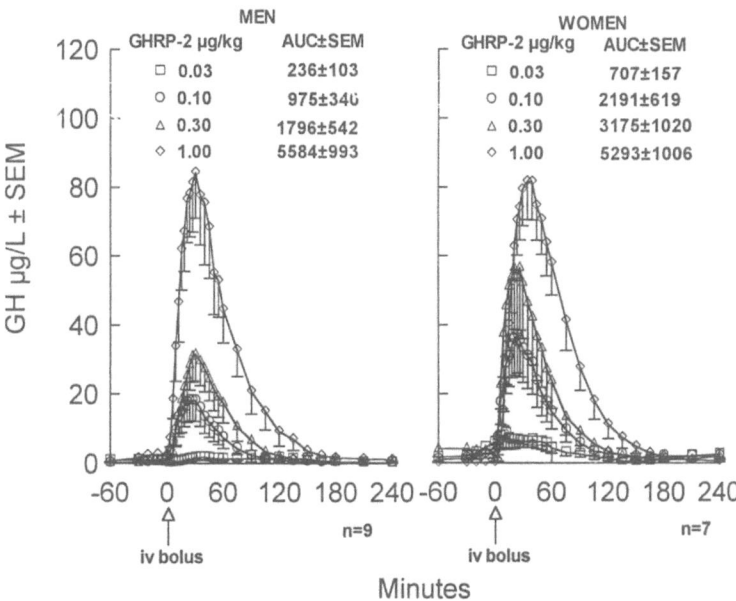

FIGURE 27.3. Dose–response curve for GHRP-2 actions in normal young men (*left panel*) and women (*right panel*). Values are the mean ± SEM. AUC (area under the curve), µg/l•4 h. Age, BMI, and IGF-I (µg/l) values for the men were 25 ± 1.7, 25 ± 0.7, and 301 ± 46, and for the women 25 ± 1.6, 22 ± 0.7, and 310 ± 52, respectively.

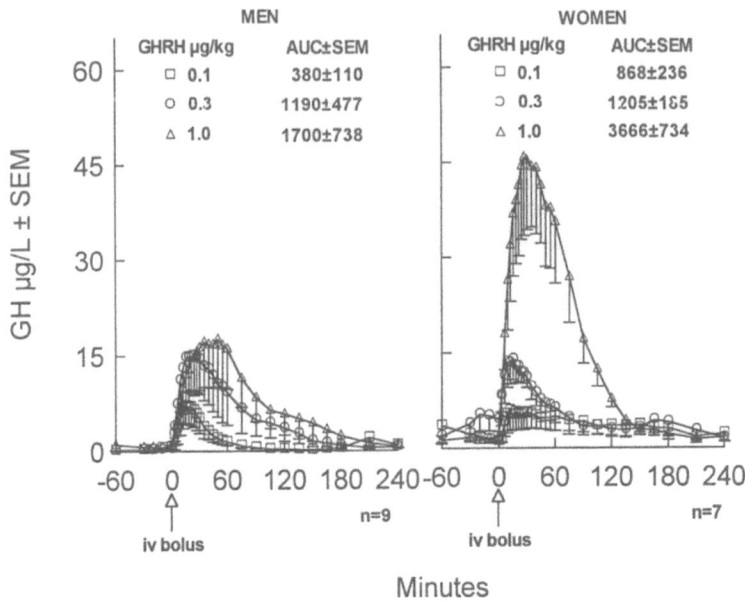

FIGURE 27.4. Dose–response curve for GHRH 1-44NH$_2$ action in normal young men (*left panel*) and women (*right panel*). Values are mean ± SEM. Subjects are the same as those reported in Figure 27.3.

lower doses of 0.1 and 0.3 µg/kg, the GH response was about twofold higher in women than men. Thus, more GH was released in normal young women than in young men by the low dosage of GHRP-2 as well as the high dosage of GHRH. In regard to the physiological regulation of GH secretion, the effects of low-dose GHRP-2 probably are more relevant. If a putative GHRP-like ligand exists and is the helper of GHRH, a greater low-dose effectiveness of GHRP-2 in women could indicate a more major role for the putative GHRP-receptor ligand in the regulation of GH secretion in young women than young men. Also implied is that if a GHRP-like ligand deficiency exists, replacement of the deficiency may be more readily achieved with a lower GHRP dosage in women.

Combined GHRP-2 and GHRH Administration in Equal Dosages

Knowing the peptide, dosage, sex, and age dependencies of the GH responses to GHRP-2 and GHRH, it is not difficult to envision that unexpected results may be revealed when the peptides are administered acutely together in equal or unequal dosage combinations. Data from these studies in normal young men and women are recorded in Figures 27.5 and 27.6. When equal i.v. bolus doses of GHRP-2 and GHRH were administered together in the low dosage of

27. Stimulated GH in Normal Humans   281

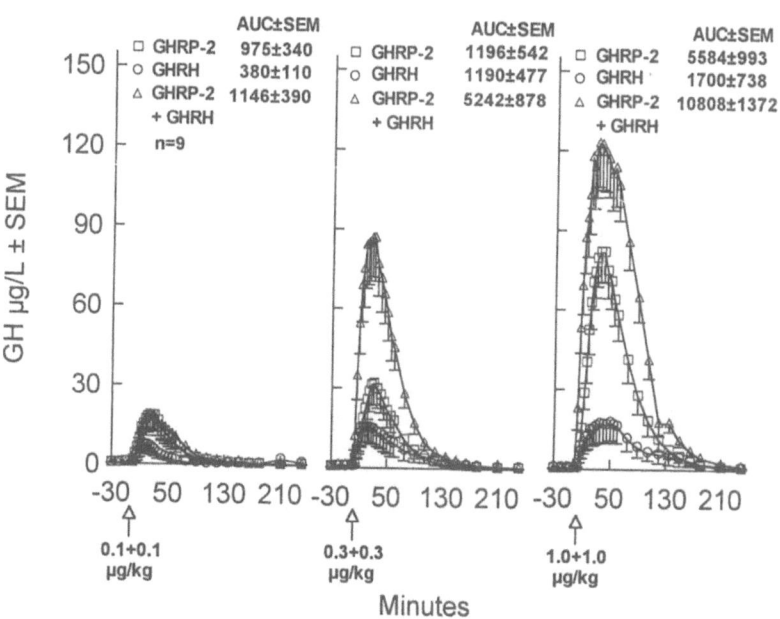

FIGURE 27.5. Serum GH responses to GHRP-2 and GHRH alone and together in equal dosages in normal young men. Values are mean ± SEM. Subjects are the same as those reported in Figure 27.3.

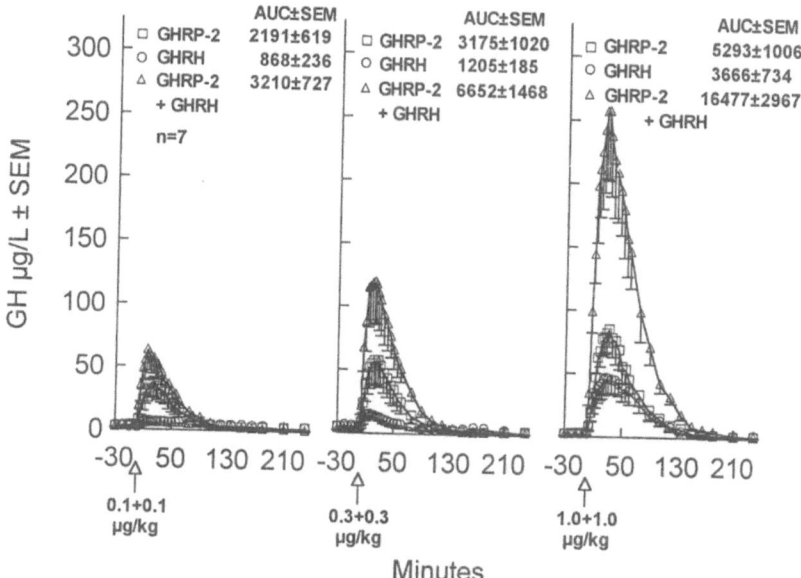

FIGURE 27.6. Serum GH responses to GHRP-2 and GHRH infusion alone and together in equal dosages in normal young women. Values are mean ± SEM. Subjects are the same as those reported (women) in Figure 27.3.

0.1 µg/kg, the GH response to the two peptides was additive and not synergistic, which is in contrast with the synergistic release of GH induced by the combined peptides at the higher dosages of 0.30 and 1 µg/kg. The combined peptides released more GH in normal young women than normal young men at all three combined equal dosages of each peptide, 0.1, 0.3, and 1 µg/kg.

Combined GHRP-2 and GHRH Administration in Unequal Dosages

Gender differences in the GH responses in normal young men and women were peptide, dosage, and sex dependent. In Figure 27.7, a marked synergistic release of GH was induced in men by 0.1 µg/kg GHRP-2 + 1 µg/kg GHRH 1-44NH$_2$. Also GH was released synergistically in the women but the degree of synergism was less in women than in men. The dosage and peptide dependencies are again underscored by the results (Fig. 27.7B when high-dose rather than low-dose GHRP-2 was administered with low-dose GHRH. There was no synergistic GH release by 1 µg/kg GHRP-2 + 0.1 µg/kg GHRH in men, while in women GH was released synergistically by this dosage combination. Sex-dependent unequal GH release was also elicited by combined 1 µg/kg GHRP-2 + 0.3 µg/kg GHRH 1-44NH$_2$ (not shown), and although GH was released synergistically in both men and women, synergistic secretion, was greater in women.

Combined Administration of Very Low Dose GHRP-2 and High-Dose GHRH

One of the most informative GH responses to GHRP-2 + GHRH was obtained by the combined dosage of 0.03 µg/kg GHRP-2 plus 1 µg/kg GHRH (3,4). As shown in Figure 27.8, this very small amount of ~2 µg GHRP-2 per subject did not release GH when administered alone; however, in combination with ~70 µg GHRH 1-44NH$_2$ this subthreshold GH-releasing dose augmented the GH response to GHRH in both normal young men and women. These results allow some new insight into the action of GHRP-2. Because the in vivo synergistic release of GH induced by GHRP + GHRH has been unexplained by a direct pituitary action of these two combined peptides in vitro, and as an i.v. bolus subthreshold GH-releasing dosage of GHRP-2 in combination with GHRH 1-44NH$_2$ induced synergistic release of GH, we suggest that GHRP acts on the hypothalamus rather than the pituitary to produce synergistic release of GH (3–5). In this study, administration of a large dose of exogenous GHRH obviated the possibility that the synergistic GHRP-2 hypothalamic action resulted from release of endogenous GHRH by GHRP-2. In addition, a series of other studies indicate that the hypothalamic action of a very low dose of GHRP-2 does not involve the inhibition of somatostatin (SRIF) release from the hypo-

27. Stimulated GH in Normal Humans   283

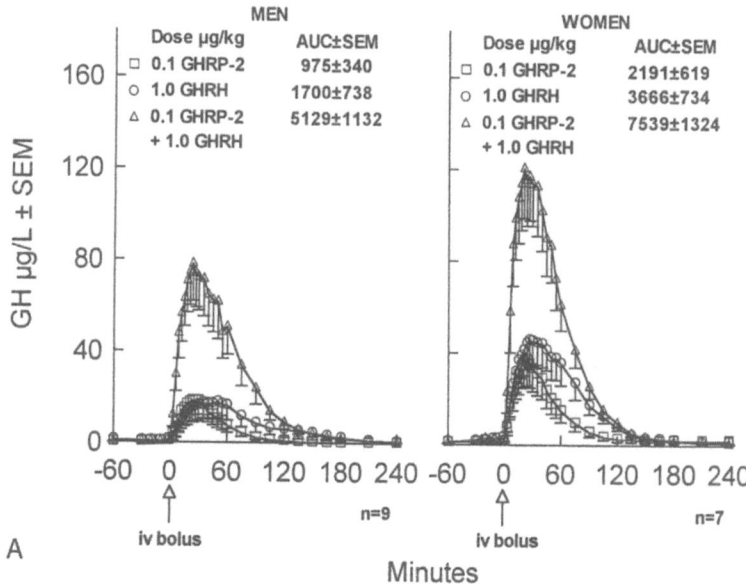

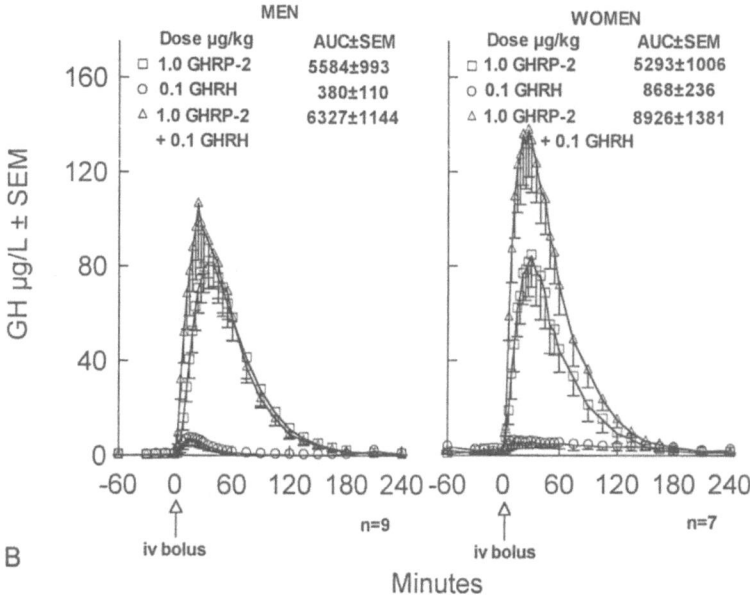

FIGURE 27.7A,B. Serum GH responses to GHRP-2 and GHRH infusion alone and together in unequal dosages in normal young men (*left panels*) and women (*right panels*). Values are the mean ± SEM. Subjects are the same as those reported in Figure 27.3.

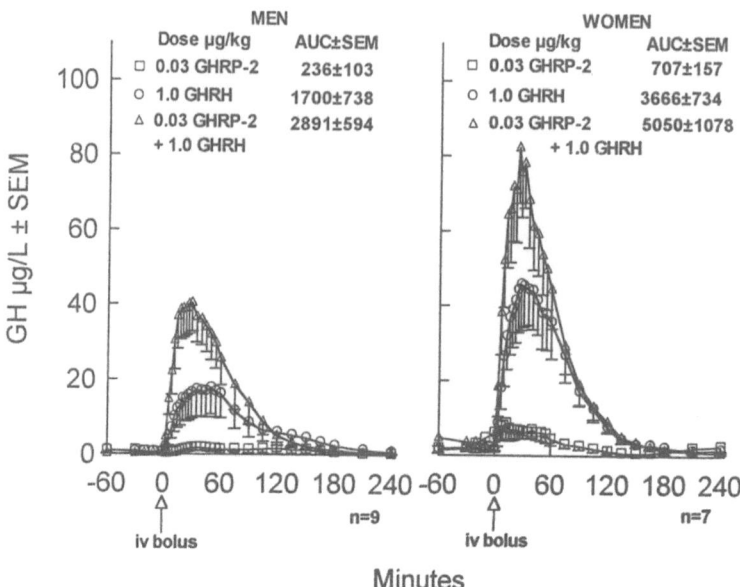

FIGURE 27.8. Serum GH responses to very low dose [mean dose, 1.7 (women) or 2.0 (men) µg/subject] of GHRP-2 or 1 µg/kg GHRH 1-44NH$_2$ infusions alone and together in normal young men (*left panel*) and women (*right panel*). Values are the mean ± SEM. Subjects are the same as those reported in Figure 27.3.

thalamus or opposition of SRIF pituitary inhibitory action on GH release. To explain these results, we have proposed that the hypothalamic action of GHRP is to release U-factor or unknown factor, and that U-factor plus GHRH act on the pituitary to release GH synergistically.

GH Responses to GHRP-2 Postmenopausal Women off and on Estrogen

The results in Figure 27.9 reveal that estrogen administration to postmenopausal women did not alter the GH response to 1µg/kg GHRP-2 or GHRH 1-44NH$_2$ alone, but enhanced their joint actions. These data are considered preliminary, because the women were different in the two groups and the type, duration, and dosage of estrogen administered varied. When the peptides were administered together, GH was released synergistically on or off estrogen, but the synergy was greater on estrogen. The results in Figure 27.10 demonstrate that when a larger dose of 3 µg/kg GHRP-2 was administered s.c. to post- or premenopausal women, neither estrogen nor oral contraceptives (premenopausal women) altered the GH response in these women. However, stimulated GH release was much greater in the pre- versus postmenopausal women.

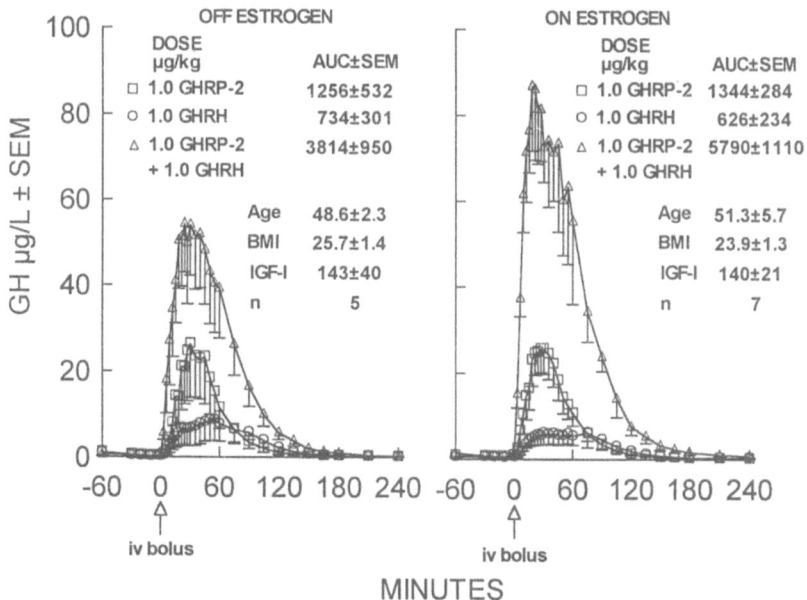

FIGURE 27.9. GHRP-2- and GHRH 1-44NH$_2$-stimulated serum GH responses to the secretagogues given alone and together in postmenopausal women off (*left panel*) and on (*right panel*) estrogen (ERT). Values are the mean ± SEM.

Endogenous GHRP Deficiency

The GHRP-2-GHRH dual-stimulation approach has been utilized to investigate the pathophysiology of decreased GH secretion in normal older men and women. The hypothesis has been proposed that in some subjects this may be caused by a deficiency of the putative hypothalamic GHRP-like hormone. Indirect evidence suggests (but does not establish that a natural GHRP-like hormone exists and is involved in the physiological regulation of GH release. Previously, other investigators have proposed that decreased GH secretion in normal older subjects is probably the result of a deficiency of GHRH, or increased release of SRIF from the hypothalamus, or an enhanced pituitary action of SRIF.

Possible evidence of decreased GH secretion indicating a deficiency of the putative hypothalamic GHRP-like hormone in older subjects is presented in Figure 27.11. This seemingly meaningful finding demonstrates that a low GH response to maximal 1 μg/kg i.v. bolus GHRH 1-44NH$_2$ is dramatically reversed by a low dose of 0.1 μg/kg GHRP-2 when given together with GHRH. Low-dose GHRP-2 alone, like high-dose GHRH, releases only a small amount of GH. In addition, when these subjects were given 1 μg/kg GHRP-2 alone, the GH response was equivalent to that of the combined peptides at this dosage. These results have been interpreted as follows. The response to GHRH

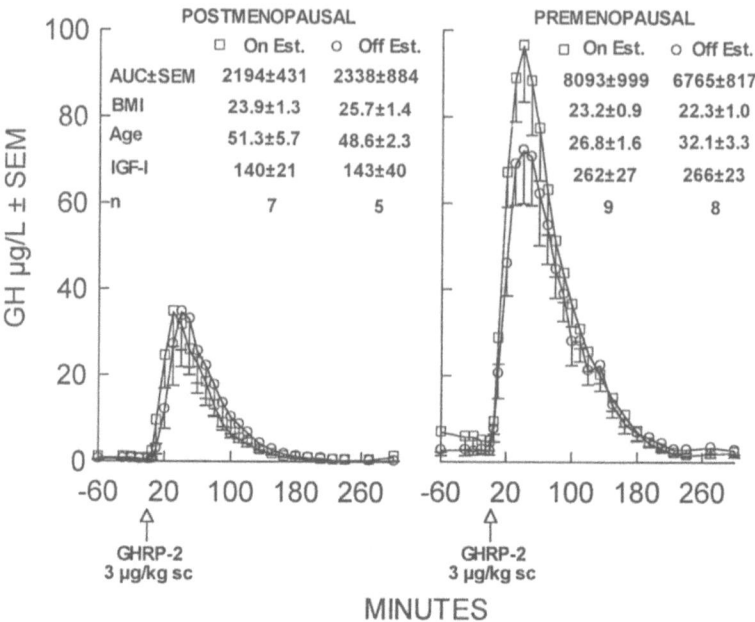

FIGURE 27.10. GHRP-2- and GHRH 1-44NH$_2$-stimulated serum GH responses for the secretagogues infused alone and together in postmenopausal women on and off estrogen (ERT) (*left panel*) and premenopausal women on and off estrogen (oral contraceptives) (*right panel*). Values are the mean ± SEM.

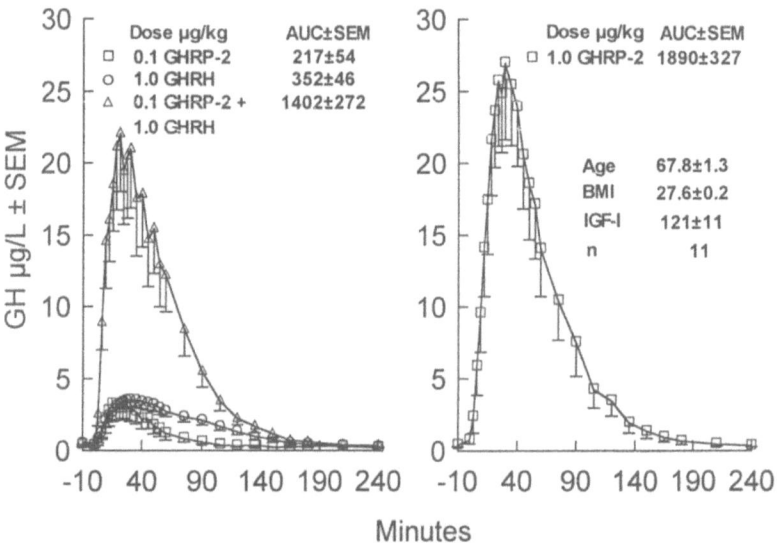

FIGURE 27.11. Serum GH responses to low-dose GHRP-2 (0.1 µg/kg) and high-dose GHRH 1-44NH$_2$ (1.0 µg/kg) infusions administered alone and together in normal older subjects with decreased GH secretion and low IGF-I levels (121 ± 11 µg/l) (*left panel*). *Right panel*, GH responses to 1 µg/kg GHRP-2 in the same 11 subjects. Values are the mean ± SEM.

could be in principle caused by low pituitary stores of GH or increased release or action of SRIF. However, the first explanation does not appear to be the reason for the impaired action of GHRH, because low-dose GHRP-2 very effectively reverses the impaired pituitary action of GHRH on GH release. A number of results have accumulated to indicate that low-dose GHRP-2 does not increase SRIF release or attenuate its pituitary action. Also, given the low responses to 1 µg/kg GHRH alone, it appears that in these subjects, even if endogenous GHRH were secreted in large amounts, it still would be ineffective in releasing GH. Thus, neither GHRH nor SRIF alone appear to be primarily responsible for the decreased GH secretion in older subjects. An additional finding which indicates that these subjects are secreting endogenous GHRH is that the GH-releasing action of GHRP is substantially dependent on the secretion of endogenous GHRH (6,7). In these older subjects, a greater release of GH was induced by 1 µg/kg GHRP-2 than by this dose of GHRH 1-44NH$_2$, again indicating that low-dose GHRP-2 does not induce synergism with a maximal GHRH dose because it releases endogenous GHRH. Because of the low GH response to a maximal dose of exogenous GHRH, it is possible to conclude that even if GHRP-2 released endogenous GHRH, it would be ineffective in releasing GH. Thus, because it does not seem possible to explain the large GH response to 1 µg/kg GHRP-2 via the release of endogenous GHRH, it is proposed that GHRP-2 induces its synergistic GH-releasing effect via release of U-factor from the hypothalamus.

Summary and Conclusion

The interactions of GHRP, GHRH, and SRIF seem convoluted and difficult to understand despite a number of clear-cut findings both in vivo and in vitro. For example, GHRP and GHRH act on different receptors and stimulate different signal transduction pathways. Regardless of the latter points, which indirectly suggest that the GHRP + GHRH synergism is induced because of a direct pituitary action of the two peptides, the GH-releasing activity of GHRP-2 + GHRH is only additive in vitro but synergistic in vivo. We have been unable to explain the in vivo synergism by a direct effect of low-dose GHRP on the release or action of GHRH or SRIF. However, this does not mean that GHRP-2 never has a direct effect on the release or action of GHRH and SRIF. A high, but not low, dose of GHRP-2 appears to release GHRH. Also, GHRP does attenuate the GH-inhibiting action of SRIF in vitro and in vivo, but because GHRP and GHRH are equally effective in attenuating SRIF, this action of GHRP is not considered unique (3,4).

In earlier studies, GHRP was designated a functional SRIF antagonist by Thorner et al. (8). However, our findings in animals and humans, that the GH response to GHRP could be increased under conditions in which SRIF is inhibited, led us to conclude that GHRP-attenuating action on SRIF release from the hypothalamus or pituitary SRIF action appear to be a minor rather than major point. In addition, it is acknowledged that postulating that a hypothalamic U-factor or unknown factor plays an integral role in the action of GHRP adds considerably to the complexity of how GHRP acts to release GH. To explicate the GHRP + GHRH synergistic release of GH, GHRP is envisioned to act on the hypothalamus to release U-factor, which in turn acts together with GHRH on the pituitary to release GH synergistically. It is unclear whether synergism reflects a pharmacological or a physiological mechanism. Nevertheless, evidence seems strong that a special GHRP hypothalamic action, such as U-factor release, is revealed by combined GHRP and GHRH stimulation. This inference has resulted in our attempt to isolate not only the putative GHRP-like ligand but also U-factor from porcine hypothalamic extracts. Also, U-factor has been incorporated into our theoretical models, which depict the actions of GHRP.

A general hypothesis proposed is that the primary physiological role of the putative GHRP-like hormone is not to stimulate or regulate the release of endogenous GHRH but rather to act together with GHRH in a complementary way to facilitate the actions of GHRH either additively or synergistically (9). At times, GHRH may need a helper to overcome the potent inhibitory action of SRIF on GH release.

The combined actions of GHRH and GHRP are envisioned to add another degree of independency to the GH regulatory network. This added aspect is considered to be comparable to the independency of the SRIF system. The complexity of the GH regulatory network could readily be viewed to incorporate two major systems: the SRIF inhibitory system, presumably being responsible for the frequency of the GH pulses as well as the level of basal secretion of GH between pulses, and the GHRH–GHRP stimulatory system responsible for the effectual release of GH.

Acknowledgments. This work was supported in part by National Institutes of Health grants AM-06164, DK40202, and PHS RR05096-09 (G.C.R.C.). Special appreciation is expressed to the technicians and fellows of the Endocrinology and Metabolism Section of the Department of Medicine, and to Robin Alexander for typing the manuscript. Special thanks are expressed to Dr. Fred Wagner at BioNebraska for supplying GHRH 1-44NH$_2$.

References

1. Bowers CY, Momany F, Reynolds GA, Hong A. On the *in vitro* and *in vivo* activity of a new synthetic hexapeptide that acts on the pituitary to specifically release growth hormone. Endocrinology 1984;114:1537–45.
2. Bowers CY, Reynolds GA, Durham D, Barrera CM, Pezzoli SS, Thorner MO. Growth hormone releasing peptide stimulates GH release in normal men and acts synergistically with GH-releasing hormone. J Clin Endocrinol Metab 1990;70:975–82.
3. Bowers CY. Xenobiotic growth hormone secretagogues. In: Bercu B, Walker R, eds. Growth hormone secretagogues. New York: Springer Verlag, 1996:9–28.

4. Bowers CY. GHRP+GHRH synergistic release of GH: scope and implication. In: Bercu B, Walker R, eds. Growth hormone secretagogues. New York: Marcel Dekker, 1998:1–25.
5. Camanni F, Ghigo E, Arvat E. Growth hormone releasing peptides and their analogs. Front Neuroendocrinol 1998;19:47–72.
6. Bowers CY, Sartor AO, Reynolds GA, Badger TM. On the action of the growth hormone releasing hexapeptide GHRP. Endocrinology 1991;128:2027–35.
7. Pandya N, DeMott-Friberg R, Bowers CY, Barkan AL, Jaffe CA. Growth hormone (GH) releasing peptide-6 requires endogenous hypothalamic GH-releasing hormone for maximal GH stimulation. J Clin Endocrinol Metab 1998;83:1186–9.
8. Thorner MO, Vance ML, Hartman ML, Holl RW, Evans WS, Veldhuis EV, et al. Physiological role of somatostatin on growth hormone regulation in man. Metabolism 1990;39:9(suppl. 2):40–2.
9. Bowers CY. GH releasing peptide (GHRPs). In: Kostyo J, Goodman H, eds. Handbook of physiology. New York: Oxford University Press 1999:267–97.

Part V

Novel Techniques in Investigating the Reproductive and GH Axis

28

Genetic and Transgenic Models to Investigate the Growth Hormone Axis and Sexual Dimorphism

LAWRENCE A. FROHMAN AND RHONDA D. KINEMAN

Investigation of the role of the GH axis in sexually dimorphic function requires model systems in which the axis can be perturbed in the absence of confounding changes in other pituitary and target organ hormones. The classical paradigm of ablation of an endocrine gland with subsequent replacement of the affected hormone cannot be applied to the pituitary, because the loss of other pituitary hormones confounds the interpretation of the observed changes.

The discovery of a series of spontaneously occurring genetic models and the creation of a broad spectrum of transgenic animals with selective alterations of individual components of the GH axis have circumvented this problem. Table 28.1 lists the models described to date that have been found useful in studying the regulation of GH secretion and the role of GH in the generation of sexual dimorphism. They have been grouped on the basis of the site at which the natural or experimentally induced disturbances occur.

The three models that disturb hypothalamic function all cause a decrease in GH-releasing hormone (GHRH) secretion and result in GH deficiency. GSH-1 is a nuclear transcription factor that is required for the expression of the GHRH gene. Mice with a GSH-1 deletion lack GHRH, are phenotypically growth retarded, and their pituitaries contain reduced numbers of somatotropes with diminished GH content (1). Partial reductions in GHRH mRNA and protein levels are seen on a secondary basis in two transgenic models in which the hGH reporter gene linked to either a tyrosine hydroxylase (2) or a GHRH promoter (3) is expressed in the hypothalamus.

The majority of models (including all the genetic models) are associated with pituitary hypofunction. Five separate spontaneous inactivating mutations that reduce GH secretion are now well characterized. The *lit* mouse has a mutation in the GHRH receptor (GHRH-R) gene that impairs ligand binding and signaling critical for GH synthesis and release (4,5). The Snell mouse has

TABLE 28.1. Genetic and transgenic mouse and rat models with altered function of the GH axis.

Site	Model	Genetic	TG / KO	Hypofunction	Hyperfunction	Component
Hypothalamus	GSH-1 deletion (m)			X		GHRH
	TH-hGH (m)		X			GH↑ in CNS
	GHRH-hGH (r)		X			GH↑ in CNS
Pituitary	lit (m)	X		X		GHRH-R gene
	dw (r)	X		X		GHRH signal transduction
	Snell (m)	X		X		Pit 1 (somatotrope)
	Ames (m)	X		X		Prop 1 (somatotrope)
	Spontaneous dwarf (r)	X		X		GH gene
	GH diphth. toxin (m)		X	X		Somatotrope
	MT-hGHRH (m)		X		X	GHRH
	MT-h/bGH (m)		X		X	GH
	MT-bGH-A (m)		X			GH
Liver/peripheral	GH-R deletion (m)		X	X		GH-R
	IGF-1 deletion (m)		X	X		IGF-I
	MT-IGF-1 (m)		X		X	IGF-I

TG, transgenic; KO, knockout; (m), mouse; (r), rat; TH tyrosine hydroxylase; CNS, central nervous system; diphth, diphtheria; MT, metallothionein; R, receptor.

a mutation in the gene for Pit-1, a transcription factor required for somatotrope, lactotrope, and thyrotrope development, while the Ames mouse carries a mutation in the Prop-1 gene, which encodes a transcription factor necessary for Pit-1 gene expression (6). Both models exhibit a dwarf phenotype. The spontaneous dwarf rat (SDR) contains a mutation in the GH gene resulting in the total absence of immunodetectable GH (7). The dwarf (dw) rat exhibits a yet to be defined mutation in the GHRH signal transduction system (8).

Two transgenic mouse models result in hyperfunction of the GH axis. The mMT-hGHRH transgenic mouse has widespread expression of the GHRH gene under the control of the promoter for mouse metallothionein, a housekeeping gene (9). Although hypothalamic expression does occur, the highest level of transgene expression is in the pituitary (10). The elevated GHRH level leads to GH hypersecretion, marked pituitary hyperplasia, and a 2- to 2.5 fold increase in body size. The same promoter has been used to overexpress both the bGH and hGH genes, resulting in a similar phenotype except for the lack of pituitary hyperplasia (11). GH deficiency is produced in mice expressing a mGH-diphtheria toxin transgene that leads to selective destruction of somatotropes (12). Reduced linear growth also occurs in the mMT-bGH antagonist mouse, in which overexpression of a GH antagonist transgene results in a competitive inhibition of GH binding to its receptor (13).

Several transgenic models have been developed to study the physiological changes associated with perturbations of the peripheral effects of GH. Deletion of the GH receptor (GH-R) gene results in elevated GH levels, but a total absence of GH action and a dwarf phenotype (14). Deletion of the IGF-I gene also leads to a dwarf phenotype, although not so severe (15), while overexpression of an IGF-I transgene results in overgrowth similar to that seen in the transgenic GH mouse (16). Three representative genetic models studied in the authors' laboratory are described here in greater detail to provide examples of the utility of this entire cohort of murine and rodent strains.

Little Mouse (*lit*)

The *lit* mouse, originally described by Eicher and Beamer (17), was shown to have an absence of GH and cyclic AMP responsiveness to GHRH. However, the mice responded normally to signal transduction probes with actions distal to the GHRH-R (18). The defect was subsequently attributed to a point mutation in the extracellular domain of the GHRH-R that abolished its ligand-binding ability (4,5). The *lit* mouse was also shown to be unresponsive in vivo and in vitro to GHRP-6 (19), a GH secretagogue that binds to a G-protein-coupled receptor distinct from GHRH-R (Fig. 28.1). Although initially unexplained, this observation is consistent with the many recent studies demonstrating a requirement for intact GHRH-R function for GH responsiveness to GHRP-6 and other GH secretagogues. Of particular interest is the observation that heterozygous +/*lit* mice,

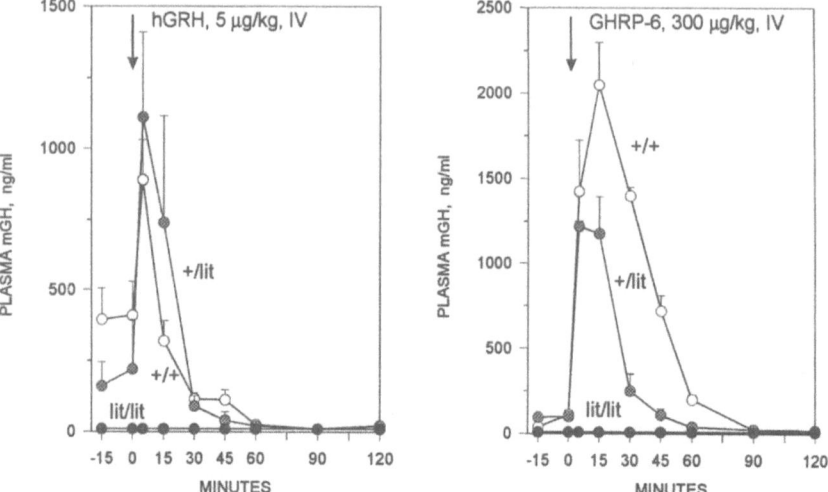

FIGURE 28.1. Growth hormone secretion in response to a single intravenous injection of GHRH (*left panel*) or GHRP-6 (*right panel*) in anesthetized wild-type (+/+), heterozygous (+/−), and homozygous (*lit/lit*) mice. Shown are the mean ± SEM. Adapted from Jansson et al. (19).

which are phenotypically normal and exhibit an intact GH secretory response to GHRH, have reduced GH responses to GHRP-6, implying that the mutated GHRH-R may exert a dominant negative effect.

The little mouse has been useful in demonstrating the role of GH secretion in the sexually dimorphic pattern of the hepatic epidermal growth factor receptor (EGF-R). The hepatic EGF-R is greater in male than in female rodents and can be increased in males by testosterone. In *lit* mice, EGF-R levels are decreased in both sexes and do not exhibit sexual dimorphism (20). Testosterone treatment of female *lit/lit* mice failed to increase EGF-R binding, in contrast to the threefold increase observed in +/*lit* mice, suggesting that androgen-induced changes in GH secretion exert a mediating role. Treatment of *lit/lit* mice with hGH (by either twice daily injections or a constant infusion) increased EGF-R binding in both males and females, but only to levels observed in +/*lit* female mice. This result is in contrast to the results in hypophysectomized mice in which twice daily injections of hGH increased EGF-R binding to the levels observed in normal male mice. The difference in the responses to GH in these two models may be explained by a relative resistance to GH in *lit* as compared to hypophysectomized mice because of the longer duration of GH deficiency or the low but continuously present levels of GH in *lit* mice that may interfere with the "masculinization" of the response.

The *lit* mouse has also been used to investigate the sexually dimorphic pattern of the mouse urinary protein (mUP) gene. This gene is expressed to a greater extent in males than in females and is inducible in females by a male pattern of GH injections, demonstrating that testosterone is not required for the sexually dimorphic effect of GH. The *lit* mouse does not express the mUP gene under basal conditions and the gene is unresponsive to testosterone. However, GH treatment does increase mUP mRNA levels to normal (21).

Dwarf Rat (*dw*)

The *dw* rat was originally discovered as a spontaneous recessive mutation by Charlton et al. (8). This model exhibits moderate growth retardation, decreased numbers of somatotropes, reduced pituitary GH content, and a partially impaired GH response to GHRH. In contrast to the *lit* mouse, the GH response in vitro is about 70%–75% that in normal somatotropes, when expressed as a fraction of pituitary cell content. GHRH-stimulated cyclic AMP generation, however, is markedly reduced to only 1%–2% of that in normals (22). The signal transduction defect is yet undefined, though it appears to be associated with impaired $G_s\alpha$ function (23). Hypothalamic GHRH mRNA levels are increased and somatostatin mRNA levels are decreased in *dw* rats, and the sexually dimorphic patterns in these two hypothalamic hormones, seen in normal rats, persists in *dw* rats (24).

Several studies have utilized the *dw* rat to examine sexually dimorphic hepatic function. Hepatic carbonic anhydrase (CA)-III levels are much greater in normal male rats, and this dimorphism is retained in *dw* rats (25). Continuous GH infusion in male *dw* rats partially "feminized" CA-III levels while IGF-I infusions, which had comparable growth-promoting effects, did not. Thus, the expression of CA-III is sensitive to the pattern of GH secretion and is independent of its effects on growth. The findings also implicate the low basal levels rather than the high peaks of GH as being responsible for the sexually dimorphic expression of this enzyme.

GH-binding protein (GHBP) is two- to threefold greater in normal female rats, and this difference is not seen in *dw* rats (26). Continuous GH infusion in male *dw* rats increases GHBP levels to those of normal females. The hepatic GH receptor (GH-R), which is derived from the same gene as is GHBP as a consequence of alternative splicing, is also greater in normal but not in *dw* female rats (27). Continuous GH infusion in female *dw* rats increased GH-R binding.

The activity of 11-beta-hydroxysteroid dehydrogenase (11-β-HSD), which converts corticosterone to the inert 11-dehydrocorticosterone, is greater in the liver of normal male compared to female rats, and estradiol suppresses 11-β-HSD gene transcription. The sex differences are less pronounced in *dw* rats, and continuous GH infusion in male *dw* rats decreases both enzyme activity and mRNA levels (28).

Spontaneous Dwarf Rat

The spontaneous dwarf rat (SDR) has a mutation in the splice acceptor site of the third intron of the GH gene that produces a frameshift leading to a premature stop codon. A low level of GH mRNA is present, but there is no detectable GH (7). Although there are no identifiable somatotropes, a comparably sized population of "null" cells are present that represent non-GH-containing somatotropes because they express greater than normal levels of the GHRH-R and exhibit enhanced cyclic AMP responses to GHRH (29). As with other models of GH deficiency, hypothalamic GHRH mRNA levels are increased and somatostatin mRNA levels are decreased. Treatment with GH and IGF-I suppresses GHRH-R mRNA levels, but only GH treatment reverses the changes in hypothalamic hormone gene expression. Pituitary GH secretagogue (GHS) receptor (GHS-R) mRNA is also increased in the SDR and is decreased by GH (although not IGF-I) treatment (30). The positive association between GHRH and GHS-R mRNA implies that GHRH is required for maintenance of GHS-R gene expression. This idea is supported by our recent observation that short-term intravenous infusion of GHRH stimulated GHS-R expression, although this effect was not observed in vitro (31). Although the precise mediation of this effect remains to be determined, the simulatory interaction of the two GH-releasing systems GHRH and GHS provides a possible explanation for the impaired responses to various GH secretagogues under conditions of reduced or absent GHRH–GHRH-R function.

The SDR has also been used as a model for studying hepatic cytochrome P450 sexual dimorphism. The predominant enzyme form in males (CYP2C11) is reduced to one-third of normal in male SDRs, while the female-specific form (CYP2C12) is not detectable in female SDRs (32). Continuous GH infusion induced CYP2C12 expression in both male and female SDRs, whereas intermittent GH injections elevated CYP2C11 to normal levels in males.

Summary

The genetic and transgenic models of disturbances of the GH axis provide unique opportunities for the study of individualized components of the GH feedback axis, somatotrope function, and effects of GH secretion and action. In particular, the importance of the GH secretory pattern in generating the sexually dimorphic expression of GH-dependent proteins has become better understood because of these models. Their use in expanding our present knowledge of extrahepatic effects of GH related to growth, metabolism, and feedback can be expected to provide new and important findings.

References

1. Li H, Zeitler PS, Valerius MT, Small K, Potter SS. *Gsh-1*, an orphan Hox gene, is required for normal pituitary development. EMBO J 1996;15:714–24.
2. Szabo M, Butz MR, Banerjee SA, Chikaraishi DM, Frohman LA. Autofeedback suppression of growth hormone (GH) secretion in transgenic mice expressing a human GH reporter targeted by tyrosine hydroxylase 5′ flanking sequences to the hypothalamus. Endocrinology 1995;136:4044–8.

3. Flavell DM, Wells T, Wells SE, Carmignac DF, Thomas GB, Robinson ICAF. Dominant dwarfism in transgenic rats by targeting human growth hormone (GH) expression to hypothalamic GH-releasing factor neurons. EMBO J 1996;15:3871–9.
4. Godfrey P, Rahal JO, Beamer WG, Copeland NG, Jenkins NA, Mayo KE. GHRH receptor of little mice contains a missense mutation in the extracellular domain that disrupts receptor function. Nat Genet 1993;4:227–32.
5. Lin SC, Lin CR, Gukovsky I, Lusis AJ, Sawchenko PE, Rosenfeld MG. Molecular basis of the little mouse phenotype and implications for cell type-specific growth. Nature (Lond) 1993;364:208–13.
6. Sornson MW, Wu W, Dasen JS, Flynn SE, Norman DJ, O'Connell SM, et al. Pituitary lineage determination by the prophet of Pit-1 homeodomain factor defective in Ames dwarfism. Nature (Lond) 1996;384:327–33.
7. Takeuchi T, Suzuki H, Sakurai S, Nogami H, Okuma S, Ishikawa H. Molecular mechanism of growth hormone (GH) deficiency in the spontaneous dwarf rat: detection of abnormal splicing of GH messenger ribonucleic acid by the polymerase chain reaction. Endocrinology 1990;126:31–8.
8. Charlton HM, Clark RG, Robinson IC, Porter-Goff AEP, Cox BS, Bugnon C, et al. Growth hormone-deficient dwarfism in the rat: a new mutation. J Endocrinol 1988;119:51–8.
9. Hammer RE, Brinster RL, Rosenfeld MG, Evans RM, Mayo KE. Expression of human growth hormone-releasing factor in transgenic mice results in increased somatic growth. Nature (Lond) 1985;315:413–6.
10. Frohman LA, Downs TR, Kashio Y, Brinster RL. Tissue distribution and molecular heterogeneity of human growth hormone-releasing factor in the transgenic mouse. Endocrinology 1990;127:2149–56.
11. Palmiter RD, Norstedt G, Gelinas RE, Hammer RE, Brinster RL. Metallothionein-human GH fusion genes stimulate growth of mice. Science 1983;222:809
12. Behringer RR, Mathews LS, Palmiter RD, Brinster RL. Dwarf mice produced by genetic ablation of growth hormone-expressing cells. Genes Dev 1988;2:453–61.
13. Chen WY, Wight DC, Wagner TE, Kopchick JJ. Expression of a mutated bovine growth hormone gene suppresses growth of transgenic mice. Proc Natl Acad Sci USA 1990;87:5061–5.
14. Zhou Y, Xu BC, Maheshwari HG, He L, Reed M, Lozykowski M, et al. A mammalian model for Laron syndrome produced by targeted disruption of the mouse growth hormone receptor/binding protein gene (the Laron mouse). Proc Natl Acad Sci USA 1997;94:13215–20.
15. Li J-L, Grinberg A, Westphal H, Sauer B, Accili D, Karas M, et al. Conditional knockout of mouse IGF-I gene using the cre/loxP system (abstract). Program 80th Annu Mtg Endocrinol Soc 1998:86.
16. Mathews LS, Hammer RE, Behringer RR, D'Ercole AJ, Bell GI, Brinster RL, et al. Growth enhancement of transgenic mice expressing human insulin-like growth factor I. Endocrinology 1988;123:2827–33.
17. Eicher EM, Beamer WG. Inherited ateliotic dwarfism in mice: characteristics of the mutation, little, on chromosome 6. J Hered 1976;67:87–91.
18. Jansson J-O, Downs TR, Beamer WG, Frohman LA. Receptor-associated resistance to growth hormone-releasing factor in dwarf "little" mice. Science 1986;232:511–2.
19. Jansson J-O, Downs TR, Beamer WG, Frohman LA. The dwarf "little" (lit/lit) mouse is resistant to growth hormone (GH)-releasing peptide (GH-RP-6) as well as to GH-releasing hormone (GRH) (abstract). Program 68th Annu Mtg Endocrinol Soc 1986:397.

20. Jansson J-O, Ekberg S, Hoath SB, Beamer WG, Frohman LA. Growth hormone enhances hepatic epidermal growth factor receptor concentration in mice. J Clin Invest 1988;82:1871–6.
21. al-Shawri R, Wallace H, Harrison S, Jones C, Johnson D, Bishop JO. Sexual dimorphism and growth hormone regulation of a hybrid gene in transgenic mice. Mol Endocrinol 1992;6:181–90.
22. Downs TR, Frohman LA. Evidence for a defect in growth hormone-releasing factor signal transduction in the dwarf (dw/dw) rat pituitary. Endocrinology 1991;129:58–67.
23. Zeitler PA, Downs TR, Frohman LA. Impaired growth hormone releasing-hormone signal transduction in the dwarf (dw) rat is independent of a generalized defect in the stimulatory G-protein, Gs-alpha. Endocrinology 1993;133:2782–6.
24. Frohman LA, Downs TR, Sato M. The use of transgenic and genetic models to study the neuroendocrine regulation of growth hormone secretion. In: Muller EE, Cocchi D, Locatelli V, eds. Berlin: Springer-Verlag, 1993:131–40.
25. Jeffery S, Carter ND, Clark RG, Robinson IC. The episodic secretory pattern of growth hormone regulates liver carbonic anhydrase. III. Studies in normal and mutant growth-hormone-deficient dwarf rats. Biochem J 1990;266:69–74.
26. Carmignac DF, Wells T, Carlsson LM, Clark RG, Robinson IC. Growth hormone (GH)-binding protein in normal and GH-deficient dwarf rats. J Endocrinol 1992;135:447–57.
27. Carmignac DF, Robinson IC, Enberg B, Norstedt G. Growth hormone receptor regulation in growth hormone-deficient dwarf rats. J Endocrinol 1993;138:267–74.
28. Low SC, Chapman.KE, Edwards CR, Wells T, Robinson IC, Seckl JR. Sexual dimorphism of hepatic 11 beta-hydroxysteroid dehydrogenase in the rat: the role of growth hormone patterns. J Endocrinol 1994;143:541–8.
29. Kamegai J, Unterman TG, Frohman LA, Kineman RD. Hypothalamic/pituitary axis of the spontaneous dwarf rat: autofeedback regulation of growth hormone (GH) includes suppression of GH releasing-hormone receptor messenger ribonucleic acid. Endocrinology 1998;139:3554–60.
30. Kamegai J, Wakabayashi I, Miyamoto K, Unterman TG, Kineman RD, Frohman LA. Growth hormone (GH)-dependent regulation of pituitary GH secretagogue receptor (GHS-R) mRNA levels in the spontaneous dwarf rat. Neuroendocrinology 1998;68:312–8.
31. Kamegai J, Wakabayashi I, Unterman TG, Frohman LA, Kineman RD. Growth hormone-releasing hormone (GHRH) stimulates pituitary GH-secretagogue receptor (GHS-R) mRNA levels in vivo (abstract). Program 80th Annu Mtg Endocrinol Soc 1998;63–4.
32. Shimada M, Murayama N, Nagata K, Hashimoto H, Ishikawa H, Yamazoe Y. A specific loss of growth hormone abolished sex-dependent expression of hepatic cytochrome P450 in dwarf rats: reversal of the profiles by growth hormone-treatment. Arch Biochem Biophys 1997;337:34–42.

29

Estrogen Receptor Expression in the Pituitary Gland

KEITH E. FRIEND AND IAN E. MCCUTCHEON

Estradiol has been demonstrated to affect the production or release of all six major anterior pituitary hormones (1–5). Although some of the effects of estradiol on hormonal secretion are primarily hypothalamic in nature, other physiologically relevant actions appear to occur directly within the anterior pituitary gland itself (6–9). According to our current understanding of the signal transduction mechanics of estradiol, the presence of an estrogen receptor (ER) is necessary for such direct actions. This has been the primary rationale for performing experiments to determine which types of anterior pituitary cells express estrogen receptors. By defining which anterior pituitary cells are ER positive, we are identifying which populations of cells possess the necessary cellular machinery to make them at least capable of responding directly to estradiol.

Two major ERs have been identified to date. ER-alpha was first isolated and characterized in 1986 (10), and ER-beta was isolated and characterized a decade later (11). Both are ligand-activated transcriptional factors that bind to specific DNA sequences, known as estrogen response elements (ERE), located in the regulatory region of estradiol-responsive genes. ER-alpha and ER-beta have 95% amino acid sequence homology in their DNA-binding domains, but are only 55% homologous in their ligand-binding domains (11). Evidence for an additional estrogen receptor located in the cell membrane, which would then be capable of mediating some of the more rapid, nontranscriptional physiological actions attributed to estradiol, has also been presented (12), but to date this receptor has not been isolated or fully characterized.

ER-Alpha Expression in the Normal Pituitary Gland

The earliest studies to examine the anterior pituitary gland for the presence of estrogen receptor were binding experiments using ^{3}H-estradiol. Several groups of investigators identified high-affinity estrogen binding in the anterior pituitary gland (13,14) as early as 1976. Although this methodology was used to determine the effects of gonadectomy (15) and of glucocorticoid (14), androgen (15), and estradiol (16) administration, it was generally not possible to define which anterior pituitary cell types accounted for the observed changes. The development of immortalized anterior pituitary cell lines (e.g., GH-3 cells) provided investigators with initial models to study the direct actions of estradiol on hormone production (17).

In 1981, using a combined autoradiographic-immunocytochemical technique, Keefer quantified the amount of nuclear estrogen uptake in different anterior pituitary cell types in the male rat. The following rank order was determined by silver grain counting: somatotropes = lactotropes > gonadotropes = corticotropes > thyrotropes. Thyrotropes were reported to contain approximately half the amount of estrogenic radioactivity found in somatotropes or lactotropes (18).

More recently, double-labeling immunohistochemistical experiments for ER-alpha protein and each of the six major anterior pituitary hormones were performed in normal human pituitary. Three hundred cells of each type were counted and the percentages of ER-alpha-containing cells were as follows: growth hormone (GH) (2.3%), follicle-stimulating hormone (FSH) (70%), luteinizing hormone (LH) (83%), prolactin (PRL) (50%), thyroid-stimulating hormone (TSH) (4%), and adrenocorticotropic (ACTH) (1%). The ER antibody used in these experiments was specific for ER-alpha. Because somatotropes and lactotropes are thought to have a common progenitor cell, the somatomammotrope, the authors suggested that one key point of divergence in the development of the two cell types might involve regulation of the ER-alpha gene, upregulation occurring as cells differentiate into PRL-producing cells or downregulation occurring as they differentiate into GH-producing cells (19). Although the percentage of ER-alpha-expressing GH cells (2.3%) in the anterior pituitary is consistent with the expected number of somatomammotropes, triple immunostaining to confirm that the ER-alpha-expressing GH cells were indeed somatomammotropes was not performed.

A combined in situ hybridization/immunocytochemistry technique identified ER mRNA in all cell types present in the anterior pituitary, with the highest levels observed in the lactotrope population (20). The synthetic oligonucleotide probe used for the in situ hybridization portion of the experiments was complementary to amino acids 1–24 of the ER-alpha molecule and would not be expected to detect ER-beta.

ER-Alpha Expression in Pituitary Adenomas

Several investigators have examined human pituitary adenomas for the presence of ER protein or mRNA. The rationale for performing these measurements have generally been twofold. First, ER expression in adenomas might provide insight into which cell types in the normal pituitary express ER (e.g., if prolactin-producing adenomas contain ER, then normal prolactin-producing cells are also likely to contain ER). As adenomas represent clonal expansions of one particular cell type and are generally well-differentiated tumors, this is not an unreasonable proposal and has been at least partially verified experimentally. Second, the pattern of ER expression might provide some insight into whether estrogen could be playing a direct role in oncogenesis in those tumors that show gender-related differences in prevalence (e.g., ACTH- and prolactin-producing tumors). An extension of this argument would be that the ER could be modulating tumor growth after the tumor has developed. In the case of prolactin-producing tumors, it does appear that many are ER positive; however, the importance of this observation has not been adequately investigated with respect to either tumorigenesis or tumor growth.

One of the earliest experiments to assess ER expression in pituitary tumors was performed by Pinchon et al. (21). Using an estradiol-binding assay, these investigators detected ER in 14 of 23 tumors. The highest concentrations were observed in prolactinomas (71% positive; mean ER level, 20.6 fmol/mg protein), followed by chromophobe adenomas (86% positive; 4.1 fmol/mg protein), and then GH-producing tumors (33% positive; 1.4 fmol/mg protein). Several years later, using an enzyme immunoassay, Nakoa et al. identified ER in 10 of 56 pituitary adenomas (22). Again, the highest levels were found in prolactinomas (43% positive; 33.5 fmol/mg), followed by gonadotropin-immunostaining tumors (50% positive; 13.5 fmol/mg protein), and then combined GH-/PRL-producing adenomas (13% positive; 10.6 fmol/mg protein). No ER was detected in GH-secreting or nonfunctioning tumors. Stefaneanu and Kovacs et al. found no detectable ER in any pituitary adenomas with a ER immunohistochemistry methodology (23).

In 1994, 41 human pituitary adenoma specimens were examined for the presence of ER-alpha mRNA using a ribonuclease protection assay (RPA). ER-alpha mRNA prevalence was high in tumors immunostaining positively for prolactin (100%), moderate in tumors immunostaining positively for both GH and prolactin (40%), and absent in tumors immunostaining positively only for GH (0%). Among the tumors immunostaining positively for both GH and prolactin, ER-alpha expression was uniformly associated with elevated serum prolactin concentrations. Among the gonadotropin-immunostaining tumors, 59% were ER-alpha positive; those with gonadotrope secretory characteristics (determined by electron microscopy) were uniformly ER-alpha positive, while those with null cell or oncocytic characteristics were uniformly ER-alpha negative. Of the tumors that were immunonegative, 27% were ER-alpha positive. ER-alpha immunohistochemistry in 14 of the aforementioned tumors revealed a 100% correlation with RPA results, while ^{3}H-estradiol binding, performed on 9 tumors, demonstrated an 87% correlation (19).

ER-Alpha Isoforms

A number of ER-alpha isoforms have been identified recently in the anterior pituitary gland. Truncated estrogen receptor product-1 (TERP-1), is an ER-alpha isoform present in the anterior pituitary gland of the rat. The expression of TERP-1 is a highly regulated process in that it is estrogen inducible, which was demonstrated by the administration of estradiol to male animals (24). The abundance of the mRNA encoding this naturally occurring isoform varies many hundredfold throughout the rat estrous cycle, reaching peak levels on the morning of proestrus (25). The wide range of expression of TERP-1 throughout the estrous cycle is probably a physiological manifestation of the E_2-inducible nature of this isoform (25). The phenomenon of a ligand potently inducing a truncated form of its own receptor is unusual among the steroid hormone receptor superfamily. Other larger but less abundant TERP isoforms have also been identified (26). The structures of the TERP isoforms isolated to date are shown in Figure 29.1. Although the TERP

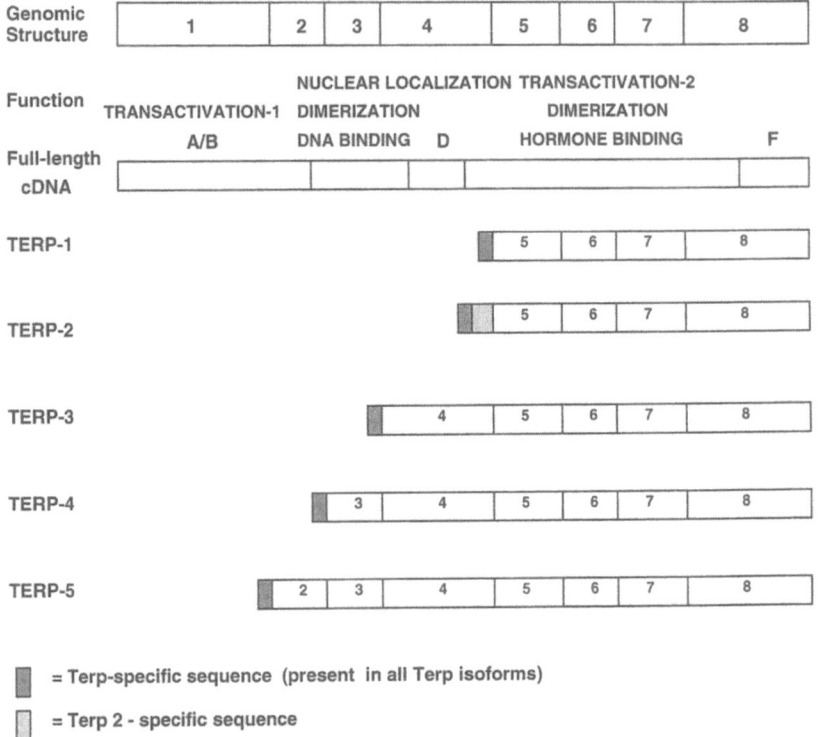

FIGURE 29.1. Structure of the truncated estrogen-receptor protein (TERP) isoforms that have been isolated to date in the anterior pituitary gland of the rat. Shown also are the genomic organization and functional domains of the receptor.

isoforms appear to be able to modulate the ability of the full-length ER-alpha to activate ER-responsive genes, the physiological role of these isoforms has not been defined.

ER mRNAs corresponding to splice variants in which either exon 4, exons 3 and 4, or exons 5 and 6 are deleted have also been identified in the anterior pituitary gland of the rat. The expression of these isoforms, however, is not pituitary specific as they were also localized to uterus, testes, heart, hypothalamus, and liver. An antibody to the ER C-terminus bound to full-length ER (64 kDa) and smaller proteins (50–55 kDa and 40–45 kDa); the smaller proteins would be consistent with the predicted size of some deletion isoforms (25).

Alternatively spliced ER variant mRNAs have also been identified in human pituitary tumors. Approximately 82% of prolactinomas expressed multiple ER variants including exons 3, 4, 5, and 7. Some tumors (55%) also expressed an exon 3 deletion isoform. Among the glycoprotein hormone-producing tumors, 57% expressed exon 2, 5, and 7 deletion isoforms. Of note, only the exon 4 and 7 deletion isoforms were identified in normal pituitary tissue. All deletion isoforms were coexpressed with normal full-length ER-alpha (27). The functional significance of the ER deletion isoforms remains uncertain also although some isoforms can modulate the function of the full-length ER-alpha (28).

ER-Beta Expression

ER-beta mRNA expression has been identified in oxytocin- and vasopressin-secreting neurons of the rat supraoptic and paraventricular nuclei (29) and in the anterior pituitary gland, along with a variant of ER-beta named ER-beta$_2$ (30). The precise distribution and functional significance of ER-beta in the pituitary gland has not been determined. This information should become available as investigators apply some of the methodologies used to define the range of ER-alpha expression to this more recently identified receptor, as is discussed further in Chapter 33.

References

1. Veldhuis JD, Samojlik E, Evans WS, Rogol AD, Ridgeway CE, Crowley WF, et al. Endocrine impact of pure estradiol replacement in postmenopausal women: alterations in anterior pituitary hormone release and circulating sex steroid hormone concentrations. Am J Obstet Gynecol 1986;155:334–9.
2. Rossmanith WG, Stabler C, Benz R, Bornstein SR, Scherbaum WA. Role of ovarian sex steroids in the regulation of thyrotropin (TSH) secretion in hypogonadal women. Acta Endocrinol 1992;127(2):131–7.
3. Wiedemann E, Schwartz E, Frantz AG. Acute and chronic effects upon somatomedin C activity, growth hormone and prolactin in man. J Clin Endocrinol Metab 1975;42: 942–50.

4. Burgess LH, Handa RJ. Chronic estrogen-induced alterations in adrenocorticotropin and corticosterone secretion, and glucocorticoid receptor-mediated functions in female rats. Endocrinology 1992;131(3):1261–9.
5. Kineman RD, Faught WJ, Frawley LS. Steroids can modulate transdifferentiation of prolactin and growth hormone cells in bovine pituitary cultures. Endocrinology 1992;130(6):3289–94
6. Ho KY, Thorner MO, Krieg, RJ Jr, Lau SK, Sinha YN, Johnson ML, et al. Effects of gonadal steroids on somatotroph function in the rat: analysis by the reverse hemolytic plaque assay. Endocrinology 1988;123:1405–11.
7. Maurer RA, Notides AC. Identification of an estrogen-responsive element from the 5'-flanking region of the rat prolactin gene. Mol Cell Biol 1987;7:4247–54.
8. Shupnik MA, Gharib SD, Chin WW. Divergent effects of estradiol on gonadotropin gene transcription in pituitary fragments. Mol Endocrinol 1989;3:474–80.
9. Shupnik MA, Weinmann CM, Notides AC, Chin WW. An upstream region of the rat luteinizing hormone beta gene binds estrogen receptor and confers estrogen responsiveness. J Biol Chem 1989;264:80–6.
10. Green S, Walter P, Kumar V, Krust A, Bornert JM, Argos P, et al. Human oestrogen receptor cDNA: sequence, expression and homology to v-erb-A. Nature (Lond) 1986;320:134–9.
11. Kuiper GGJM, Enmark E, Pelto-Huikko M, Nilsson S, Gustafsson JA. Cloning of a novel estrogen receptor expressed in rat prostate and ovary. Proc Natl Acad Sci USA 1996;93:5925–30.
12. Pappas TC, Gametchu B, Watson CS. Membrane estrogen receptors identified by multiple antibody labeling and impeded-ligand binding. FASEB J 1995;9(5):404–10.
13. Ginsburg M, Maclusky NJ, Morris ID, Thomas PJ. The specificity of oestrogen receptor in brain, pituitary and uterus. Br J Pharmacol 1977;59(3):397–402.
14. Lisk RD, Reuter LA. Dexamethasone: increased weights and decreased [^{3}H] estradiol retention of uterus, vagina and pituitary in the ovariectomized rat. Endocrinology 1976;99(4):1063–70.
15. Barley J, Ginsburg M, MacLusky NJ, Morris ID, Thomas PJ. Sex differences in the distribution of cytoplasmic oestrogen receptors in rat brain and pituitary: effects of gonadectomy and neonatal androgen treatment. Brain Res 1997;129(2):309-18.
16. Morris ID. Changes in brain, pituitary and uterine cytoplasmic oestrogen receptors induced by oestradiol-17 beta in the ovariectomized rat. J Endocrinol 1976;71(3):343–9.
17. Tashjian AH Jr, Yasumura Y, Levine L, Sato GH, Parker ML. Establishment of clonal strains of rat pituitary tumor cells that secrete growth hormone. Endocrinology 1968;82(2):342–52.
18. Keefer DA. Quantification of in vivo ^{3}H-estrogen uptake by individual anterior pituitary cell types of male rat: a combined autoradiographic-immunocytochemical technique. J Histochem Cytochem 1981;29(suppl 1A):167–74.
19. Friend KE, Chiou YK, Lopes MBS, Laws ER Jr, Hughes KM, Shupnik MA. Estrogen receptor expression in human pituitary: correlation with immunohistochemistry in normal tissue, and immunohistochemistry and morphology in macroadenomas. J Clin Endoclinol Metab 1994;78:1497–504.
20. Stefaneanu L, Kovacs K, Horvath E, Lloyd RV, Buchfelder M, Fahlbusch R, Smyth H. In situ hybridization study of estrogen receptor messenger ribonucleic acid in human adenohypophysial cells and pituitary adenomas. J Clin Endocrinol Metab 1994;78(1):83–8.

21. Pinchon MF, Bression D, Peillon F, Milgrom E. Estrogen receptors in human pituitary adenoma. J Clin Endocrinol Metab 1980;51:897–902.
22. Nakoa H, Koga M, Arao M, Nakao M, Sato B, Kishimoto S, Saitoh Y, Arita N, Mori S. Enzyme immunoassay for estrogen receptors in human pituitary adenomas. Acta Endocrinol 1989;120:233–8.
23. Stefaneanu L, Kovacs K. Immunocytochemical approach to demonstrate the estrogen receptor in human adenohypophyses and pituitary adenomas with monoclonal antibody. Med Sci Res 1988;16:449–50.
24. Friend KE, Ang LW, Shupnik MA. Estrogen regulates the expression of several different estrogen receptor mRNA isoforms in rat pituitary. Proc Natl Acad Sci USA 1995;92:4367–71.
25. Friend KE, Resnick EM, Ang LW, Shupnik MA. Specific modulation of estrogen receptor mRNA isoforms in rat pituitary throughout the estrous cycle and in response to steroid hormones. Mol Cell Endocrinol 1997;131:147–55.
26. Zhu Z, Ang LE, Friend KE. Identification of two novel truncated estrogen receptor isoforms in rat pituitary. In: 6th International Conference on Cell Biology, 1996, San Francisco, CA.
27. Chaidarun SS, Klibanski A, Alexander JM. Tumor-specific expression of alternatively spliced estrogen receptor messenger ribonucleic acid variants in human pituitary adenomas. J Clin Endocrinol Metab 1997;82(4):1058–65.
28. Zhang QX, Hilsenbeck SG, Fuqua SA, Borg A. Multiple splicing variants of the estrogen receptor are present in individual human breast tumors. J Steroid Biochem Mol Biol 1996;59(3–4):251–60.
29. Hrabovszky E, Kallo I, Hajszan T, Shughrue PJ, Merchenthaler I, Liposits Z. Expression of estrogen receptor beta messenger ribonucleic acid in oxytocin and vasopressin neurons of the rat supraoptic and paraventricular nuclei. Endocrinology 1998;139(5):2600–4.
30. Petersen DN, Tkalcevic GT, Kozataylor PH, Turi TG, Brown TA. Identification of estrogen receptor beta(2), a functional variant of estrogen receptor beta expressed in normal rat tissues. Endocrinology 1998;139(3):1082–92.

30

Metabolic Effects of GH/IGF-I, Testosterone, and Estrogen: Studies in Children and Adults

NELLY MAURAS

During no other time in postnatal human development are there such massive changes in body composition as in puberty. GH and IGF-I production are greatly enhanced, and insulin and sex steroidal hormones increase during this period, allowing the child's transformation into a fully grown adult with reproductive maturity. Some of the changes observed are directly related to the enhancement of whole-body protein synthetic rates and the increase of specific protein pools as well as to the increased calcium absorption and retention, leading to increased bone mineralization, net bone accrual, and eventual achievement of peak bone mass. Safe and reproducible investigative techniques are now available that permit the assessment of protein, calcium, and glucose kinetics, as well as body compositional changes. These tools are uniquely useful in human studies, particularly pediatric studies, as they do not involve radioactive materials, and are relatively noninvasive and nontoxic. We have studied the relative contributions of these anabolic hormones to the metabolic changes of puberty and assessed the effects of these hormones in chronic catabolic states.

Stable Isotopes Techniques

Protein

The basic principles of the techniques used in these experiments have been extensively reviewed (1) and are summarized in Figure 30.1. An essential amino acid, leucine, with a labeled carbon, L-[1-^{13}C]-Leu, is infused at fixed rates into the plasma compartment. It diffuses inside the cell where, in the fasted state, it has two potential fates: either it is transaminated into its α-keto acid (αketoisocaproic acid, α-KIC) and eventually oxidized, or it is incorporated into protein. In the oxidative process, because the ^{13}C label is in the first

Leucine Kinetic Model

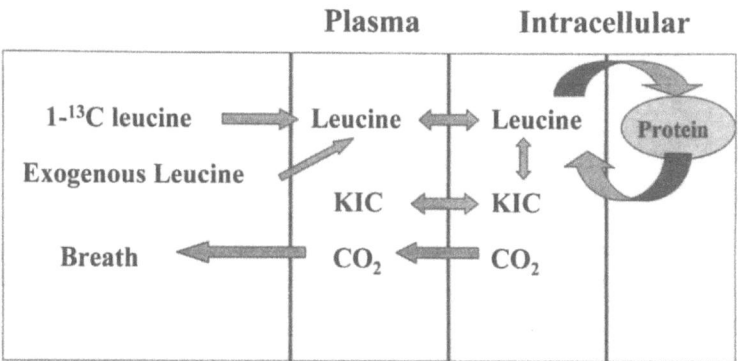

FIGURE 30.1. A stable tracer of leucine (^{13}C-Leu) is infused at fixed rates in plasma, diffuses into the intracellular space where it is either incorporated into protein (which can also break down and contribute free leucine back to the cell), or is transaminated into its α-keto acid (KIC) and eventually oxidized, generating CO_2. KIC diffuses into the plasma compartment. The rates of appearance and disappearance of the tracer allow estimates of whole-body protein degradation and synthesis, respectively (1,2).

carbon of leucine, $^{13}CO_2$ is generated, captured in exhaled breath. Using gas chromatography/mass spectrometry (GCMS) and isotope ratio mass spectrometry methods, we can measure the enrichment of the tracer in plasma and exhaled air. We assume that, at steady state, the rate of appearance (Ra) of the tracer equals its rate of disappearance; hence, Leucine Ra = [(Ei/Ep)-1]*F, where Ei is the enrichment of the infusate, Ep is the enrichment of plasma, and F is the infusion rate of the tracer (2,3). Leucine oxidation rate (Leu Ox) is calculated using the enrichment of ^{13}C-KIC, corrected for CO_2 recovery, and the expired rate of $^{13}CO_2$ production as previously described (4). Then nonoxidative leucine disposal or NOLD = Ra–Leu Ox. The leucine Ra thus serves as an index of whole-body proteolysis and NOLD as an index of whole-body protein synthesis. Further compartmentalization of which specific protein pools are affected by hormonal interventions or disease states can be achieved by measuring the fractional biosynthetic rates of specific plasma proteins, e.g., fibrinogen and albumin (5,6), and contractile muscle proteins (e.g., myosin heavy chain) (7); however, the latter technique is invasive, requiring muscle biopsies, and is not particularly applicable to studies in children.

Glucose

Similar principles as already described are used to estimate hepatic glucose output in the postabsorptive state. An infusion of [6,6, $^2H_2$] glucose is given, and the enrichment of the tracer in plasma is measured by GCMS as previously described (8).

Calcium

Stable labeled tracers of calcium (Ca), ^{44}Ca, and ^{42}Ca were given orally (p.o.) and intravenously (i.v.) respectively, and serum and urine samples were collected to measure the enrichment of both tracers using a dual-filament thermal ionization quadrupole mass spectrometer (9). The fractional absorption of Ca (α) was calculated from the ratio of the recovery of the p.o./i.v. tracer in urine: $\alpha = \int {}^{44}\text{Ca} / \int {}^{42}\text{Ca}$. Analysis of the Ca kinetics was performed using a three-pool compartmental model (10) and the simulation and analysis modeling program (11), (Fig. 30.2).

GH and IGF-I Effects

Extensive data suggest that the mechanism of action of GH is mediated by IGF-I in many, but not all, tissues in vivo. IGF-I mediates the linear growth-promoting actions of GH, as it improves the growth of both GH-deficient and IGF-I deficient (Laron syndrome) patients. Acute GH administration can be associated with a mild decrease in insulin sensitivity and compensatory hyperinsulinemia (12). However, IGF-I administration has dichotomous effects on carbohydrate metabolism, with either insulin-like (glucose-lower-

Calcium kinetics: multicompartmental model

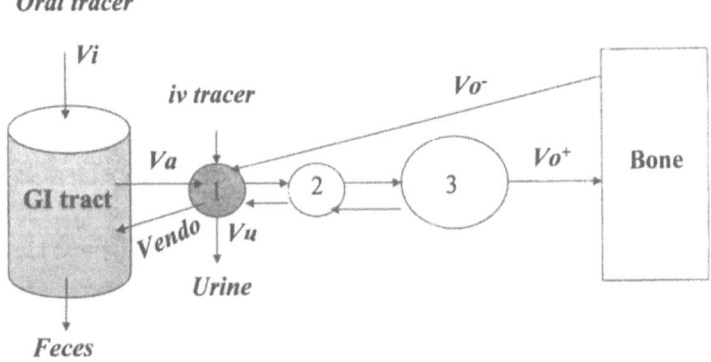

FIGURE 30.2. A three-compartment model is used to estimate changes in calcium (Ca) fluxes after ^{44}Ca is given orally and ^{42}Ca i.v. (9,10). Vi, dietary Ca intake; Va, Ca absorption from the gastrointestinal tract; Vu. urinary Ca loss; $Vendo$, endogenous fecal excretion; Vo^+ is the net forward flow of Ca into bone; Vo^-, net bone Ca resorption. Reproduced with permission from Abrams et al. (30).

ing) effects at higher doses (13,14), or maintenance of normal glucose homeostasis at lower doses (15).

The action of GH on whole-body protein pools is, however, indistinguishable from that of IGF-I. When given to healthy volunteers, GH and IGF-I each selectively stimulated whole-body protein synthesis without any effects on proteolysis (3,15), contrary to the antiproteolytic effects observed with insulin (16). Given in combination to calorically deprived volunteers, GH and IGF-I were observed to cause greater nitrogen retention than either compound given separately (17). However, when given to normally fed volunteers, the combination was not more anabolic on whole-body protein than each compound given separately (18). Taken in aggregate, these data strongly suggest that the protein-anabolic effects of GH are mediated through IGF-I (and the type 1 IGF-I receptor) and that by inference, in puberty, the marked enhancement of whole-body protein-anabolic responses to GH is mediated through IGF-I generation.

The direct lipolytic effects of GH are well characterized; however, these do not appear to be mediated through IGF-I as there are no functional type 1 IGF-I receptors in adipocytes (19). The indirect effects of IGF on lipolysis can be biphasic, with suppression of triglycerides, free fatty acids, and ketone body production, during both low-dose and high-dose infusions, similar to an insulin-like effect (20). However, enhanced lipolysis and increased free fatty acid concentrations have been observed with more prolonged IGF-I administration (19). The latter may be caused by the chronic insulinopenia caused by IGF-I. Because in puberty there tends to be a net gain in lean body mass in both sexes, yet a concomitant increase in adipose mass in the female, the data suggest that there may be differential sensitivity to the protein-anabolic and -lipolytic effects of GH between the sexes. Further studies are needed to characterize these changes throughout all stages of puberty.

We have studied the route of delivery of peptide hormones with these metabolic tools, and it is apparent that there are significant differential effects on metabolic fuels depending on the mode of administration. For example, IGF-I, when administered as 200 µg kg^{-1}·day^{-1} as a continuous subcutaneous infusion for 16 h/day had no detectable effects on rates of protein synthesis or oxidation, whereas when given as twice daily bolus injections for 5 days it generated a robust protein-anabolic response (21). Carbohydrate oxidation and glucose production rates were similar regardless of route of delivery, but lipid oxidation decreased during bolus injections but not continuous s.c. administration. Free IGF-I concentrations were higher after bolus injections, suggesting that there may be a dose-dependent effect of free IGF-I at the tissue level and that carbohydrate metabolism may be more sensitive to IGF-I than protein anabolism.

Testosterone Effects

The anabolic effects of androgenic hormones have been previously studied in a variety of situations in experimental animals and humans. The effects of testosterone administration in the prepubertal child are particularly interesting, as the rate of change in sex-steroid hormones is dramatically altered at this time. We studied a group of prepubertal normal boys (mean age, 12.9 ± 0.6 years) treated with exogenous intramuscular testosterone enanthate injections for 1 month utilizing infusions of ^{13}C-Leu and ^{42}Ca and ^{44}Ca as described (22); ^{15}N-glutamine was also infused to measure glutamine production rates. Glutamine is the most abundant amino acid in the body and provides a large reservoir of endogenous nitrogen, necessary for protein synthesis. We found that measures of proteolysis increased 18% ($p = 0.036$), protein oxidation decreased 49% ($p = 0.004$), with a net gain in whole-body protein synthesis by 34% ($p = 0.009$), as well as an increase in glutamine de novo synthesis (+32%; $p < 0.05$). We also found a net increase in calcium absorption and retention and in the rate of the net calcium flow into bone. Mean and peak serum GH levels, as measured by frequent blood sampling, as well as plasma IGF-I concentrations markedly increased during these experiments. Thus, these effects may represent a pure androgen effect, an amplification of GH action, or a combination of the two.

More recently (23), we studied the opposite model; i.e., we evaluated, in normal eugonadal males between 18 and 25 years of age, a multiplicity of kinetic, metabolic, and hormonal factors, as well as body composition both before and after the induction of hypogonadism with the GnRH analogue Lupron given twice in 3 weeks. Studies were paired, 10 weeks apart. The induction of hypogonadism was severe, with mean testosterone concentrations decreasing into the prepubertal range. The leucine kinetics inversely mirrored those found in the boys given testosterone, with a 13% decline in proteolysis ($p = 0.01$) and in protein synthesis (-13%; $p = 0.01$) without any plasma amino acid concentration changes (Fig. 30.3A). All subjects also showed a marked decrease in fat-free mass (FFM) as measured by dual-emission X-ray absorptiometry (DEXA) and an increase in fat mass (FM). Rates of lipid oxidation, as measured by indirect calorimetry, decreased (-31%; $p = 0.05$), as well as resting energy expenditure (-9%; $p = 0.05$). Muscle strength of the leg extensors markedly decreased, as measured by isokinetic dynamometry (Fig. 30.3B). Interestingly, deconvolution analysis of the GH concentration profiles showed no decrease in pulsatile GH secretion but an increase in the basal rate of production ($p < 0.02$) with invariant plasma IGF-I concentrations (Fig. 30.4). These data support a direct effect of androgens on whole-body lipid and protein metabolism, independently of peripheral GH and IGF-I production.

In an attempt to examine the possible synergistic effect of testosterone and GH on measures of protein metabolism and body composition, we are presently studying young boys with GH deficiency, before and after the administration of testosterone, alone and in combination with GH. Each study is paired, 1 month apart, with the order of the intervention arms randomized (each subject is studied three times). Preliminary results show that testosterone and GH exert synergistic effects on body composition, protein anabo-

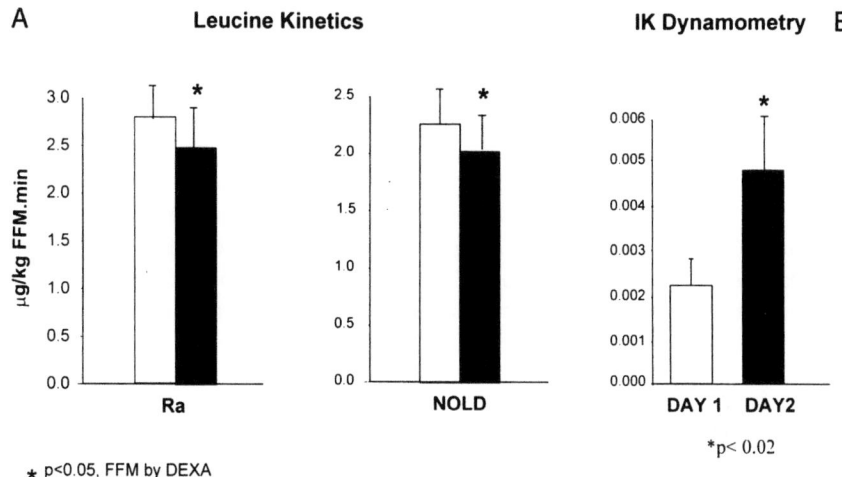

FIGURE 30.3. *A* Changes in rates of whole-body proteolysis (Ra) and synthesis (NOLD) after induction of hypogonadism in normal men. *B* Changes in muscle strength in the leg quadriceps muscle in the same men (*right*). *Open bars*, baseline data; *solid bars*, after GnRHa treatment. Reproduced with permission from Mauras et al. (23).

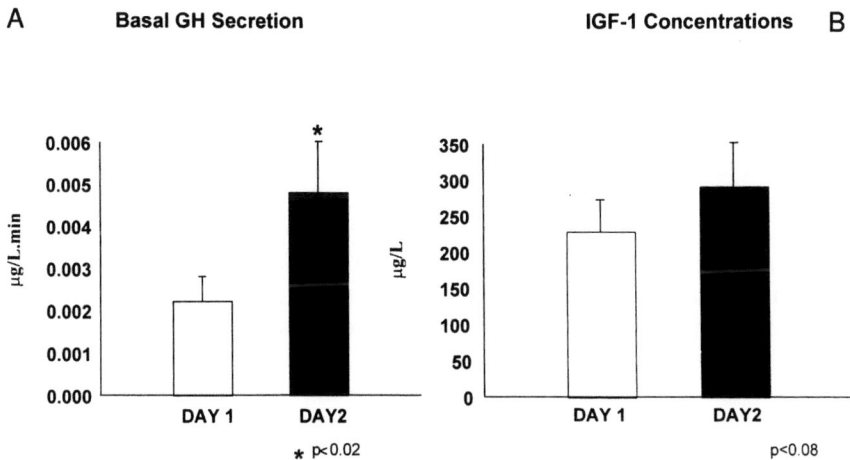

FIGURE 30.4. *A* Changes in GH basal secretory rate in young men rendered hypogonadal by the administration of a GnRHa. *B* Changes in plasma IGF-I concentrations in the same subjects. Day 1, baseline data; Day 2, after treatment. Data from Mauras et al. (23).

lism, and IGF-I generation (24), again suggesting that GH and testosterone can independently affect body composition in males.

Using stable calcuim (Ca) tracer infusions, we have examined the effects of androgens on measures of Ca metabolism in men with Leuprulide induced hypogonadism but maintained GH secretion. We observed an increase in urinary Ca losses after 4 weeks and 10 weeks of hypogonadism (+43%; $p = 0.007$; +73%, $p = 0.003$, respectively), and a 21% decrease in bone calcium deposition ($p < 0.05$) (25). These data suggest that short-term, severe deficiency in gonadal steroids can have profoundly negative effects on calcium and bone metabolism, underscoring the critical importance of normal androgen production in male calcium–bone metabolism.

Estrogen Effects

The role of endogenous estrogens in the body compositional changes of normal puberty are not completely understood. In the male, for example, the endogenous rise in testosterone is also accompanied by an estrogen increase, making it difficult to dissect the contribution of each to the overall changes observed. We studied the impact of estrogen on measures of whole-body protein anabolism by performing whole-body protein kinetic analysis, using ^{13}C-leucine tracer infusions, in a group of young girls (12.2 ± 0.3 years) with either Turner syndrome or hypogonadotropic hypogonadism, before and after oral or intramuscular estrogen administration for 1 month (26). Contrary to the effects observed in the boys given testosterone, girls had no detectable changes in protein metabolism after estrogen treatment (Fig. 30.5). Even though these were not normal girls, and hence the model may not be directly applicable to physiological puberty, the data nonetheless suggest inferentially that the anabolic effect of androgens on protein may be a direct androgen effect and not secondary to aromatization. Further studies selectively blocking aromatization of androgens in the male may offer further insight into this question.

The girls with Turner syndrome were also studied with Ca tracers during estrogen treatment (27). Similar to prepubertal boys given androgens, they had a significant increase in oral Ca absorption ($p = 0.03$), Ca retention ($p = 0.04$), and total Ca balance ($p = 0.04$) after treatment. However, contrary to boys, total bone turnover decreased during sex hormone treatment. As these girls have primary gonadal failure, and many already have severe osteopenia at a young age (28), our results indicate the need for further studies to assess the role of dietary supplementation with calcium and vitamin D, especially in those girls whom estrogen replacement is intentionally delayed in an attempt to increase final height. Using a highly sensitive estradiol assay, Klein et al. (29) have shown significant differences in the circulating concentrations of estradiol in prepubertal boys and girls, with much higher concentrations in the female. These recent data suggest that before puberty the prepubertal gonad actively produces small amounts of sex steroids, putatively necessary for the suppression of gonadotropin production, and potentially important in the maintenance of normal bone mineralization.

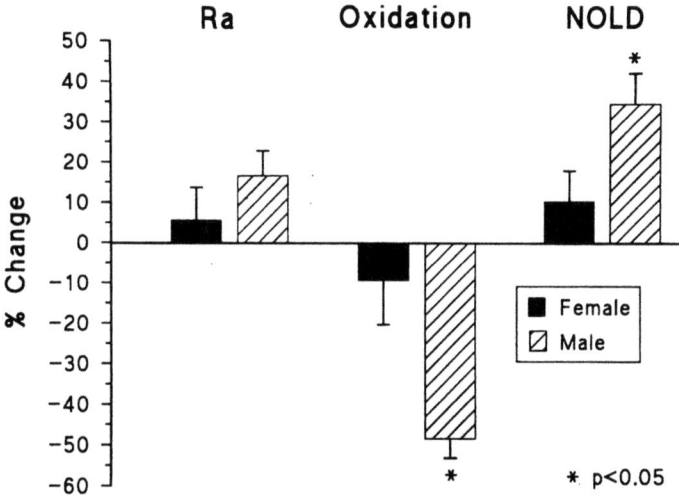

FIGURE 30.5. Comparison of changes in whole-body leucine kinetics in hypogonadal girls treated with estrogen vs. prepubertal boys treated with testosterone for 1 month. Ra is an estimate of proteolysis, and NOLD is an estimate of protein synthesis. There were no significant changes in protein kinetics in the girls as compared to the boys. Reproduced with permission from Mauras (26).

Summary

Safe and reproducible investigative techniques, particularly useful in pediatric studies, have allowed us to examine the metabolic effects of complex anabolic hormones. GH anabolic effects are mediated through IGF-I, and testosterone and GH appear to operate synergistically to enhance the anabolic explosion that occurs in puberty. Estrogen treatment in the female on the other hand exerts marked effects on calcium–bone metabolism but less pronounced effects at the whole-body protein level. However, it is possible that different estrogen doses or treatment in individuals with intact gonads may be effective or that different protein pools may be affected by estrogens. Much more work is needed in this area, in as much as the more we understand the anabolic effects desired in bone, muscle, or at the whole-body level, the more specific and targeted our therapeutic interventions can become.

Acknowledgments. This author is grateful to Annie Rini and Susan Welch, Brenda Sager, Alfred Yergey and Nancy Vieira, Randall Urban, Kimberly O'Brien, Morey W. Haymond, and Johannes D. Veldhuis for their support and collaboration.

References

1. Wolfe RR. Calculation of substrate kinetics: single-pool model. In: Wolfe RR, ed. Radioactive and stable isotope tracers in biomedicine. New York: Wiley-Liss, 1992: 119–44.
2. Steele R. Influences of glucose loading and of injected insulin on hepatic glucose output. Ann NY Acad Sci 1959;82:420–30.
3. Horber FF, Haymond MW. Human growth hormone prevents the protein catabolic side effects of prednisone in humans. J Clin Invest 1990;86:265–72.
4. Schwenk WF, Beaufrere B, Haymond MW. Use of reciprocal pool specific activities to model leucine metabolism in humans. Am J Physiol 1985;249:E646–50.
5. DeFeo P, Gaisano MG, Haymond MW. Differential effects of insulin deficiency on albumin and fibrinogen synthesis in humans. J Clin Invest 1991;88:833–40.
6. Fu AZ, Morris JC, Ford GC, Nair KS. Sequential purification of human apolipoprotein B-100, albumin, and fibrinogen by immunoaffinity chromatography for measurement of protein synthesis. Anal Biochem 1997;247:228–36.
7. Balagopal P, Rooyackers OE, Adey DB, Ades PA, Nair KS. Effects of aging on in vivo synthesis of skeletal muscle myosin heavy-chain and sarcoplasmic protein in humans. Am J Physiol 1997;273:E790–800.
8. Bier DM, Leake RD, Haymond MW, Arnold KJ, Bruenke LD, Sperling MA, Kipnis DM. Measurement of "true" glucose production rates in infancy and childhood with 6,6-dideuteroglucose. Diabetes 1997;26:1016–23.
9. Yergey AL, Abrams SA, Vieria NE, Eastell R, Hillman LS, Covell DG. Recent studies of human calcium metabolism using stable isotope tracers. Can J Physiol Pharmacol 1990;68:973–6.
10. Neer R, Berman N, Fisher L, Rosenberg R. Multicompartmental analysis of calcium kinetics in normal adult males. J Clin Invest 1967;46:1364–79.
11. Berman M, Weissberger AJ. SAAM manual (simulation, analysis and modeling). DHEW Publ 78–180. Bethesda: National Institutes of Health, 1990.
12. Saenger P, Attie KM, Dimartino-Nardi J, Hintz R, Frahm L, Frane JW. Metabolic consequences of 5-year growth hormone (GH) therapy in children treated with GH for idiopathic short stature. Genentech Collaborative Study Group. J Clin Endocrinol Metab 1998;83:3115–20.
13. Zenobi PD, Jaeggi-Groisman SE, Riesen WF, Roder ME, Froesch ER. Insulin-like growth factor I improves glucose and lipid metabolism in type 2 diabetes mellitus. J Clin Invest 1992;90:2234–41.
14. Guler HP, Zapf J, Froesch ER. Short-term metabolic effects of recombinant human insulin-like growth factor I in healthy adults. N Engl J Med 1987;317:137–40.
15. Mauras N, Beaufrere B. Recombinant human IGF-I enhances whole body protein anabolism and significantly diminishes the protein-catabolic effects of prednisone in humans, without a diabetogenic effect. J Clin Endocrinol Metab 1995;80: 869–74.
16. Fukagawa NK, Minaker KL, Rowe JW, Goodman MN, Matthews DE, Bier DM, Young VR. Insulin-mediated reduction of whole body protein breakdown. Dose-response effects on leucine metabolism in post-absorptive men. J Clin Invest 1985;76:2306–11.
17. Kupfer SR, Underwood LE, Baxter RC, Clemmons DR. Enhancement of the anabolic effects of growth hormone and insulin-like growth factor I by use of both agents simultaneously. J Clin Invest 1993;91:391–6.

18. Mauras N. Combined recombinant human growth hormone and recombinant human insulin-like growth factor I: lack of synergy on whole body protein anabolism in normally fed subjects. J Clin Endocrinol Metab 1995;80:2633–7.
19. Zapf J, Schoenle E, Waldrogel M, Sand I, Froesch ER. Effect of trypsin treatment on rat adipocytes on biological effects and binding of insulin and IGFs: further evidence for the action of IGFs through the insulin receptor. Eur J Biochem 1981;113:605–9.
20. Turkalj I, Keller U, Ninnis R, Vosmeer S, Stauffacher W. Effect of increasing doses of recombinant human insulin-like growth factor I on glucose, lipid, and leucine metabolism in man. J Clin Endocrinol Metab 1992;75:1186–91.
21. Mauras N, Martha PM, Quarmby V, Haymond MW. Recombinant human IGF-I administration in man: Differential sensitivity to the metabolic effects of subcutaneous (SC) bolus vs. continuous delivery. Am J Physiol 1997;272E349–55.
22. Mauras N, Haymond MW, Darmaun D, Vieira NE, Abrams SA, Yergey AL. Calcium, and protein kinetics in prepubertal boys. Positive effects of testosterone. J Clin Invest 1994;93:1014–9.
23. Mauras N, Hayes V, Welch S, Rini A, Helgeson K, Dokler M, Veldhuis JD, Urban R. Testosterone deficiency in young men: marked alterations in whole body protein kinetics, strength and adiposity. J Clin Endocrinol Metab 1998;83:1886–92.
24. Mauras N. Additive effects of GH and testosterone on the anabolic changes of puberty: a metabolic study. Pediatr Res 1998;43(abstract #459).
25. Mauras N, Hayes V, Yergey AL. Profound hypogonadism has significant negative effects on calcium balance in males: a calcium kinetic study. In: Proceedings, 80th Annual Meeting Endocrine Society 1998:3(abstract #88).
26. Mauras N. Estrogens do not affect whole-body protein metabolism in the prepubertal female. J Clin Endocrinol Metab 1985;80:2842–5.
27. Mauras N, Vieira NE, Yergey AL. Estrogen therapy enhances calcium absorption and retention and diminishes bone turnover in young girls with Turner's syndrome: a calcium kinetic study. Metabolism 1997;46:908–13.
28. Mora S, Weber G, Guarneri MP, Nizzoi G, Pasolini D, Chiumello G. Effect of estrogen replacement therapy on bone mineral content in girls with Turner's syndrome. Obst Gynecol 1992;79:747–51.
29. Klein KO, Baron J, Colli MJ, McDonnell DP, Cutler GB. Estrogen levels in childhood determined by an ultrasensitive recombinant cell bioassay. J Clin Invest 1994;94:2475–80.
30. Abrams SA, Schanler RJ, Yergey AL, Vieira NE, Bronner F. Compartmental analysis of calcium metabolism in very low-birth-weight infants. Pediatr Res 1994;36:424–8.

31

Innovative Quantitative Neuroendocrine Techniques

MICHAEL L. JOHNSON AND MARTIN STRAUME

Many neuroendocrine hormones exhibit dramatic changes in serum concentrations, which occur many times per day. Figure 31.1 presents two typical examples of a 24-h time series that show the pulsatile nature of the luteinizing hormone (LH) concentration. It is obvious, by inspection, that these patterns are somewhat different. However, we are interested in a more quantitative (i.e., mathematical and statistical) description of how hormone times series differ. Specifically, we wish to derive information about the number of pulses, the interpulse intervals, the concentration pulse magnitudes and widths, the secretion pulse magnitudes and shapes, basal secretion, and the variability of these quantities. To answer these questions, our group has developed or applied several mathematical and statistical techniques for the analysis of neuroendocrine hormone pulsatility and regularity. These methods include cluster analysis (1), deconvolution analysis (2,3), copulsatility analysis (4), and approximate entropy (5).

In considering methods for the analysis of hormone time series, it is informative to consider some unique features of this type of data and in particular the properties of the experimental uncertainties contained within the data. For example, hormone assays are expensive and thus there are relatively few data points. For various reasons, there will be missing data points and because the assays are expensive these incomplete data sets must be analyzed. In addition, many assays will have high levels of experimental uncertainties, i.e., experimental noise. As is seen below, these characteristics determine the statistical validity of the methods of analysis.

Cluster Analysis

Our first attempt at developing a pulse detection method was the cluster analysis method (1). This method uses a sliding pooled t-test to identify data points within the hormone time series that correspond to statistically significant increases and decreases in the hormone concentrations. A nadir is assumed to be a significant decrease followed by a significant increase. All regions that are not within a nadir are considered to be a part of a hormone pulse.

The results of the cluster analysis of the data in Figure 31.1 are shown as the horizontal dashes at an LH concentration of unity. Clearly, the cluster method can identify peaks within a series, but does not identify peaks at the edges of a time series. Thus, quantities such as height and numbers of pulses and interpulse intervals can be evaluated. However, the cluster program does not provide information about the secretion of the hormone into the serum and the elimination of the hormone from the serum. This information must be obtained from a deconvolution analysis (2,3).

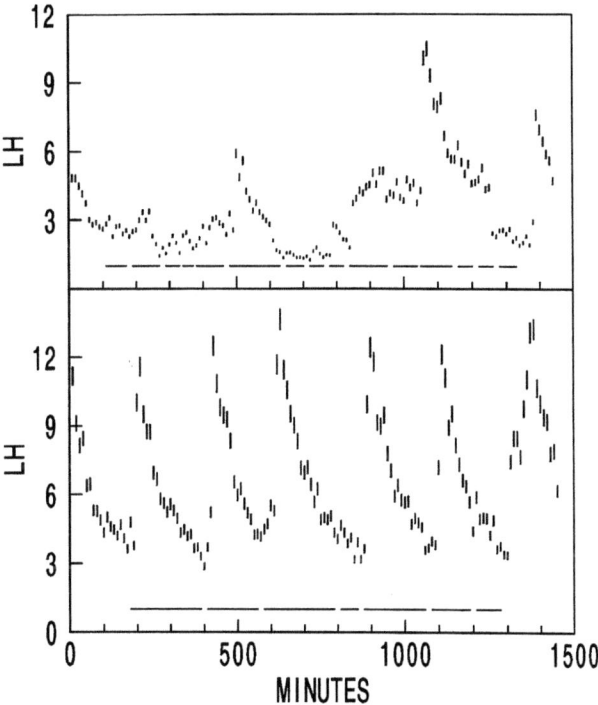

FIGURE 31.1. Two typical human serum luteinizing hormone times series. The vertical lines correspond to the ± 1 SD uncertainties of the concentrations sampled at 10-min intervals. The horizontal dashes at LH concentrations of unity indicate regions where the cluster analysis method located "peaks" within the data series (cluster size = 2; nadir size = 2; T-scores = 2.0).

Deconvolution Analysis

The concentration of a hormone in the serum is a balance between two processes: secretion into the serum and elimination from the serum. To obtain the concentration as a function of time the two processes are combined via a convolution integral. A graphical explanation of a convolution integral is shown in Figure 31.2. Assume for the moment that the upper panel of Figure 31.2 depicts the shape of a secretion event. Now consider a rectangular approximation of the same secretion event (the second panel from the top in Fig. 31.2). For this case, the time axis is simply divided into a series of very small time increments and the secretion during each time interval is approximated as a rectangle. Of course, for the actual integral there is an infinite number of infinitely small time increments but it is easier to visualize the five discrete rectangles shown in Figure 31.2. Once the secretion within each of these rectangles (time points) takes place, the concentration increases by that amount, and then immediately begins to be eliminated according to its particular kinetics (as indicated in the third panel from the top in Fig. 31.2). What we observe as the final concentration profile is the sum of all these individual eliminations as in the bottom panel. Note that the lower panel of Figure 31.2 depicts a rapid increase in concentration followed by a slow decrease in the concentration, as can be seen in actual experimental data such as shown in Figure 31.1.

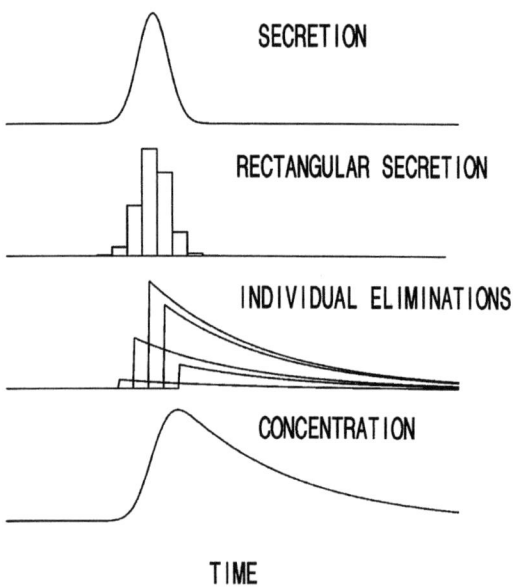

FIGURE 31.2. A diagrammatic depiction of the convolution process.

31. Quantitative Neuroendocrinology

It is comparatively easy to calculate the total concentration if the forms of the secretion and elimination are known. However, we are concerned with the inverse process, that is, starting with the concentration as a function of time and calculating both the secretion and the elimination that generated the concentration. This inverse process is known as a deconvolution. Deconvolutions are computationally complex, time consuming, and error prone. However, it is the only way to obtain the information from limited data.

Our approach to the deconvolution of hormone time series data is to derive a mathematical model for the form of a hormone concentration pulse, and then by nonlinear least-squares methods, fit the actual experimental data to a series of these mathematical forms (secretory bursts) occurring at various times. Other deconvolution methods, require comparatively low amounts of experimental noise and equally spaced data (i.e., no missing data). The least-squares approach was chosen because of its statistical validity. The general mathematical form that we have utilized is a convolution integral of the secretion and the elimination processes (functions):

$$C(t) = \int_{\infty}^{\pm} S(z) H(t-z) dz \qquad (1)$$

where H is a heavyside step function of $t-z$ (the time since the secretion):

$$H(t-z) = \begin{bmatrix} 1, & \text{if } t-z \geq 0 \\ 0, & \text{if } t-z \leq 0 \end{bmatrix} \qquad (2)$$

The heavyside function is used to ensure that clearance from the serum will not occur before hormone secretion into the serum. E is a function that characterizes the elimination function, typically a one- or two-exponential decay:

$$E(t-z) = e^{-\left[\frac{\ln 2(t-z)}{HL}\right]} \qquad (3)$$

or,

$$E(t-z) = (1-f_2) e^{-\left[\frac{\ln 2(t-z)}{HL_1}\right]} + f_2 e^{-\left[\frac{\ln 2(t-z)}{HL_2}\right]} \qquad (4)$$

where f_2 is the fraction of the second elimination component and the HL's are the elimination half-lives of the components (i.e., compartments). S is a function that describes the secretion process at any time z. We have utilized several forms for the secretion profile (3). Initially, we assumed that the secretion was simply the sum of Gaussian-shaped secretion events (2):

$$S(z) = S_0 + \sum_{i=1}^{n} e^{lnH_i - \frac{1}{2}\left(\frac{z-PP_i}{SD}\right)^2} \tag{5}$$

where lnH_i is the natural logarithm of the height of the ith secretion event, PP_i is the secretion peak position of the ith secretion event, and SD is the standard deviation of the secretion events. We also developed a waveform-independent method that utilizes known low-exponential kinetics and a tabulation of the sample secretion profile.

The analysis using the waveform-independent secretion profiles has shown that in a large number of cases a Gaussian form is a good approximation. Figure 31.3 presents an analysis of the data in the lower panel of Figure 31.1,

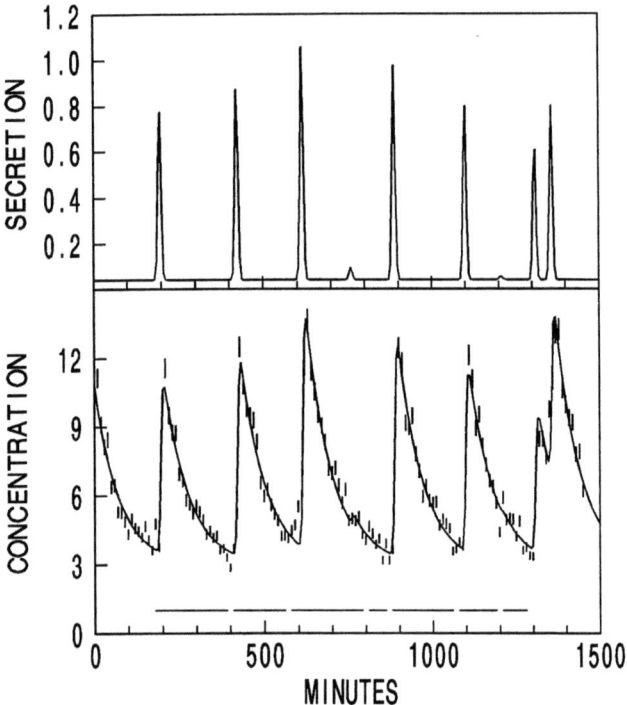

FIGURE 31.3. A model-based deconvolution analysis of the data shown in the lower panel of Figure 31.1. The *upper panel* of this figure presents the derived secretion profile and the *lower panel* includes the calculated concentration profile. The assumptions are that the secretion is a series of Gaussian-shaped events and that there is single-compartment elimination.

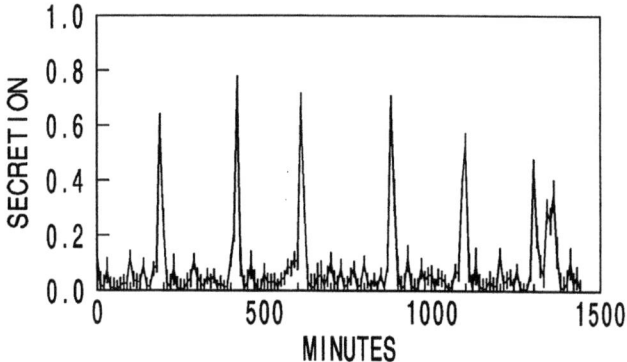

FIGURE 31.4. An alternative (waveform-independent) deconvolution analysis of the data shown in the lower panel of Figure 31.1. The assumptions for this figure are that the secretion is positive at every time point of the series, but is not required to be of Gaussian shape, and there is a single-compartment elimination half-time of 50.9 min.

assuming that the secretion pattern is given by the sum of Gaussian curves, i.e., Eq. 5, and a single compartment elimination model. The lower panel is the same as the lower panel of Figure 31.1 with the addition of the "best-fit" calculated concentration curve. The upper panel plots the secretion rate as a function of time. This analysis estimated the elimination half-life to be 50.9 ± 2.8 min. It appears that there are eight significant secretion events; the small peak at 1209 min is not significant. Note the agreement with the cluster analysis, as shown in Figures 31.1 and 31.3.

Figure 31.4 presents an analogous analysis of the same data without the assumption that the secretion events are of Gaussian shape. In this case, the assumptions are that a small positive amount of secretion occurs at each time point and a single compartment elimination of 50.9 minutes. While these assumptions lead to a somewhat noisier secretion profile it is clear that the general shape of the secretion events are approximately Gaussian or bell-shaped curves.

Copulsatility Analysis

It is of interest to know how the timing of one series of hormone pulses relates to the timing of one, or more, other hormone series. We used combinatorial statistics to assign a probability that the observed timing of two, or more, series of secretion pulses occurs randomly (4).

Approximate Entropy

Approximate entropy (*ApEn*) was developed and formulated by Pincus to statistically discriminate time series by quantifying their unequal regularity (5). Of particular significance is the property of *ApEn* to reliably quantify the regularity of finite-length time series even in the presence of noise and measurement inaccuracy, a property unique to *ApEn* and not shared by other methods common to nonlinear dynamical (chaotic) systems theory (5).

More specifically, *ApEn* measures the logarithmic likelihood that runs of patterns in a time series that are close for m consecutive observations remain close when considered as $m + 1$ consecutive observations. Greater pattern regularity (i.e., higher probability of remaining close) yields smaller *ApEn* values, whereas greater independence among sequential values of an ordered time series yields larger *ApEn* values.

Calculation of *ApEn* requires prior definition of the two parameters m and r. The parameter m is the length of run to be compared (as alluded to earlier) and r is a filter (the magnitude that will discern "close" and "not close," as described next). *ApEn* is thus formally defined as $ApEn(m,r,N)$, where N is the number of data points in the time series being considered. *ApEn* values can only be validly compared when computed for the same m, r, and N values (5). For optimal statistical validity, *ApEn* is typically implemented in hormone time series by using m values of 1 or 2 and r values of approximately 0.2 SD of the series being considered (5).

ApEn is calculated (5) according to the following. Given N data points in a time series, $u(1), u(2),...,u(N)$, the set of $N-m+1$ possible vectors, $x(i)$, are formed with m consecutive u values such that $x(i)=[u(i),...,u(i+m-1)]^T$, $i=1,2,...,N-m+1$. The distance between vectors $x(i)$ and $x(j)$, $d[x(i),x(j)]$, is defined as the maximum absolute difference between corresponding scalar elements of the respective vectors. For each of the $N-m+1$ vectors $x(i)$, a value for $C_i^m(r)$ is computed by comparing all $N-m+1$ vectors $x(j)$ to vector $x(i)$ such that

$$C_n^m(r) = \frac{\text{number of } x(j) \text{ for which } d[x(i),x(j)] \leq r}{N-m+1} \qquad (6)$$

These $N-m+1$ $C_i^m(r)$ values measure the frequency that patterns were encountered that are similar to the pattern given by $x(i)$ of length m within tolerance r. Note that for all i, $x(i)$ is always compared relative to $x(i)$ (i.e., to itself), so that all values $C_i^m(r)$ are positive. Now, define

$$\Phi^m(r) = \frac{\sum_{i=1}^{N-m+1} \ln C_i^m(r)}{N-m+1} \qquad (7)$$

from which the approximate entropy $ApEn(m,r,N)$ is given by

$$ApEn(m,r,N) = \Phi^m(r) - \Phi^{m+1}(r) \qquad (8)$$

ApEn defined in this way can be interpreted (with $m = 1$, for example) as a measure of the difference between (a) the probability that runs of length 1 will recur within tolerance r and (b) the probability that runs of length 2 will recur within that tolerance (5).

For the lower time series in Figure 31.1 *ApEn(1,0.2,144)* is 1.101 ± 0.043, while for the upper time series *ApEn(1,0.2,144)* it is 0.951 ± 0.028. Here, the $\pm$ SEM is estimated from the within-sample error. Clearly, it appears that based upon this statistical metric, these two time series are different. We have also utilized *ApEn* to distinguish pulsatile growth hormone in normal and acromegalic individuals (6).

It is also possible to evaluate the *ApEn* for the time series if they were truly random. This is done by literally shuffling the values and repeating the *ApEn* calculation. By repeating this shuffling process hundreds of times, statistical uncertainty of the truly random *ApEn(1,0.2,144)* for the upper panel of Figure 31.1 is 1.730 ± 0.043 and 1.789 ± 0.043 for the lower panel. Clearly, both these profiles are highly ordered, as neither yields an *ApEn* value that is close to that expected for a random time series.

Conclusions

We have developed or applied several mathematical and statistical techniques for the analysis of neuroendocrine hormone pulsatility and regularity: cluster analysis (1), deconvolution analysis (2,3), copulsatility analysis (4), and approximate entropy (5). These techniques have proven to be useful in the analysis of hormone pulsatility data. They are also useful in other fields of science, such as in studying the patterns of growth in children (7), in fluorescence spectroscopy (8), and in studying the cooperativity of oxygen binding to human hemoglobin (9).

Acknowledgments. The authors acknowledge the support of the National Science Foundation Science and Technology Center for Biological Timing at the University of Virginia (NSF DIR-8920162), the Clinical Research Center at the University of Virginia (NIH RR-00847), the University of Maryland at Baltimore Center for Fluorescence Spectroscopy (NIH RR-08119), and National Institutes of Health grant GM-35154.

References

1. Veldhuis JD, Johnson ML. Cluster analysis: a simple, versatile and robust algorithm for endocrine pulse detection. Am J Physio 1986;250:E486–93.

2. Veldhuis JD, Carlson ML, Johnson ML. The pituitary gland secretes in bursts: appraising the nature of glandular secretory impulses by simultaneous multiple-parameter deconvolution of plasma hormone concentrations. Proc Natl Acad Sci USA 1987;84:7686–90.
3. Johnson ML, Veldhuis JD. Evolution of deconvolution analysis as a hormone pulse detection method. Methods Neurosci 1995;28:1–24.
4. Veldhuis JD, Johnson ML, Seneta E. Analysis of the co-pulsatility of anterior pituitary hormones. J Clin Endocrinol Metab 1991;73:569–76.
5. Pincus SM. Quantification of evolution of order to randomness in practical time series analysis. Methods Enzymol 1994;240:68–89.
6. Hartman ML, Pincus SM, Johnson ML, Matthews DH, Faunt LM, Vance ML, et al. Enhanced basal and disorderly growth hormone secretion distinguish acromegalic from normal pulsatile growth hormone release. J Clin Invest 1994;94:1277–88.
7. Lampl M, Veldhuis JD, Johnson, ML. Saltation and stasis: a model of human growth. Science 1992;258:801–3.
8. Szmacinski H, Lakowicz JR, Johnson, ML. Fluorescence lifetime imaging microscopy (FLIM). Methods Enzymol 1994;240:723–48.
9. Johnson ML. Statistical thermodynamic modeling of hemoglobin cooperativity. Adv Biophys Chem 1995;5:179–231.

32

Sexual Dimorphism of Liver Cytochrome P-450 Gene Expression: GH Pulse-Activated STAT Signaling Mechanisms

DAVID J. WAXMAN

Steroid Hydroxylase Liver P-450 Enzymes

Humans and other mammals are exposed to a large number of toxic foreign chemicals, many of which are lipophilic and thus have a tendency to persist in fatty tissues. In response to this environmental challenge, mammals have evolved a large number of genes that encode cytochrome P-450 enzymes (CYPs), integral membrane proteins of the endoplasmic reticulum that oxygenate lipophilic compounds and thereby facilitate their elimination. More than two dozen individual P-450 enzymes have been purified from rat liver, a widely studied animal tissue (1,2). Several of these enzymes exhibit dual specificities: they oxygenate structurally diverse foreign chemicals, while they exhibit strikingly narrow stereoselectivities for hydroxylation of testosterone and other endogenous steroids (2). In addition to serving as substrates, steroid hormones play a key role in regulating the expression in liver of sex-specific P-450 enzymes.

Sex-Specific Expression and Hormonal Regulation of Liver P-450s

The physiological requirements with respect to steroid hormone hydroxylation differ between the sexes, and accordingly the steroid hydroxylase liver P-450s from gene families *CYP2* and *CYP3* are expressed in a sex-dependent manner and subject to unique postnatal developmental control (3,4). In the rat, a brief exposure to androgen during the first 3 days of life is sufficient to permanently program the male-specific pattern of liver P-450 gene expression that emerges several weeks later at adolescence, while the same exposure conversely suppresses the adult expression of those P-450s normally restricted to females (5,6). This imprinting, or programming action of early androgen on the liver is analogous to the imprinting effect of neonatal androgen on the development of mammalian sexual behavior and other brain functions. It is not caused by a direct action of androgen on the liver, but rather is mediated by the temporal patterns of circulating growth hormone (GH), which are sex dependent and under gonadal control (7,8). In the female rat, plasma GH levels are nearly time invariant, whereas in males GH is present in plasma intermittently (plasma GH peaks detectable every 3.5 h, and separated by a GH-free interval of 2–2.5 h). GH can induce two mutually exclusive patterns of liver gene expression, depending on whether the target cell is stimulated by GH in a continuous manner (female liver) or intermittently (male liver).

Studies of the underlying cellular and molecular mechanisms that govern the sexually dimorphic effects of plasma GH profiles have primarily focused on two liver P-450 genes: (1) *CYP2C11*, which encodes an adult male-specific testosterone 2α- and 16α-hydroxylase that is activated by intermittent GH pulses in male rats; and (2) *CYP2C12*, a steroid sulfate 15β-hydroxylase, which is activated in a female-specific manner by the female pattern of continuous plasma GH (2,3). Other GH-regulated, male-specific rat liver P-450s include the steroid 6β-hydroxylase CYP3A2, the testosterone 15α-hydroxylase CYP2A2, and the fatty acid ω-hydroxylase CYP4A2 (9). Although the regulation of these other P-450s differs from that of 2C11, as judged from their GH-independent expression in hypophysectomized rat liver, recent studies using the more selective monosodium glutamate model for GH deficiency establish that these male-specific liver P-450s are also inducible by plasma GH pulses (10).

GH pulse induction thus corresponds to a general mechanism that is of fundamental importance for the male-specific effects of plasma GH pulses on liver gene expression. Although many endocrine factors are secreted in a pulsatile manner, GH represents a unique example in which alternative temporal patterns of hormone secretion lead to distinct biological responses. These patterns of GH regulation are not unique to the rat: corresponding patterns of sex-dependent GH regulation characterize several mouse liver steroid hydroxylase P-450s (7,11) and several non-P-450 liver enzymes as well (8,12). Human liver P-450 levels and their associated drug-metabolic activities are also determined in part by age, sex, or GH status (13,14). Studies of the regulation by GH of the prototypic rat P-450s 2C11 and 2C12 are therefore likely to be applicable to the regulation of liver-expressed genes in general.

Transcriptional Effects of GH on Sex-Specific *Cyp* Genes

GH regulates the sex-specific expression of the P-450 genes *2C11* and *2C12* at the level of transcriptional initiation, as shown by nuclear RNA (hnRNA) analysis (15) and by nuclear run-on transcription assay (15,16). In vitro characterizations of the 5′-flank of the *2C11* gene and the *2C12* gene have identified specific DNA sequences that interact in a sex-dependent and GH-regulated manner with DNA-binding proteins (putative transcription factors) that are differentially expressed in male versus female rat liver nuclei (15,17). As descibed here, two factors appear to be of particular importance (Fig. 32.1): these are GH-regulated nuclear factor, GHNF, which binds to five sites along the *2C12* promoter and at one site in the proximal *2C11* promoter (17), and the male GH pulse-activated STAT5b (18), which we have recently found binds to at least one site within the *2C11* gene. These *trans*-acting factors are hypothesized to contribute to the sex-specific transcription of the *2C11* and *2C12* genes working in conjunction with several liver-enriched transcription factors. Several such factors contribute to basal gene expression in the case of *2C11* (19) and *2C12* (17,20,21). More detailed molecular studies are necessary to establish the functional significance of these and other basal factors that may contribute to or amplify GH-regulated transcriptional events.

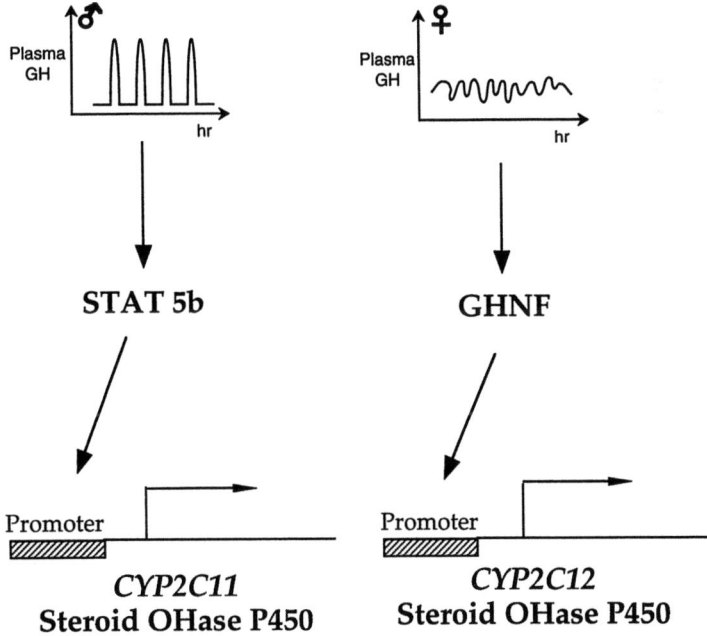

FIGURE 32.1. Distinct intracellular signaling pathways lead to male-versus-female pattern transcriptional activation of *CYP2C11* and *CYP2C12* by GH.

Role of STAT5b in GH Pulse Regulation of Male-Specific Gene Transcription

The mechanistic basis for the discrimination by hepatocytes between male and female plasma GH profiles leading to sex-specific liver gene transcription is only partially understood. This discrimination is likely to occur at the cell surface, where GH binds to and activates its plasma membrane-bound receptor, GH receptor (GHR), via a sequential dimerization mechanism (22). GH pulse frequency is the most critical determinant for GH stimulation of a male pattern of liver P-450 expression, with hepatocytes requiring a minimum GH off-time of ~2.5 h to respond to GH pulses by induction of *2C11* (23). This requirement for a 'recovery period' to achieve *2C11* expression is not met in the case of female rats, where hepatocytes are exposed to GH continuously.

Recent studies carried out in this laboratory strongly support the hypothesis that GH activates two independent intracellular signaling pathways in hepatocytes, one stimulated by intermittent GH pulses (male pattern), leading to *2C11* activation, and the other induced by continuous GH exposure (female pattern), leading to *2C12* expression. Because GH binding to GHR activates the receptor-associated, tyrosine-specific kinase JAK2 (24), we initially investigated whether the marked sex difference in liver P-450 gene expression is associated with a sex-dependent pattern of tyrosine phosphorylated proteins that signal to the nucleus. We discovered that a latent cytoplasmic transcription factor termed STAT5b (25,26) is intermittently, and repeatedly, activated in adult male rat liver in direct response to each incoming GH pulse (18). The activation of liver STAT5b by male-like GH pulses is now known to involve (sequential) tyrosine phosphorylation of STAT5b in the cytosol (18), STAT5b phosphorylation on serine or threonine (27,28), STAT5b nuclear translocation (18,29), and ultimately the binding of STAT5b to its target sites in GH pulse-responsive genes, such as *CYP2C11* (Fig. 32.2).

The proposal that GH pulse-activated STAT5b is a key, global determinant of male-specific liver gene transcription (18) is given strong support by studies of *STAT5b* gene knockout mice (30). Male *STAT5b*$^{-/-}$ mice exhibit two striking phenotypes not seen in *STAT5b*$^{-/-}$ females. First, a whole-body growth rate deficiency emerges in the STAT5b-deficient male mice at puberty. Growth deficiency and several other physiological characteristics of these mice are reminiscent of Laron-type dwarfism, a genetic disease involving GH tissue resistance that occurs in humans (31). Second, in the knockout male mice, there is a selective loss of male-specific liver P-450 gene expression that is coupled to a derepression of those P-450 genes whose expression is normally limited to adult females (30). Surprisingly, the closely related (90% identical) STAT5a protein (25,26) is unable to substitute for STAT5b (30), despite its intrinsic responsiveness to GH activation by the same tyrosine phosphorylation mechanism (32). Further study is required to verify that these *STAT5b*$^{-/-}$ mice are, indeed, GH pulse resistant, rather than displaying a phenotype which results from a perturbation of their endogenous pituitary GH secretory patterns.

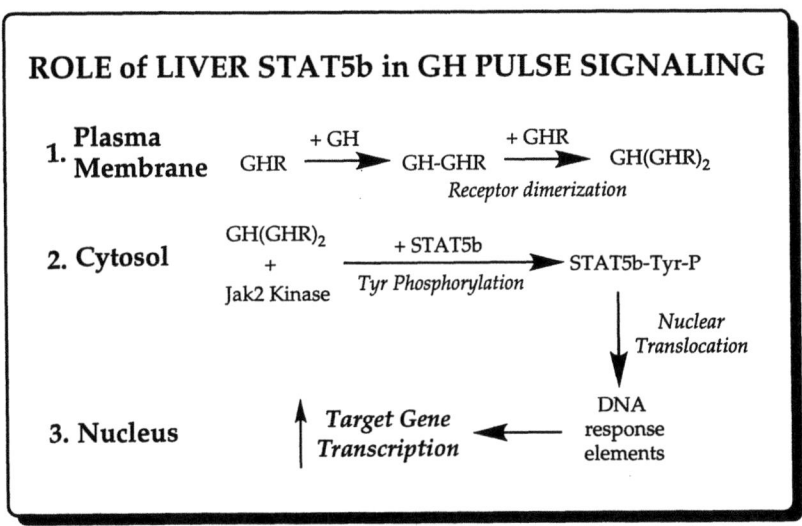

FIGURE 32.2. Role of liver STAT5b in GH pulse signaling.

Responses of STAT5b and GHNF to Continuous GH

In contrast to the repeated activation of STAT5b by male GH pulses, continuous GH exposure desensitizes the STAT5b activation pathway by a mechanism that may involve enhanced dephosphorylation of both STAT5b and the GHR-JAK kinase signaling complex (33). Consequently, little or no activated STAT5b transcription factor is present in the nucleus of adult female rats (18). Rather, continuous GH treatment of male rats leads to expression of a novel non-STAT nuclear factor, GHNF, which, as noted earlier, binds to multiple sites along the 5'-flank of *CYP2C12*, and is greatly enriched in female as compared to male liver nuclei (17). The molecular nature of GHNF, and the precise role that it plays in the sexually dimorphic expression of *2C12* and other continuous GH-regulated liver genes, is presently unknown. Further study will be required to elucidate the mechanisms whereby continuous GH regulates GHNF activity or protein levels in the nucleus of female rats.

GH Pulse-Activated STAT5b Cycle

Data obtained by several laboratories studying GH signaling or related cytokine or growth factor signaling pathways (34) supports the following proposed sequence of events for the activation of STAT5b by plasma GH pulses. (1) GHR dimerizes at the cell surface on binding GH. This leads to the activation of JAK2 tyrosine kinase (24). (2) JAK2 phosphorylates GHR at tyrosine residues 534, 566, and 627, each of which can serve as a docking site for STAT5b (35). (3) STAT5b is recruited to these GHR phosphotyrosine docking sites via the STAT protein SH2 domain. Once bound to GHR in this manner, STAT5b itself then becomes phosphorylated on tyrosine residue 699 (36) in a reaction that is believed to be catalyzed by GHR-associated JAK2 tyrosine kinase. (4) STAT5b becomes phosphorylated on a serine or threonine residue(s) in a step that alters its DNA-binding and perhaps also its transcriptional activity (27,28). (5) The tyrosine-phosphorylated STAT5b homodimerizes via SH2 domain interactions in a process that appears to be coupled to nuclear translocation of the activated STAT transcription factor (36). It is uncertain whether the serine/threonine phosphorylation reaction (step 4) precedes or follows STAT5b dimerization. (6) The homodimeric STAT complex binds to $TTCN_3GAA$-containing DNA response elements and stimulates target gene transcription (37–39). (7) Signaling from GHR-JAK2 to STAT5b is subsequently terminated by a process that involves inhibition, inactivation, or internalization of the GHR-JAK2 signaling complex (40). (8) Nuclear STAT5b is deactivated by a process that involves phosphotyrosine dephosphorylation as a key first step (28). This dephosphorylation reaction may be catalyzed by the SH2 domain-containing phosphotyrosine phosphatase SHP1 (29). (9) Key signaling components, likely including STAT5b and perhaps also GHR and JAK2, reactivate and recycle back to the plasma membrane in time for the hepatocyte to respond to a subsequent GH pulse (28,40).

Steps 1–3, 5, and 6 of the foregoing pathway are fairly well understood at present. Step 4, GH-induced serine (or threonine) phosphorylation of STAT5b, was demonstrated for GH-activated STAT5b in liver tissue and in cultured liver cells by our laboratory (27,28), and has been confirmed by others for interleukin-2-activated STAT5 in T cells (41). This secondary phosphorylation of STAT5b is analogous to the serine phosphorylation of STAT3 at its consensus MAP kinase site, $PMS^{727}P$, which is conserved in STAT5, and which greatly enhances the STAT transcription activation potential (42). Less well understood are the cellular and biochemical events that underlie the termination of the GHR-JAK2 signaling (step 7), STAT5b dephosphorylation (step 8), and the resetting of GH signaling components during the interpulse interval (step 9). These last three steps are key for the physiological requirement that STA5b signaling undergo a rapid downregulation at the conclusion of each GH pulse.

Conclusion

Recent studies have elucidated key features of the cellular and molecular mechanisms through which GH regulates hepatic expression of cytochrome P-450 and other genes that respond in a sexually dimorphic manner to plasma GH patterns. Because the effects of GH on liver P-450 gene expression are distinguished by their dependence on the pulsatility and frequency of pituitary GH release, these studies have provided fundamental new knowledge about the underlying cellular and molecular mechanisms whereby the temporal patterns of hormonal stimulation regulate gene expression. Knowledge gained from studies of GH actions on liver enzyme expression in the rat model may provide basic insight into analogous processes that occur in humans, in which GH is also secreted by the pituitary gland in an episodic and sex-dependent manner (43–45) and GH pulsatility is an important determinant of long bone growth rates (46,47), as it is in the rat (23,47). Future studies on the STAT5b-dependent signaling pathways induced by physiological GH pulses may shed light on factors that contribute to the pathophysiological conditions associated with postreceptor Laron syndrome defects in GH signaling (48) and to relative GH resistance associated with aging, diabetes, malnutrition, and renal failure (49).

Acknowledgments. Studies carried out in the author's laboratory were supported in part by NIH grant DK33765.

References

1. Ryan DE, Levin W. Purification and characterization of hepatic microsomal cytochrome P-450. Pharmacol Ther 1990;45:153–239.
2. Waxman DJ. Interactions of hepatic cytochromes P-450 with steroid hormones. Regioselectivity and stereospecificity of steroid metabolism and hormonal regulation of rat P-450 enzyme expression. Biochem Pharmacol 1988;37:71–84.
3. Waxman DJ. Regulation of liver-specific steroid metabolizing cytochromes P450: cholesterol 7-alpha-hydroxylase, bile acid 6-beta-hydroxylase, and growth hormone-responsive steroid hormone hydroxylases. J Steroid Biochem Mol Biol 1992;43:1055–72.
4. Mode A, Tollet P, Strom A, Legraverend C, Liddle C, Gustafsson JA. Growth hormone regulation of hepatic cytochrome P450 expression in the rat. Adv Enzyme Regul 1992;32:255–63.
5. Waxman DJ, Dannan GA, Guengerich FP. Regulation of rat hepatic cytochrome P-450: age-dependent expression, hormonal imprinting, and xenobiotic inducibility of sex-specific isoenzymes. Biochemistry 1985;24:4409–17.
6. Waxman DJ, LeBlanc GA, Morrissey JJ, Staunton J, Lapenson DP. Adult male-specific and neonatally programmed rat hepatic P-450 forms RLM2 and 2a are not dependent on pulsatile plasma growth hormone for expression. J Biol Chem 1988;263:11396–406.
7. Shapiro BH, Agrawal AK, Pampori NA. Gender differences in drug metabolism regulated by growth hormone. Int J Biochem Cell Biol 1995;27:9–20.
8. Jansson J-O, Ekberg S, Isaksson O. Sexual dimorphism in the control of growth hormone secretion. Endocr Rev 1985;6:128–50.
9. Waxman DJ, Chang TKH. Hormonal regulation of liver cytochrome P450 enzymes. In: Ortiz de Montellano PR, ed. Cytochrome P450: structure, mechanism, and biochemistry, 2nd ed. New York: Plenum Press, 1995:391–417.

10. Waxman DJ, Ram PA, Pampori NA, Shapiro BH. Growth hormone regulation of male-specific rat liver P450s 2A2 and 3A2: induction by intermittent growth hormone pulses in male but not female rats rendered growth hormone deficient by neonatal monosodium glutamate. Mol Pharmacol 1995;48:790–7.
11. Noshiro M, Negishi M. Pretranslational regulation of sex-dependent testosterone hydroxylases by growth hormone in mouse liver. J Biol Chem 1986;261:15923–7.
12. Srivastava PK, Waxman DJ. Sex-dependent expression and growth hormone regulation of class alpha and class mu glutathione-S-transferases in adult rat liver. Biochem J 1993;294:159–65.
13. Cheung NW, Liddle C, Coverdale S, Lou JC, Boyages SC. Growth hormone treatment increases cytochrome P450-mediated antipyrine clearance in man. J Clin Endocrinol Metab 1996;81:1999–2001.
14. Levitsky LL, Schoeller DA, Lambert GH, Edidin DV. Effect of growth hormone therapy in growth hormone-deficient children on cytochrome P-450-dependent 3-N-demethylation of caffeine as measured by the caffeine $^{13}CO_2$ breath test. Dev Pharmacol Ther 1989;12:90–5.
15. Sundseth SS, Alberta JA, Waxman DJ. Sex-specific, growth hormone-regulated transcription of the cytochrome P450 2C11 and 2C12 genes. J Biol Chem 1992;267:3907–14.
16. Legraverend C, Mode A, Westin S, Strom A, Eguchi H, Zaphiropoulos PG, et al. Transcriptional regulation of rat P-450 2C gene subfamily members by the sexually dimorphic pattern of growth hormone secretion. Mol Endocrinol 1992;6:259–66.
17. Waxman DJ, Zhao S, Choi HK. Interaction of novel sex-dependent, growth hormone-regulated liver nuclear factor with *CYP2C12* promoter. J Biol Chem 1996;271:29978–87.
18. Waxman DJ, Ram PA, Park SH, Choi HK. Intermittent plasma growth hormone triggers tyrosine phosphorylation and nuclear translocation of a liver-expressed, Stat 5-related DNA binding protein. Proposed role as an intracellular regulator of male-specific liver gene transcription. J Biol Chem 1995;270:13262–70.
19. Strom A, Equchi H, Mode A, Legraverend C, Tollet P, Stromstedt PE, et al. Characterization of the proximal promoter and two silencer elements in the CYP2C11 gene expressed in rat liver. DNA Cell Biol 1994;13:805–19.
20. Tollet P, Lahuna O, Ahlgren R, Mode A, Gustafsson JA. CCAAT/enhancer-binding protein-alpha-dependent transactivation of CYP2C12 in rat hepatocytes. Mol Endocrinol 1995;9:1771–81.
21. Lahuna O, Fernandez L, Karlsson H, Maiter D, Lemaigre FP, Rousseau GG, et al. Expression of hepatocyte nuclear factor 6 in rat liver is sex-dependent and regulated by growth hormone. Proc Natl Acad Sci U S A 1997;94:12309–13.
22. de Vos AM, Ultsch M, Kossiakoff AA. Human growth hormone and extracellular domain of its receptor: crystal structure of the complex. Science 1992;255:306–12.
23. Waxman DJ, Pampori NA, Ram PA, Agrawal AK, Shapiro BH. Interpulse interval in circulating growth hormone patterns regulates sexually dimorphic expression of hepatic cytochrome P450. Proc Natl Acad Sci U S A 1991;88:6868–72.
24. Argetsinger LS, Campbell GS, Yang X, Witthuhn BA, Silvennoinen O, Ihle JN, et al. Identification of JAK2 as a growth hormone receptor-associated tyrosine kinase. Cell 1993;74:237–44.
25. Mui AL, Wakao H, O'Farrell AM, Harada N, Miyajima A. Interleukin-3, granulocyte-macrophage colony stimulating factor and interleukin-5 transduce signals through two STAT5 homologs. EMBO J 1995;14:1166–75.

26. Liu X, Robinson GW, Gouilleux F, Groner B, Hennighausen L. Cloning and expression of Stat5 and an additional homologue (Stat5b) involved in prolactin signal transduction in mouse mammary tissue. Proc Natl Acad Sci U S A 1995;92:8831–5.
27. Ram PA, Park SH, Choi HK, Waxman DJ. Growth hormone activation of Stat 1, Stat 3, and Stat 5 in rat liver. Differential kinetics of hormone desensitization and growth hormone stimulation of both tyrosine phosphorylation and serine/threonine phosphorylation. J Biol Chem 1996;271:5929–40.
28. Gebert CA, Park SH, Waxman DJ. Regulation of STAT5b activation by the temporal pattern of growth hormone stimulation. Mol Endocrinol 1997;11:400–14.
29. Ram PA, Waxman DJ. Interaction of growth hormone-activated STATs with SH2-containing phosphotyrosine phosphatase SHP-1 and nuclear JAK2 tyrosine kinase. J Biol Chem 1997;272:17694–702.
30. Udy GB, Towers RP, Snell RG, Wilkins RJ, Park SH, Ram PA, et al. Requirement of STAT5b for sexual dimorphism of body growth rates and liver gene expression. Proc Natl Acad Sci U S A 1997;94:7239–44.
31. Laron Z. Disorders of growth hormone resistance in childhood. Curr Opin Pediatr 1993;5:474–80.
32. Smit LS, Vanderkuur JA, Stimage A, Han Y, Luo G, Yu-Lee LY, et al. Growth hormone-induced tyrosyl phosphorylation and deoxyribonucleic acid binding activity of Stat5A and Stat5B. Endocrinology 1997;138:3426–34.
33. Gebert CA, Park SH, Waxman DJ. Down-regulation of liver JAK2-STAT5b signaling by the female plasma pattern of continuous growth hormone stimulation. Mol Endocrinol 1999;13:213–27.
34. Darnell JEJ. STATs and gene regulation. Science 1997;277:1630–5.
35. Hansen JA, Hansen LH, Wang X, Kopchick JJ, Gouilleux F, Groner B, et al. The role of GH receptor tyrosine phosphorylation in Stat5 activation. J Mol Endocrinol 1997;18:213–21.
36. Gouilleux F, Pallard C, Dusanter-Fourt I, Wakao H, Haldosen LA, Norstedt G, et al. Prolactin, growth hormone, erythropoietin and granulocyte-macrophage colony stimulating factor induce MGF-Stat5 DNA binding activity. EMBO J 1995;14:2005–13.
37. Bergad PL, Shih HM, Towle HC, Schwarzenberg SJ, Berry SA. Growth hormone induction of hepatic serine protease inhibitor 2.1 transcription is mediated by a Stat5-related factor binding synergistically to two gamma-activated sites. J Biol Chem 1995;270:24903–10.
38. Ganguly TC, O'Brien ML, Karpen SJ, Hyde JF, Suchy FJ, Vore M. Regulation of the rat liver sodium-dependent bile acid cotransporter gene by prolactin. Mediation of transcriptional activation by Stat5. J Clin Invest 1997;99:2906–14.
39. Subramanian A, Wang J, Gil G. STAT 5 and NF-Y are involved in expression and growth hormone-mediated sexually dimorphic regulation of cytochrome P450 3A10/lithocholic acid 6beta-hydroxylase. Nucleic Acids Res 1998;26:2173–8.
40. Gebert CA, Park SH, Waxman DJ. Termination of growth hormone pulse-induced STAT5b signaling. Mol Endocrinol 1999;13:38–56.
41. Beadling C, Ng J, Babbage JW, Cantrell DA. Interleukin-2 activation of STAT5 requires the convergent action of tyrosine kinases and a serine/threonine kinase pathway distinct from the Raf1/ERK2 MAP kinase pathway. EMBO J 1996;15:1902–13.
42. Wen Z, Zhong Z, Darnell JE Jr. Maximal activation of transcription by Stat 1 and Stat3 requires both tyrosine and serine phosphorylation. Cell 1995;82:241–50.

43. Pincus SM, Gevers EF, Robinson ICAF, Van Den Berg G, Roelfsema F, Hartman ML, et al. Females secrete growth hormone with more process irregularity than males in both humans and rats. Am Physiol Soc 1996;270:E107–15.
44. Weltman A, Weltman JY, Hartman ML, Abbott RD, Rogol AD, Evans WS, et al. Relationship between age, percentage body fat, fitness, and 24-hour growth hormone release in healthy young adults: effects of gender. J Clin Endocrinol Metab 1994;78: 543–8.
45. Winer LM, Shaw MA, Baumann G. Basal plasma growth hormone levels in man: new evidence for rhythmicity of growth hormone secretion. J Clin Endocrinol Metab 1990;70:1678–86.
46. Keret R, Ashkenazi IE, Wasserman M, Bauman B, Pertzelan A, Ticher A, et al. Two types of growth hormone rhythms in boys with constitutional short stature. J Pediatr Endocrinol Metab 1996;9:599–607.
47. Clark RG, Robinson IC. Growth induced by pulsatile infusion of an amidated fragment of human growth hormone releasing factor in normal and GHRF-deficient rats. Nature (Lond) 1985;314:281–3.
48. Freeth JS, Silva CM, Whatmore AJ, Clayton PE. Activation of the signal transducers and activators of transcription signaling pathway by growth hormone (GH) in skin fibroblasts from normal and GH binding protein-positive Laron syndrome children. Endocrinology 1998;139:20–8.
49. Maheshwari H, Sharma L, Baumann G. Decline of plasma growth hormone binding protein in old age. J Clin Endocrinol Metab 1996;81:995–7.

33

Intracellular Signaling Networks

JAN-ÅKE GUSTAFSSON

Several years ago our laboratory demonstrated that the sexually dimorphic liver metabolism in rodents is secondary to the sexually dimorphic growth hormone (GH) secretory pattern (1–3). The sexual differences with reference to GH secretory pattern result from both neonatal imprinting mechanisms and steroid hormone effects in the adult period. The nature of the neonatal imprinting that irreversibly programs aspects of the GH secretory pattern needs to be further elucidated, but seems to involve participation of both androgens and estrogens (4).

As biomarkers for sexually differentiated liver metabolism, we cloned and studied the GH regulated female-specific cytochrome P-450 isoform 2C12 as well as the male-specific isoform 2C11. 2C12 is induced by the female-type GH secretory pattern which is experimentally mimicked by GH administration via osmotic minipumps; however, 2C11 is turned on by the male GH secretory pattern as mimicked by intermittent injections of GH (5). These effects of GH on hepatic cytochrome P-450 isoforms are mediated via changes in the rate of transcription of these two genes. In the case of 2C11, the GH effects appear to be direct on the 2C11 gene and do not require ongoing protein biosynthesis, whereas the GH regulation of 2C12 seems to be indirect as the effect of GH is blocked by protein biosynthesis inhibitors. It is possible that GH regulates the expression of an upstream gene product, which then secondarily affects the expression of 2C12. In addition to these two model genes, a plethora of other genes are under the control of the sexually differentiated GH secretory pattern. It appears as if most, if not all, sex-steroid control of liver cytochrome P-450 is mediated via effects of the GH secretory pattern; i.e., the sex steroids do not act directly on the liver cytochrome genes but rather via the hypothalamopituitary system affecting the GH secretory pattern.

Work in our laboratory is concerned with various aspects of GH mechanism of action with particular focus on the mechanisms behind the different effects of continuous GH and intermittent GH, respectively. Based upon available studies, it appears quite clear that GH may utilize several signal transduction pathways to exert its effects on hepatic genes. GH induction of 2C12 involves several transcription factors such as HNF-3, HNF-6, IRE-ABP, and C/EBPα, as well as possibly some orphan nuclear receptors such as HNF-4 (6–10). On the other hand, 2C11 seems to be under the control of the JAK-STAT pathway, possibly in cross talk with the MAP kinase pathway, which leads to serine phosphorylation of STAT5A in addition to the JAK-2 catalyzed tyrosine phosphorylation of STAT5A and STAT5B. In this context, we have shown the existence of a preformed complex between inactive ERK and STAT5A, an interesting biochemical expression of cross talk between two signal transduction pathways. We have also obtained indications that specific GH-regulated proteins participate in desensitization of GH signaling via the JAK-STAT5 pathway. We are currently attempting to characterize the specific nature of these negative-feedback proteins.

Recent opportunities to address neonatal programming mechanisms are dramatically enhanced by the discovery of a novel estrogen receptor, estrogen receptor-β (ER-β) (11). This ER seems to convey most of the effects of estrogens in the organism, although the classical estrogen receptor-α (ER-α) is a major regulator of reproductive physiology. Expression of ER-β is very widespread in the body and is especially evident in the prostate, the ovaries, the urogenital tract, bone, vessels, immune system, lung, and intestinal tract (12). In the brain, ER-β seems to be somewhat more broadly expressed than ER-α and is found from the olfactory lobe to the brainstem. In the paraventricular nucleus, ER-β appears to be the only estrogen receptor present and is colocalized with vasopressin and oxytocin-containing neurons. In some parts of the brain, ER-α and ER-β appear to be expressed in the same neurons, for instance, in the bed nucleus of the striae terminalis as well as in the medial preoptic area. Coexistence of ER-α and ER-β in certain cells is of significant interest because ER-α and ER-β have been shown to heterodimerize, and it is believed that the heterodimer is the preferred state over ER-α or ER-β homodimers. Furthermore, the occurrence of ER-β in the brain probably explains the preservation of specific estrogen uptake as well as estrogenic regulation of gene expression in brains of so-called ERKO animals, i.e., ER-α knockout mice. It will be of great interest to investigate whether any estrogen receptor colocalizes with GH-releasing hormone or somatostatin in appropriate neurons to start to dissect the pathways mediating sex-hormone effects on GH biosynthesis and release.

As nuclear receptors are of paramount importance in intracellular signaling some aspects of these transcription factors are discussed next. Nuclear receptors are ligand-activated transcription factors that regulate the expression of a plethora of genes. The nuclear receptor superfamily currently encompasses close to 100 members, but it is expected that this group of important regulatory molecules will continue to expand. The first nuclear receptors to be identified were the steroid hormone receptors, and the first steroid receptor to become known was the estrogen receptor, now renamed estrogen receptor-α (ER-α). Another intensively studied steroid receptor, is the glucocorticoid receptor which represents the first-described mammalian transcription factor

(13). The glucocorticoid receptor (GR) was the first member of the family to be cloned, first as a partial clone (14) and later as a full-length clone (15). Following these breakthroughs, remaining known steroid receptors were cloned during the 1980s. Genomic work has confirmed what had already been determined during the biochemical era of steroid receptor research (1960s and 1970s); namely, that these transcription factors have a modular structure encompassing three domains: an N-terminal domain, a central DNA-binding domain, and a C-terminal ligand-binding domain (16–18). The DNA-binding domain was characterized by two so-called zinc finger structures, each with a zinc atom coordinated by four cysteincs. This very typical characteristic of steroid receptors made it possible to use molecular biology techniques to clone further members of this family by searching for genes with homologous DNA-binding domains. Consequently a large number of novel members of the superfamily were cloned, most of which appear not to bind steroid hormones. Consequently, "nuclear receptors" is nowadays a more appropriate name, which refers to the preferred localization of these transcription factors inside the cell.

Because it turned out to be a difficult task to identify ligands for most of the novel members of the nuclear receptor family, these new receptors were termed orphan receptors (19). Subsequently some of them have been further characterized and shown to be activated by both steroidal or nonsteroidal ligands. The peroxisome proliferator-activated receptors (PPARs), for instance, are activated by a wide range of synthetic or naturally occurring compounds, notably fatty acids (20) and their derivatives, such as prostaglandins and leukotrienes (19). Other examples are RLD-1 (LXR-α) and OR-1 (UR, NER, LXR-β) (21), which are structurally closely related but nevertheless show distinct biological properties. In their case, side-chain-oxygenated sterols appear to be the preferred ligands (22) and, like PPARs (of which three isoforms exist), they are involved in the control of lipid homeostasis. Recently, other orphan receptors have turned out to have a very broad ligand-binding specificity. Examples of such receptors are PXR (23) and FXR (24), and it may be argued that at least PXR is a variant of a steroid hormone receptor. However, whereas the classical steroid receptors bind their ligands with very high affinity (K_d values, between 0.1 and 10 nM), these new "steroid receptors" need much higher steroid concentrations to be activated (K_d values, 100 nM–10 µM). One cannot excluded the possibility that their "true" ligands will yet be found and that such physiological ligands will display higher affinity for their receptors.

During this exciting development, it has become quite apparent that nuclear receptors constitute an extremely important regulatory network of transcriptional factors with the ability to control a huge number of genes including many that are essential for life. For instance, knockouts of many of these receptors have proven to be lethal, such as HNF-4, COUP TF-1, and TF-2. In most cases, nuclear receptors are apparently ligand regulated, but at least some seem to be activated via other pathways or are constitutively active.

Nuclear receptors represent one of the important mechanisms of the organism to respond to both internal and external signals. Important hormones such as estrogens, androgens, glucocorticoids, mineralocorticoids, progestins, and vitamin D mainly exert their control over physiological systems via these receptors. Furthermore, products of intermediary metabolism such as fatty acids and their metabolites can exert feed-forward and feed-back control over various metabolic networks via members of the nuclear receptor superfamily.

Acknowledgments. This work was supported by a grant from the Swedish Medical Research Council (No.13X-2819).

References

1. Mode A, Gustafsson J-Å, Jansson J.-O, Edén S, Isaksson O. Association between plasma level of growth hormone and sex differentiation of hepatic steroid metabolism in the rat. Endocrinology 1982;111:1692–7.
2. Gustafsson J-Å, Mode A, Norstedt G. Central control of prolactin and estrogen receptors in rat liver—expression of a novel endocrine system, the hypothalamo-pituitary-liver axis. Annu Rev Pharmacol Toxicol 1983;23:259–78.
3. Gustafsson J-Å, Mode A, Norstedt G, Skett P. Sex steroid induced changes in hepatic enzymes. Annu Rev Physiol 1983;45:51–60.
4. Gustafsson J-Å, Stenberg Å. Specificity of neonatal, androgen-induced imprinting of hepatic steroid metabolism in rats. Science 1976;191:203–4.
5. Zaphiropoulos PG, Mode A, Norstedt G, Gustafsson J-Å. Regulation of sexual differentiation in drug and steroid metabolism. Trends Pharmacol Sci 1989;10:149–53.
6. Legraverend C, Eguchi H, Ström A, Lahuna O, Mode A, Tollet P, Westin S, Gustafsson J-Å. Transactivation of the rat CYP2C13 gene promoter involves HNF-1, HNF-3, and members of the orphan receptor subfamily. Biochemistry 1994;33:9889-97.
7. Tollet P, Lahuna O, Ahlgren R, Mode A, Gustafsson J-Å. CCAAT/enhancer-binding protein-α-dependent transactivation of CYP2C12 in rat hepatocytes. Mol Endocrinol 1995;9:1771–81.
8. Lahuna O, Fernandez L, Karlsson H, Maiter D, Lemaigre FP, Rousseau GG, Gustafsson J-Å, Mode A. Expression of hepatocyte nuclear factor-6 in rat liver is sex-dependent and regulated by growth hormone. Proc Natl Acad Sci USA 1997;94:12309–13.
9. Buggs C, Tollet P, Zhao HF, Nasrin N, Mode A, Gustafsson J-Å, Alexander-Bridges M. IRE-ABP, an SRY-like protein, inhibits C/EBPα-stimulated expression of the sex-specific cytochrome P450 2C12 gene. Mol Endocrinol 1998;12:1294–309.
10. Ström A, Westin S, Eguchi H, Gustafsson J-Å, Mode A. Characterization of orphan nuclear receptor binding elements in sex-differentiated members of the CYP2C gene family expressed in rat liver. J Biol Chem 1995;270:11276–81.
11. Kuiper GGJM, Enmark E, Pelto-Huikko M, Nilsson S, Gustafsson J-Å. Cloning of a novel estrogen receptor expressed in rat prostate and ovary. Proc Natl Acad Sci 1996;93:5925–30.
12. Kuiper GGJM, Carlquist M, Gustafsson J-Å. Estrogen is a male and female hormone. Sci Med 1998;5:36–45.

13. Payvar F, Wrange Ö, Carlstedt-Duke J, Okret S, Gustafsson J-Å, Yamamoto KR. Purified glucocorticoid receptors bind selectively *in vitro* to a cloned DNA fragment whose transcription is regulated by glucocorticoids *in vivo*. Proc Natl Acad Sci 1981;78:6628–32.
14. Miesfeld R, Okret S, Wikström A-C, Wrange Ö, Gustafsson J-Å, Yamamoto K. Characterization of a steroid receptor gene and mRNA in wild-type and mutant cells. Nature (Lond) 1984;312:779–81.
15. Hollenberg SM, Weinberger C, Ong ES, Cerelli G, Oro A, Lebo R, Thompson EB, Rosenfeld MG, Evans RM. Primary structure and expression of a functional human glucocorticoid receptor cDNA. Nature (Lond) 1985;318:635–41.
16. Wrange Ö, Gustafsson J-Å. Separation of the hormone- and DNA-binding sites of the hepatic glucocorticoid receptor by means of proteolysis. J Biol Chem 1978;253:856–65.
17. Wrange Ö, Carlstedt-Duke J, Gustafsson J-Å. Purification of the glucocorticoid receptor from rat liver cytosol. J Biol Chem 1979;254:9284–90.
18. Carlstedt-Duke J, Okret S, Wrange Ö, Gustafsson J-Å. Immunochemical analysis of the glucocorticoid receptor: identification of a third domain separate from the steroid-binding and DNA-binding domains. Proc Natl Acad Sci USA 1982;79:4260–4.
19. Enmark E, Gustafsson J-Å. Orphan nuclear receptors—the first eight years. Mol Endocrinol 1996;10:1293–307.
20. Göttlicher M, Widmark E, Li Q, Gustafsson J-Å. Fatty acids activate a chimera of the clofibric acid-activated receptor and the glucocorticoid receptor. Proc Natl Acad Sci 1992;89:4653–7.
21. Teboul M, Enmark E, Li Q, Wikström AC, Pelto-Huikko M, Gustafsson J-Å. OR-1, a member of the nuclear receptor superfamily that interacts with the 9-*cis*-retinoic acid receptor. Proc Natl Acad Sci USA 1995;92:2096–100.
22. Janowski BA, Willy PJ, Devi TR, Falch JA, Mangelsdorf DJ. An oxysterol signaling pathway mediated by the nuclear receptor LXRα. Nature (Lond) 1996;358:728–31.
23. Kliewer SA, Moore JT, Wade L, Staudinger JL, Jones MA, McKee DD, Oliver BM, Willson TM, Zetterstrom RH, Perlmann T, Lehmann J. An orphan nuclear receptor activated by pregnanes defines a novel steroid signaling pathway. Cell 1998;92:73–82.
24. Forman BM, Goode E, Chen J, Oro AE, Bradley DJ, Perlmann T, Noonan DJ, Burka LT, McMorris T, Lamph WW, Evans RM, Weinberger C. Identification of a nuclear receptor that is identified by farnesol metabolites. Cell, 1995;81:687–93.

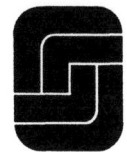

Author Index

A

Anderson, Stacey M., 93–121
Arvat, Emanuela, 249–260
Attie, Kenneth M., 235–240

B

Batson, Judy M., 172–182
Bellantoni, Michele F., 44–53
Bengtsson, Bengt-Åke, 219–226
Blackman, Marc R., 44–53
Boeppple, Paul A., 122–130
Bowers, Cyril Y., 261–276, 277–289
Bray, Megan J., 93–121
Broglio, Fabio, 249–260

C

Cacicedo, Lucinda, 153–159
Camanni, Franco, 249–260
Carr, Bruce R., 74–81
Carro, Eva, 243–248
Carter, H. Ballentine, 44–53
Casanueva, Felipe F., 243–248
Cassorla, Fernando, 3–10
Castaño, Justo Pastor, 183–191
Ceda, Gian Paolo, 202–208
Christiansen, Jens Sandahl, 67–73
Christmas, Colleen, 44–53

Clifton, Donald K., 144–152
Copeland, Kenneth C., 160–171

D

Desenzani, Paolo, 20–31
Dieguez, Carlos, 243–248

E

Edén, Staffan, 195–201
Evans, William S., 93–121, 172–182

F

Fernandez, Gumersindo, 153–159
Friend, Keith E., 301–307
Frohman, Lawrence A., 293–300
Fryburg, David A., 82–92

G

Garrido-Gracia, José Carlos, 183–191
Ghigo, Ezio, 249–260
Gianotti, Laura, 249–260
Giordano, Roberta, 249–260
Giustina, Andrea, 20–31
Gracia-Navarro, Francisco, 183–191

343

Granda-Ayala, Ramona, 277–289
Gustafsson, Jan-Åke, 337–341

H

Harman, S. Mitchell, 44–53
Hintz, Raymond L., 82–92
Ho, Ken K.Y., 54–66
Hoffman, Andrew R., 202–208
Houchin, Lisa D., 11–19

J

Jahn, Linda A., 82–92
Janssen, Yvonne Johanna Henrica, 209–218
Johannsson, Gudmundur, 219–226
Johnson, Michael L., 318–326
Jorgensen, Jens Otto Lunde, 67–73

K

Kerrigan, James R., 32–43, 172–182
Kineman, Rhonda D., 293–300
Kraus, Steven J., 172–182
Krieg, Richard J. Jr., 172–182

L

Lago, Francisca, 243–248
Lopez, Judith, 153–159

M

Martha, Paul M. Jr., 172–182, 235–240
Matt, Dennis W., 172–182
Mauras, Nelly, 308–317
McCutcheon, Ian E., 301–307
Metter, E. Jeffrey, 44–53
Metzger, Daniel L., 32–43
Muccioli, Giampiero, 249–260
Münzer, Thomas, 44–53

N

Neely, E. Kirk, 227–234

O

O'Connor, Kieran G., 44–53
Oscarsson, Jan, 195–201
O'Sullivan, Anthony J., 54–66
Ottosson, Malin, 195–201
Ovesen, Per, 67–73

P

Pabst, Katherine, 44–53
Palacios, Nuria, 153–159
Pazos, Fernando, 153–159
Peñalva, Angela, 243–248
Pombo, Manuel, 243–248
Popovic, Vera, 243–248

R

Ramirez, José Luis, 183–191
Roelfsema, Ferdinand, 209–218
Rogol, Alan D., 11–19
Román, Rossana, 3–10

S

St. Clair, Carol, 44–53
Samojlik, Eugene, 82–92
Sanchez-Franco, Franco, 153–159
Sayles, Timothy E., 172–182
Scalvini, Tiziano, 20–31
Señaris, Rosa Maria, 243–248
Shah, Nikhita, 93–121, 261–276
Steiner, Robert A., 144–152
Stevens, Thomas E., 44–53
Story, E. Shannon, 93–121, 261–276
Straume, Martin, 318–326

T

Tannenbaum, Gloria S., 133–143

V

Vahl, Nina, 67–73
Valenti, Giorgio, 202–208

Veldhuis, Johannes D., 67–73, 82–92, 93–121, 261–276

W

Waxman, David J., 327–336
Weltman, Arthur L., 82–92, 261–276
Weltman, Judy Y., 82–92, 261–276
Wideman, Laurie, 261–276

Subject Index

ACTH-releasing activity, endocrine responses, growth hormone secretagogues, 253, 255–256
Adenomas, pituitary gland, estrogen receptor expression, ER-α expression in, 302–303
Adipose tissue metabolism, growth hormone, sex-steroids, interactions, lipid metabolism and, 196–197
Androgen
  receptor blockade, somatotropic axis, sex-steroid hormone, normal male puberty, 34, 35, 36, 37
    luteinizing hormone, 34, 35
  somatostatin, growth hormone-releasing hormone, sex-steroid interaction with, somatotropic axis, 155–157
    during aging, 157–158
Androgenic modulation, growth hormone/insulin-like growth factor axis, metabolic outcomes, 82–92
  clinical research center, admission phase, 83–84
  experimental manipulations, hormone responses to, 85–86
  gonadal steroid manipulation, metabolic responses to, 85–86
  growth hormone secretory dynamics, 86–87

  insulin-like growth factor-I concentrations, 87–89
  manipulation, hormonal, 83
  secretion analysis, hormone, 84
  stimulated growth hormone secretion, 87
Approximate entropy, 319, 324–325

Basal growth hormone release, growth hormone, exercise, gender impact on, 262–263
β estradiol, sex hormone-binding protein, 24
Body composition, postmenopausal estrogen, growth hormone and, 60–62
Bone
  growth hormone action, sex steroid modulation of, 221–222
  postmenopausal estrogen, growth hormone and, 59
  remodeling, gender, response to growth hormone substitution therapy, 214–215

Calcium, growth hormone/insulin-like growth factor-I, testosterone, estrogen, metabolic effects, 310
Cluster analysis, 318–319
Copulsatility analysis, 323
CPP. *See* Precocious puberty

Deconvulution analysis, 319–323
Dimorphism, sexual, liver P-450 gene expression, growth hormone pulse-activated STAT signaling mechanisms, 327–336
Dwarf rat genetic model, growth hormone axis, sexual dimorphism, 297

Elderly men, testosterone, growth hormone/insulin-like growth factor-I axis in, 46–50
Electrolytes, gender, response to growth hormone substitution therapy, 214
Endocrine, growth hormone secretagogues, 251–256
  ACTH-releasing activity, 253, 255–256
  age, 249–260
  cortisol-releasing activity, 253, 255–256
  growth hormone-releasing activity, 251–255
  prolactin-releasing activity, 255
  receptors, 249–251
  sex, 249–260
Endogenous growth hormone releasing peptide, deficiency, 285–287
Entropy, approximate, 319, 324–325
Estradiol effects, androgen insensitivity, 7–8
Estrogen
  growth hormone, postmenopausal women, 54–66
    biological effects, 58–63
    insensitivity, estradiol effects, 7–8
    insulin-like growth factor-I, 99–105
    menopause, effect in, 54–57
    oral estrogen formulations, 57–58
    metabolic effects, 308–317
      calcium, 310
      glucose, 309–310
      protein, 308–309
      stable isotopes techniques, 308–310
    receptor blockade, somatotropic axis, sex-steroid hormone, normal male puberty, 35, 38
    growth hormone secretory, elimination dynamics, 38–40
    luteinizing hormone, 35, 38
    somatotrophs, sex-steroids effects on, dose effect, 165–167
Estrogen receptor expression, pituitary gland, 301–307
  adenomas, ER-α expression in, 302–303
  ER-α isoforms, 304–305
  ER-β expression, 305
  normal pituitary gland, ER-α expression in, 301–302
Estrogen replacement
  growth hormone and, 202–208
  Turner syndrome, 237–238
Exercise, gender impact on growth hormone response to, 261–276
  arginine, 261–262
  basal growth hormone release, 262–263
  growth hormone release, 263–266
    acute exercise, 263–264
    mechanisms, exercise-induced growth hormone release, 265–266
    training studies, 264
  growth hormone-releasing peptides, 261–262

Flutamide, somatotropic axis, sex-steroid hormone, normal male puberty, 33, 34–35

Gender, response to growth hormone substitution therapy, 209–218
  bone remodeling, 214–215
  causes and diagnosis of growth hormone deficiency in adults, 210
  electrolytes, 214
  glucose homeostasis, 214
  hematology, 213
  insulin-like growth factor-binding protein, 212, 213
  kidney function, 214

lipids, 212
patients and design of study, 211–212
physiological growth hormone secretion in adults, 209–210
plasma insulin-like growth factor, 212, 213
serum proteins, 214
thyroxine, 212
triiodothyronine, 212
Genetic models, growth hormone axis, sexual dimorphism
dwarf rat, 297
little mouse, 295–297
spontaneous dwarf rat, 297–298
GHRH. *See* Growth hormone-releasing hormone
Glucose
growth hormone/insulin-like growth factor-I, testosterone, estrogen, metabolic effects, 309–310
homeostasis, gender, response to growth hormone substitution therapy, 214
metabolism, postmenopausal estrogen, growth hormone and, 59
Gonadal function, growth hormone secretion, 243–248
nonandrogenic testicular factors, 245–247
Gonadal steroid manipulation, androgenic modulation, metabolic responses to, 85–86
Gonadotropin-releasing hormone administration, growth hormone axis, precocious puberty, 124–126
Growth hormone, 308–317
calcium, 310
estrogen, postmenopausal women, 54–66
biological effects, 58–63
estrogen replacement therapy and, 202–208
exercise, gender impact on, 261–276

basal growth hormone release, 262–263
exercise-induced growth hormone release, 263–266
acute exercise, 263–264
mechanisms, exercise-induced growth hormone release, 265–266
training studies, 264
glucose, 309–310
growth hormone-releasing peptides, 261–262
insulin-like growth factor-I, sex-steroid hormone neuromodulation, 93–121
hypothalamic actions, 98–99
nonpubertal contexts, 95–96
paradigms, sex-steroid-growth hormone axis neuroendocrine interactions, 93–96
puberty, 93–95
sex-steroid receptors, 96–97
somatostatin actions, growth hormone-releasing hormone, 97–98
menstrual cycle and, 67–73
modulation, on target tissues, sex steroids, 193–240
protein, 308–309
sex steroid
androgen insensitivity, estradiol effects, testosterone effects, 7–8
axis interactions, normal female puberty, 3–10
estradiol, growth hormone and, in Turner syndrome, 6–7
interactions, lipid metabolism and, 195–201
modulation of, 219–226
stable isotopes techniques, 308–310
stimulated release of, 277–289
age, 277–279
endogenous growth hormone releasing peptide, deficiency, 285–287
estrogen, growth hormone releasing peptide-2, 284–286
growth hormone releasing peptide-2, 279–283

Growth hormone axis
  precocious puberty, 122–130
    during active puberty, 122–124
    during gonadotropin-releasing hormone administration, 124–126
    growth velocity correlates, 127
    pituitary-gonadal suppression, 126–127
  sexual dimorphism, genetic, transgenic models, 293–300
  superovulation and, 74–81
    estradiol production, 74–75
    menopausal gonadotropins, gonadotropin-releasing hormone analogs, 76–80
Growth hormone/insulin-like growth factor axis, metabolic outcomes, androgenic modulation, 82–92
  clinical research center, admission phase, 83–84
  experimental manipulations, hormone responses to, 85–86
  gonadal steroid manipulation, metabolic responses to, 85–86
  growth hormone secretory dynamics, 86–87
  hormonal manipulation, 83
  hormone secretion analysis, 84
  serum insulin-like growth factor-I concentrations, 87–89
  stimulated growth hormone secretion, 87
Growth hormone/insulin-like growth factor-I
  sex-steroid hormone neuromodulation, 99–105
  testosterone, 44–53
    estrogen, metabolic effects, 308–317
    replacement, 20–31
Growth hormone pulse-activated STAT signaling mechanisms, sexual dimorphism, liver P-450 gene expression, 327–336
Growth hormone-releasing hormone, somatostatin, sex steroid, 153–159

androgens, somatotropic axis, 155–157
  during aging, 157–158
  hypothalamic cells, 153–155
Growth hormone releasing peptide-2, 279–280
  growth hormone-releasing hormone, combined
    equal doses, 280–282
    unequal dosages, 282
  postmenopausal women, 284–286
Growth hormone secretagogues, 241–289
  endocrine responses
    age, 249–260
    endocrine activities, 251–256
    receptors, 249–251
    sex, 249–260
Growth hormone secretion
  spontaneous, testosterone replacement and, 24–25
  stimulated, testosterone replacement and, 25
Growth spurt, pubertal, growth hormone, sex-steroid, interactions, 14–17

Hepatic lipoprotein metabolism, growth hormone, sex-steroids, interactions, lipid metabolism and, 197–198
Hydroxylase liver P-450 enzymes, steroid, 327
Hypothalamic actions
  insulin-like growth factor-I, growth hormone, sex-steroid hormone neuromodulation, 98–99
  sex-steroid actions, growth hormone axis, feed-forward, feedback network, 104–108
Hypothalamic cells, somatostatin, 153–155
Hypothalamic control mechanisms, sexually dimorphic growth hormone secretion, 133–143
  sex-steroid modulation, sexually dimorphic growth hormone secretory patterns, 138–140

sexually dimorphic patterns of hypothalamic somatostatin, 133–134
somatostatin receptor subtypes, sexually dimorphic expression, 134–137

Insulin-like growth factor
  gender, response to growth hormone substitution therapy, 212, 213
  growth hormone, metabolic outcomes, androgenic modulation, 82–92
    clinical research center, admission phase, 83–84
    experimental manipulations, hormone responses to, 85–86
    gonadal steroid manipulation, metabolic responses to, 85–86
    growth hormone secretory dynamics, 86–87
    hormonal manipulation, 83
    hormone secretion analysis, 84
    serum insulin-like growth factor-I concentrations, 87–89
    stimulated growth hormone secretion, 87
Insulin-like growth factor-binding protein, gender, response to growth hormone substitution therapy, 212, 213
Insulin-like growth factor-I
  growth hormone
    sex-steroid hormone neuromodulation, 93–121
    testosterone, 44–53
  growth hormone action, sex steroid modulation of, 219–220
  insulin-like growth factor-1 BP-3, testosterone replacement and, 25–26
  metabolic effects, 308–317
    calcium, 310
    glucose, 309–310
    protein, 308–309
    stable isotopes techniques, 308–310

Intracellular signaling networks, 337–341

Kidney function, gender, response to growth hormone substitution therapy, 214

Lipids, gender, response to growth hormone substitution therapy, 212
Lipoprotein
  growth hormone action, sex steroid modulation of, 222–223
  metabolism, hepatic, growth hormone, sex-steroids, interactions, lipid metabolism and, 197–198,
Little mouse genetic model, growth hormone axis, sexual dimorphism, 295–297
Liver P-450 enzymes, hydroxylase, steroid, 327
Liver P-450 gene expression, sexual dimorphism, growth hormone pulse-activated STAT signaling mechanisms, 327–336
Luteinizing hormone
  androgen-receptor blockade, somatotropic axis, sex-steroid hormone, normal male puberty, 34, 35
  estrogen receptor blockade, somatotropic axis, sex-steroid hormone, normal male puberty, 35, 38

Menopause
  estrogen, growth hormone and, 54–66
    biological effects, 58–63
    estrogen effect, 54–57
    oral estrogen formulations, 57–58
  growth hormone, estrogen, 54–66
Menstrual cycle, growth hormone axis, 67–73

Networks, signaling, intracellular, 337–341

Neuroendocrine techniques, quantitative, 318–326
   approximate entropy, 319, 324–325
   cluster analysis, 318–319
   copulsatility analysis, 323
   deconvulution analysis, 319–323
Nocturnal growth hormone secretion, 21
Nonandrogenic testicular factors, gonadal function, growth hormone secretion, 245–247

Oral estrogen formulations, 57–58
Oxandrolone, Turner syndrome, 236–237

P-450 enzymes, liver, hydroxylase, steroid, 327
Perifused pituitary, sex-steroid effects on, 172–182
Pituitary gland
   estrogen receptor expression, 301–307
      adenomas, ER-α expression in, 302–303
      ER-α isoforms, 304–305
      ER-β expression, 305
      normal pituitary gland, ER-α expression in, 301–302
   perifused, sex-steroid effects on, 172–182
Pituitary-gonadal suppression, and precocious puberty, 126–127
Postmenopausal women
   estrogen, growth hormone, 54–66
      biological effects, 58–63
      estrogen effect, in menopause, 54–57
      oral estrogen formulations, 57–58
   growth hormone, estrogen, 54–66
Precocious puberty, growth hormone axis, 122–130
   during active puberty, 122–124
   during gonadotropin-releasing hormone administration, 124–126
   growth velocity correlates, 127

   pituitary-gonadal suppression, during gonadotropin-releasing hormone administration, 126–127
Prolactin-releasing activity, endocrine responses, growth hormone secretagogues, 255
Protein
   gender, response to growth hormone substitution therapy, 214
   growth hormone/insulin-like growth factor-I, testosterone, estrogen, metabolic effects, 308–309,
Puberty
   growth hormone, sex-steroid, interactions, 14–17
   insulin-like growth factor-I, growth hormone, sex-steroid hormone neuromodulation, 93–95
   precocious, growth hormone axis, 122–130
   somatotrophs, sex-steroid effects on, nonhuman primate models for, 160–164

Quantitative neuroendocrine techniques, 318–326
   approximate entropy, 319, 324–325
   cluster analysis, 318–319
   copulsatility analysis, 323
   deconvulution analysis, 319–323

Remodeling, bone, gender, response to growth hormone substitution therapy, 214–215

Serum proteins, gender, response to growth hormone substitution therapy, 214
17-β estradiol, sex hormone-binding protein, 24
Sex differences, growth hormone-releasing, somatostatin neurons, 144–152
   androgen receptors, testosterone acting on, 146–147

growth hormone-releasing hormone neurons, 147–149
neuropeptide mRNAs, somatostatin, growth hormone-releasing hormone neurons, 145–146
somatostatin, growth hormone-releasing hormone neurons, sex differences in, 149–150
Sex-steroid growth hormone interactions, normal male puberty, 11–19
physiologic changes, during normal male puberty, 11–14
pubertal growth spurt, growth hormone, sex-steroid, interactions, 14–17
Sexual dimorphism
growth hormone axis, genetic, transgenic models, 293–300
dwarf rat, 297
little mouse, 295–297, 296
spontaneous dwarf rat, 297–298
liver P-450 gene expression, growth hormone pulse-activated STAT signaling mechanisms, 327–336
Sexually dimorphic growth hormone secretion, hypothalamic control mechanisms, 133–143
Signaling networks, intracellular, 337–341
Somatostatin
growth hormone-releasing hormone, sex-steroid interaction with, 153–159
androgens, somatotropic axis, 155–157
during aging, 157–158
hypothalamic cells, 153–155
insulin-like growth factor-I, 97–98
Somatostatin neurons, sex differences, 144–152
androgen receptors, testosterone acting on, 146–147
neuropeptide mRNAs, somatostatin, growth hormone-releasing hormone neurons, 145–146

somatostatin, growth hormone-releasing hormone neurons, sex differences in, 149–150
somatostatin neurons, androgen receptors, 147–149
Somatotrope heterogeneity, growth hormone regulation, 183–191
aging, somatotrope subpopulations during, 184–186
GRF, subpopulations induced by, 186
molecular signaling, 186–188
porcine somatotrope subpopulations, 186–188
postnatal growth, somatotrope subpopulations during, 184–186
somatotrope subpopulations, 184
SRIF, subpopulations induced by, 186
Somatotrophs, sex-steroids effects on, 160–171
developmental stage, 167
estrogen, dose effect, 165–167
interventional studies, 161–167
pubertal growth, nonhuman primate models for, 160–164
route of administration, 167
Somatotropic axis, sex-steroid hormone, normal male puberty, 32–43
androgen receptor blockade, 35, 36, 37
growth hormone-related proteins, 34, 35
luteinizing hormone, 34, 35
sex-steroid hormones, 34, 35
deconvolution analysis, 34
estrogen receptor blockade
growth hormone-related proteins, 35, 38
growth hormone secretory, elimination dynamics, 38–40
luteinizing hormone, 35, 38
sex-steroid hormones, 35, 38
flutamide protocol, 33, 34–35
statistical analysis, 34
tamoxifen protocol, 33, 35–38

Spontaneous dwarf rat genetic model, sexual dimorphism, 297–298
Steroid hydroxylase liver P-450 enzymes, 327
Stimulated growth hormone secretion, testosterone replacement and, 21–22
Substrate oxidation, postmenopausal estrogen, growth hormone and, 59–61

Tamoxifen, somatotropic axis, male puberty, 33, 35–38
Testosterone
  adolescents, growth hormone-I axis, 45–46
  androgen insensitivity, 7–8
  boys, growth hormone-I axis, 45–46
  elderly men, growth hormone/insulin-like growth factor-I axis in, 46–50
  estradiol and, 7–8
  growth hormone, insulin-like growth factor-I, 99–105
  growth hormone-I axis, 45–46
  growth hormone/insulin-like growth factor-I axis, 20–31, 44–53
    assays, 22
    boys, 45–46
    correlations, 26
    data analysis, 22–23
    experimental design, 21
    insulin-like growth factor-I, insulin-like growth factor-1 BP-3, 25–26
    nocturnal growth hormone secretion, 21
    serum testosterone, 23
    spontaneous growth hormone secretion, 24–25
    stimulated growth hormone secretion, 21–22, 25
  metabolic effects, 308–317
    calcium, 310
    glucose, 309–310
    protein, 308–309
    stable isotopes techniques, 308–310
  middle-aged men, growth hormone-I axis, 45–46
  young adult, growth hormone-I axis, 45–46
Thyroxine, gender, response to growth hormone substitution therapy, 212
Transgenic models, growth hormone axis, sexual dimorphism
  dwarf rat, 297
  little mouse, 295–297
  spontaneous dwarf rat, 297–298
Triiodothyronine, gender, response to growth hormone substitution therapy, 212
Turner syndrome
  estradiol, growth hormone and, 6–7
  growth hormone therapy, 227–234, 235–240
    efficacy of, 227–229
    estrogen replacement, 237–238
    growth failure, 229–232
    oxandrolone, 236–237

PROCEEDINGS IN THE SERONO SYMPOSIA USA SERIES

Continued from page ii

GHRH, GH, AND IGF-I: *Basic and Clinical Advances*
    Edited by Marc R. Blackman, S. Mitchell Harman, Jesse Roth, and
    Jay R. Shapiro

IMMUNOBIOLOGY OF REPRODUCTION
    Edited by Joan S. Hunt

FUNCTION OF SOMATIC CELLS IN THE TESTIS
    Edited by Andrzej Bartke

GLYCOPROTEIN HORMONES: *Structure, Function, and Clinical Implications*
    Edited by Joyce W. Lustbader, David Puett, and Raymond W. Ruddon

GROWTH HORMONE II: *Basic and Clinical Aspects*
    Edited by Barry B. Bercu and Richard F. Walker

TROPHOBLAST CELLS: *Pathways for Maternal-Embryonic Communication*
    Edited by Michael J. Soares, Stuart Handwerger, and Frank Talamantes

IN VITRO FERTILIZATION AND EMBRYO TRANSFER IN PRIMATES
    Edited by Don P. Wolf, Richard L. Stouffer, and Robert M. Brenner

OVARIAN CELL INTERACTIONS: *Genes to Physiology*
    Edited by Aaron J.W. Hsueh and David W. Schomberg

CELL BIOLOGY AND BIOTECHNOLOGY:
Novel Approaches to Increased Cellular Productivity
    Edited by Melvin S. Oka and Randall G. Rupp

PREIMPLANTATION EMBRYO DEVELOPMENT
    Edited by Barry D. Bavister

MOLECULAR BASIS OF REPRODUCTIVE ENDOCRINOLOGY
    Edited by Peter C.K. Leung, Aaron J.W. Hsueh, and Henry G. Friesen

MODES OF ACTION OF GnRH AND GnRH ANALOGS
    Edited by William F. Crowley, Jr., and P. Michael Conn

FOLLICLE STIMULATING HORMONE: *Regulation of Secretion and Molecular
Mechanisms of Action*
    Edited by Mary Hunzicker-Dunn and Neena B. Schwartz

SIGNALING MECHANISMS AND GENE EXPRESSION IN THE OVARY
    Edited by Geula Gibori

GROWTH FACTORS IN REPRODUCTION
    Edited by David W. Schomberg

UTERINE CONTRACTILITY: *Mechanisms of Control*
    Edited by Robert E. Garfield

NEUROENDOCRINE REGULATION OF REPRODUCTION
    Edited by Samuel S.C. Yen and Wylie W. Vale

PROCEEDINGS IN THE SERONO SYMPOSIA USA SERIES

Continued

FERTILIZATION IN MAMMALS
    Edited by Barry D. Bavister, Jim Cummins, and Eduardo R.S. Roldan

GAMETE PHYSIOLOGY
    Edited by Ricardo H. Asch, Jose P. Balmaceda, and Ian Johnston

GLYCOPROTEIN HORMONES: Structure, Synthesis, and Biologic Function
    Edited by William W. Chin and Irving Boime

THE MENOPAUSE: Biological and Clinical Consequences of Ovarian Failure: Evaluation and Management
    Edited by Stanley G. Korenman

MIX
Papier aus verantwortungsvollen Quellen
Paper from responsible sources
FSC® C105338

If you have any concerns about our products,
you can contact us on
ProductSafety@springernature.com

In case Publisher is established outside the EU,
the EU authorized representative is:
**Springer Nature Customer Service Center GmbH
Europaplatz 3, 69115 Heidelberg, Germany**

Printed by Libri Plureos GmbH
in Hamburg, Germany